# Technological Risk Assessment

# NATO ASI Series

## Advanced Science Institutes Series

*A Series presenting the results of activities sponsored by the NATO Science Committee, which aims at the dissemination of advanced scientific and technological knowledge, with a view to strengthening links between scientific communities.*

The Series is published by an international board of publishers in conjunction with the NATO Scientific Affairs Division

| | | |
|---|---|---|
| A | Life Sciences | Plenum Publishing Corporation |
| B | Physics | London and New York |
| C | Mathematical and Physical Sciences | D. Reidel Publishing Company Dordrecht and Boston |
| D | Behavioural and Social Sciences | Martinus Nijhoff Publishers The Hague/Boston/Lancaster |
| E | Applied Sciences | |
| F | Computer and Systems Sciences | Springer-Verlag Berlin/Heidelberg/New York |
| G | Ecological Sciences | |

Series E: Applied Sciences – No. 81

# Technological Risk Assessment

edited by

## Paolo F. Ricci
## Leonard A. Sagan
## Chris G. Whipple

Electrical Power Research Institute
Palo Alto, CA 94303
USA

1984 **Martinus Nijhoff Publishers**
The Hague / Boston / Lancaster
Published in cooperation with NATO Scientific Affairs Division

Proceedings of the NATO Advanced Study Institute on Technological Risk
Assessment, Erice, Sicily, Italy, May 20-31, 1981

**Library of Congress Cataloging in Publication Data**

NATO Advanced Study Institute on Technological Risk
   Assessment (1981 : Erice, Italy)
   Technological risk assessment.

   (NATO advanced science institutes series. Series E,
Applied sciences ; no. 81)
   "Proceedings of the NATO Advanced Study Institute on
Technological Risk Assessment, Erice, Sicily, Italy,
May 20-31, 1981"--T.p. verso.
   1. Technology assessment--Congresses.  2. Risk--
Congresses.  I. Ricci, Paolo F.  II. Sagan, Leonard A.
III. Whipple, Chris G.  IV. Title.  V. Series.
T174.5.N37  1981      363.1      84-4105
ISBN-13: 978-94-009-6157-9        e-ISBN-13: 978-94-009-6155-5
DOI: 10.1007/ 978-94-009-6155-5

Distributors for the United States and Canada: Kluwer Boston, Inc., 190 Old Derby
Street, Hingham, MA 02043, USA

Distributors for all other countries: Kluwer Academic Publishers Group, Distribution
Center, P.O. Box 322, 3300 AH Dordrecht, The Netherlands

TABLE OF CONTENTS

Page

I. Health Risk Assessment

Problems in Health Measurements for the
Risk Assessor, L. A. Sagan 1

Issues Related to Carcinogenic Risk Assessment from Animal Bioassay Data, K. S. Crump 31

II. Engineering Risk Assessment

Engineering Risk Analysis, W. E. Vesely 49

Quantities of Hazardous Materials,
F. K. Farmer 85

The Wider Implications of the Canvey Island
Study: A Discussion, A. V. Cohen and
B. G. Davies 101

Benefits and Risks of Industrial Development
with Special Regard to the Field of Power
Production, A. M. Angelini 111

III. Risk Evaluation and Management

Principles for Saving and Valuing Lives,
R. Zeckhauser and D. S. Shepard 133

Eight Frameworks for Regulation, L. Lave 169

A Comparison of Air Pollution Control
Standards as Adopted in Various Industrialized Countries, A. Fontanella and G. Pinchera 191

Actual and Perceived Risk: A Review of the
Literature, V. T. Covello 225

VI

Psychological Aspects of Risk: The Assessment of Threat and Control, P. J. M. Stallen
and A. Tomas                                    247

IV.  References

Risk Analysis and Technological Hazards:
A Policy-Related Bibliography, V. T. Covello
and M. Abernathy                                233

# PREFACE

This volume describes the proceedings of a NATO Advanced Study Institute on Technological Risk Assessment, held May 20-31, 1981, at Erice, Sicily. The approach taken at the Erice meeting was to explore the multidisciplinary nature of risk assessment, rather than to focus on any single aspect or approach to risk assessment. This approach provided an overview of many different types of risk assessment work, as well as insight into how these different approaches are combined to provide estimates of risk. Additionally, understanding of the perception and public reaction to risk, and market and regulatory methods that lead to risk management were major topics of the meeting.

All but two of the papers in this volume were presented at that meeting by lecturers or participants; the papers by Drs. Lave and Vesely were added by the editors because they are excellent complements to the other papers. Also included is a comprehensive bibliography on the policy-related risk literature; this was developed by Covello and Abernathy of the U.S. National Science Foundation program on Technology Assessment and Risk Analysis.

In brief, the scientific questions addressed were:

o  How may technological risks be identified and their magnitudes estimated?

o  How do people react to exposure to technological risk? By what means is risk evaluated by individuals?

o What are the societal processes for managing technological risk? How and how well do these processes work?

It is intended that these proceedings represent a broad survey of risk assessment work, rather than a more detailed study of any particular aspect of the subject. Even with this approach, several important risk assessment topics were unavoidably omitted. Had more time and resources been available, analysis of the transport and transformation of hazardous materials through air, water, soil, and the food chain would be priority items; as it is, this topic receives a small bit of attention in the case history report on the Canvey Island study. Similarly, the role of liability and insurance in risk management was not explored. Little attention was given to environmental and other impacts of technology (including property damage) that do not relate to human health risk, yet clearly these impacts are of importance. Even though the focus of this meeting was broad, the scope and variety of risk assessment work is broader.

These qualifications notwithstanding, most of the major research areas in risk analysis were addressed to some degree. The material is roughly divided between the methods for risk estimation and for risk evaluation, and this reflects the equal weights given to these topics in the NATO meeting. The estimation of risk is addressed in the front of the book. Health risk assessment is the first addressed; L. Sagan provides a descriptive overview of the methods applied to assess risks to human health, and K. Crump describes the mathematical approaches with which to model risk from exposure to harmful agents.

The engineering and physical sciences approaches to assess accident probabilities and consequences are described in W. Vesely's article, and illustrated by the Canvey Island (UK) analysis reported by F. R. Farmer and by A. Cohen and B. Davies. Professor Angelini's paper describes how electricity production choices between coal and nuclear power and siting decisions are addressed in Italy, as well as the difficulty and importance of trying to include the risk of energy supply disruptions into such decisions.

The economics of risks are explored by L. Lave and by R. Zeckhauser and D. Shepard. Dr. Lave addresses various risk decision frameworks, and the role of economic analysis in these frameworks. Drs. Zeckhauser and Shepard address the difficult and sensitive question of how one considers analytically the question of determining resource allocations applied to reducing risks to life. In a related area, A. Fontanella and G. Pinchera provide an international comparison of approaches to control air pollution.

There is much more to risk evaluation than economics and regulatory policy, as the two papers on the psychological aspects or risk indicate. V. Covello gives an overview and a guide to this literature; Drs. Stallen and Tomas describe their research to explore the key determinants in individual reactions to various types of risk. These papers help one understand that for most people, the relevant measure of risk is far more complex than an estimate of expected fatalities.

The NATO Institute and this volume would not have been possible without the support and assistance of many people. Drs. Mario di Lullo and Craig Sinclair of NATO provided funding for the meeting, as well as helpful advice regarding organization and conduct of the meeting and its arrangements. Dr. Alberto Gabriele and his assistant, Pinola Savelli, directed the local operations of the Centre for Scientific Culture "Ettore Majorana" in Erice, and any success accorded this NATO ASI is in large share due to his generous hospitality and tireless attention to the many details in hosting such a meeting. Secretarial assistance for the meeting arrangements was provided by Nancy Lambert, and these proceedings were typed by Krista Frisch, Carol Levy, and Lenore Bartyzel; we thank them for their fine work. We also thank our contributors for their papers, both at the meeting and in this volume. While their papers do not indicate the colleagial atmosphere or lively discussions of the meeting, they do provide an overview of the risk analysis field by a notable list of experts. Finally, we thank the Brookings Institution for allowing us to use Dr. Lave's article, it is a chapter from his book, The Strategy of Social Regulation, published by Brookings in 1981.

Paolo Ricci
Leonard Sagan
Chris Whipple

Palo Alto, California
January, 1983

PROBLEMS IN HEALTH MEASUREMENTS FOR THE RISK ASSESSOR

L. A. Sagan

Electric Power Research Institute
Palo Alto, California   94303

## 1.   INTRODUCTION

When in the past a new product, industry or agent was introduced into widespread use, there was little effort or felt need to
predict or anticipate possible consequences to health that might
be associated with that innovation.  One wonders how the automobile might have been accepted if it had been known..that its
use would be at the price of 50,000 lives per year in the United
States alone, nevermind the much larger number of injuries, both
transient and permanent.  We no longer are willing to accept new
products on a "wait and see" basis.  The public wants to know
about such risks before hand.  In an affluent society such as ours,
the benefits of technology are no longer so irresistable that we
are willing to overlook possible costs, particularly when those
costs are to health and when such costs are likely to be subtle
and overlooked unless assiduously examined.

I have, in the following short paper, attempted the difficult task of assessing our ability to estimate the impact of
technology on human health.  I ask the difficult questions of
not only what the qualitative nature of such effects may be, (e.g.
cancer? heart disease? birth defects?) but also the quantitative
probability of such an outcome, i.e., how likely are such effects
to occur?  These are issues of singular importance for the professional engaged in technological risk assessment, but they are
also of importance to anyone who wants to know how to interpret
the almost daily report of a newly discovered threat to health.

The material which follows draws upon the fields of toxicology, pathology, epidemiology and genetics.  It is not written

for specialists in these fields but rather for nonbiologists who wish to understand the methods and limitations of assessing the risks to health from a given technology. If there is any conclusion, it is that everyone, nonbiologists and biologists alike, should be cautious about the difficulties that confront us in attempting such assessments.

I shall address those questions which I believe the assessor will need to ask as he creates a health model for the technology or agent under examination. Unless the questions regarding health are properly drawn, the answers will not have meaning. I shall point out the limitations of the various data bases that exist and of the various measures of health that exist. I shall describe some of the models which have been adopted to estimate human health risks. Finally, as an example, I shall examine briefly the risk model that has been developed for ionizing radiation. It is not chosen because it is typical of available risk estimates, but because it is the best developed risk model which exists.

2.    TECHNOLOGY RISK ASSESSMENT:  WHAT IS A "TECHNOLOGY",
      WHAT IS "RISK" AND WHAT IS AN "ASSESSMENT"?

The "technology" may be complex, and involve an entire industry, e.g., nuclear power; or may be relatively simple, as with a single chemical agent, e.g., a new drug, a food additive, or industrial chemical. The agents which the assessor may have to consider as hazards to human populations may include synthetic or natural agents. Examples are shown in Table 1. In this paper, no attempt will be made to deal with each of these agents individually. Rather, attention will be focused on effects, their identification and measurement.

"Risk" has two components, the probability and the magnitude of the consequences. That is, one wants to know how frequently consequences can occur, and also how serious those consequences are. The risk of rabies in the United States is extremely small, since it is only an exceptional year when a case is recognized. On that basis, then, the risk could be said to be small. On the other hand, a case of rabies is almost invariably fatal; on that basis, the risk or probability of death can be said to be very large. In the former case, the risk is viewed from the point of view of the population, and in the latter case, the risk is viewed from the point of view of the affected individual. The risk assessor attempts to incorporate both of these parameters.

Assessment refers to the systematic evaluation of the technology under examination. "Systematic" is the key world here; the assessor wishes to be as comprehensive as possible. He is like a

Table 1

ENVIRONMENTAL AGENTS POTENTIALLY
HAZARDOUS TO HEALTH

## Physical

Mechanical (trauma, as with auto accidents)
Acoustic (noise and vibration)
Electromagnetic (ionizing radiation, electric fields, radar)
Temperature (heat, cold)
Atmospheric pressure
Dusts and particulates (silica, asbestos)

## Chemical

Vapors
Gases
Nutrients and preservatives
Pharmaceuticals
Pesticides
Explosives

## Biologicals

Viruses (vaccines, genetic engineering)
Bacteria

detective, pursuing every clue to uncover effects to health that may in any way be consequent to the use of the suspect technology, whether it be to those who manufacture the material in some distant country, or to those as yet unborn children who may be affected by their parents exposure to a new agent.

A few recent examples of technological risk assessments help to clarify the concept. Within the past year the National Academy of Sciences has published two studies of health risks associated with a specific technology. The titles are self-explanatory:

Potential Risk of Lung Cancer from Diesel Engine Exhausts[1]

Evaluation of the Health Risks of Ordnance Disposal Waste in Drinking Water[2]

## 3.    CREATING THE RISK ASSESSMENT HEALTH MODEL

Before beginning a risk assessment, the risk assessor must create a "model"; he must decide which health effects are to be included, and the limits or boundaries of the technology to be assessed. In order to make such choices, he must have some understanding of human disease, its reporting (and the possible errors therein) as well as the limitations of our means of studying health and disease.

In the generation of the model the assessor may wish to consider a few useful distinctions regarding human disease. One is the health-safety distinction, a second the morbidity-mortality distinction, the third the somatic-genetic distinction and the fourth, the occupational-general public distinction. A consideration of each of these will be instructive and will serve to introduce most of the concepts and problems which the risk assessor must address.

As the reader goes through the following exposition, it would be well to keep in mind the strategy which he/she will eventually employ in constructing a risk assessment model. For whatever health effect is identified as a consequence of the technology at interest, four essential elements of information must be available. The more complete the information base for each of these, the more accurate is the assessment. To the extent that his knowledge of these is deficient, to that extent his assessment will over- or underestimate the health impact.

1) Effects-What specific health effects are likely to result from the emissions or altered environment produced by a technology under examination? (These possible effects will be discussed in some detail in the following paragraphs.) In selecting health

measures or effects likely to be altered by exposure to a given technology, the assessor must have a sound knowledge of the technology, its operation, manufacture, emissions and potential for accident. He must also have familiarity with the appropriate clinical, epidemiological, and toxicological literature which will, with expert judgement, allow a consideration of the appropriate human health effects for the final assessment.

2) The "dose-response model," i.e., the quantitative relationship between extent of exposure and effect. Models which relate exposure with subsequent effects in humans have two purposes. They provide insights into the basic mechanisms underlying the appearance of disease, and more importantly for our purposes here, they permit a quantitative estimate of the frequency of disease or other effect to be expected following a known exposure. Frequently, the exposures which are experimentally used in the laboratory for observations in animals are very much higher than those which will be experienced by human populations. The mathematical model permits extrapolation from those higher doses to lower doses. The selection of the appropriate model is a matter of careful judgement and often becomes controversial. This issue is discussed in some depth elsewhere in this book (see Crump) as well as in a recent review[3]. The difficulty arises from the fact that, whereas the various models often are in good agreement at high doses where data exists, they diverge widely at low doses where there is little data. As a consequence, it is often not possible to discriminate among competing models when an attempt is made to assess the effect at relatively low doses.

3) In addition to the question of estimating toxicity, there will often be a problem of estimating the previous level of exposure. Rarely is good dosimetry available. Data from a single air-pollution monitoring station has sometimes been used to estimate exposure for the population of the entire city. In addition to the problem of estimating the total exposure of a populaion aggregated over time, it will often not be clear whether the total exposure is the significant parameter or whether peak values are also important.

It is well to keep in mind the important distinction between dose and dose-rate. The latter is often more important than the former, i.e., a dose which may be lethal if delivered acutely is more likely to be much less harmful, or even harmless, if the exposure is protracted. The usual explanation of this latter phenomenon is the existence of biological repair mechanisms which permit the individual to cope with low dose-rate exposures. These repair mechanisms are rate-limited and become overwhelmed with high exposures, allowing toxic concentrations to occur.

Having the above information on toxicity and exposure, the

risk assessor can calculate the aggregate health impact of his
technology from airborne, foodborne, or waterborne agents.

### 3.1  Health and/or Safety, or Acute vs. Chronic Effects

Acute injuries resulting from mechanical or chemical trauma
are treated as safety issues whereas illness resulting from
chronic, low dose exposures are treated as health issues. Accidents generally result in acute injuries or illnesses which follow closely upon the causative event and therefore are easily
identified as such. Accidents are therefore a fairly easy
exercise for the risk assessor.

An example may be helpful: acute radiation exposures resulting from accidents and producing doses in the range of half
of a lethal dose result in a characteristic illness known as
"the acute radiation syndrome" characterized by diarrhea, infection, bleeding, bone marrow depression and resulting in
death, or in complete recovery. The cause is rarely in doubt
since:

1.  The accidental exposure is generally known.
2.  The illness follows shortly after the exposure.
3.  The illness itself is characteristic and recognizable.

Recognition of the effects of chronic, low dose exposures present
more difficult problems in their recognition and quantification
for the following reasons:

1.  The disease appears long after the exposure. The
    interval, known as the latent period, may be measured
    in years or decades. The causative exposure may not
    have been recognized at the time of exposure, may be
    forgotten, and/or records no longer in existence.

2.  The disease is not characteristic but identical to
    that occuring spontaneously. As an example, the
    leukemia occurring after radiation exposure is in-
    distinguishable from that due to other causes. As
    a consequence, statistical and epedemiological tools
    must be resorted to in order to identify late effects.

This problem will crop up again and again throughout our discussion, i.e., the problem of establishing causation in chronic
disease suspected of having a relationship to environmental agents.
Since most mortality in an industrialized society is from chronic
disease, the problem of relating the health effects of chronic
exposures to the emissions of a technology is one which plagues
the risk assessor.

## 3.2 Occupational Health and/or Public Health

There are good reasons for considering occupational impacts in the assessment of a given technology. First of all, they are likely to be large. Just how large, no one knows, partly because much occupationa illness almost certainly remains unrecognized. Only recently have a number of chemical carcinogens been recognized. Among them are asbestos, benzidine, bis-chloromethyl ether, cadmium oxide, vinyl chloride and others.

There are two reasons why it is so difficult to arrive at estimates of occupational illness. One has been discussed above, namely, the difficulty of distinguishing between occupationally-induced illness and that which results from other causes. The other reason for our difficulty in estimating occupational illness is the growing understanding that chronic disease is often multi-factorial in origin, that is to say, arising from personal factors, both biological and behavioral, as well as from the occupational exposure. By personal I mean both such practices as cigarette smoking and diet as well as genetically inherited susceptibility to disease. What I am getting at is the notion that division of illness into two categories, occupational and nonoccupational, may be quite an unrealistic distinction. I might point out paren-thetically that this creates an enormous problem for the workers compensation system in trying to sort out claims for occupational illness.

I have been discussing the difficulty in reaching estimates of occupational illness. This difficulty can be strongly con-trasted with the relative ease of quantitating occupational injury. In the United States as well as in many other nations, lost-time injuries are required by law to be reported. Therefore, data re-garding the frequency and severity of accidents are readily avail-able. Problems do exist, however. First of all there are dif-ferences in reporting requirements among different jurisdictions so that data are not always comparable. For example, some states require that all injuries be reported, whereas others require only that injuries involving lost time (from work) be reported. A second problem relates to aggregation of data. Frequently data are available only by gross categories, e.g., manufacturing, whereas the investigator's interest may be much more specific.

There is some reason to treat occupational illness separately from risks to the public since the public and their decision-makers treat the two quite differently. There may be good reasons for this. First and foremost, occupational risks (when they are recognized) are voluntary risks and as Chauncey Starr first pointed out, people are more willing to accept voluntary risks than those which are involuntary[4].

A second reason for separate treatment of occupational risks is the opportunity for compensation of risk. The risk can be compensated directly, as in hazard pay, or for those consequences (injuries) which result, e.g., as in workers compensation. In any case, maintaining separate categories of occupational and public risks is convenient, and they can always be aggregated by interested parties.

### 3.3  Morbidity and/or Mortality

Morbidity refers to illness, mortality to death. Obviously, the two are related but they are not congruent. Some illness, such as the common cold, rarely results in death, yet takes a considerable toll in human welfare. Other illness, a conspicuous example being cancer, often results in death and here the distinction between morbidity, or incidence, and mortality may be less important. When the health consequences of a particular technology result principally in cancer, as with ionizing radiation, the differences in consequences measured by morbidity and mortality are small.

Collecting morbidity data in the assessment of a particular technology has obvious advantages. To ignore nonfatal illness is to ignore a significant burden and the risk assessor may doom his assessment if he does so. For example, eye irritation, upper respiratory infection, psychological effects, and asthmatic attacks associated with photochemical smog are thought to be major consequences of such exposure, whereas an increase in mortality from this cause is difficult to measure and in fact may be nonexistent with ambient concentrations currently found in the U.S.

Unfortunately, morbidity data are often non-existent or have been collected in a manner which makes them difficult to interpret. One reason for this is that there are rarely legal reporting requirements. These exist only for communicable diseases where public health requirements justify a need for reporting. Hospital admissions, insurance records and other secondary sources sometimes are useful, but these suffer from a second problem, namely, absence of standadized diagnostic criteria. Physicians differ considerably in their diagnostic skills and training. Differences also exist over time. In the absence of objective criteria, incidence data over time and space must be viewed with considerable caution.

It must also be recognized that not all persons have equal access to physicians nor do all persons employ the same criteria for consulting physicians. The point is that patterns of morbidity as reported by physicians may be badly biased because of failure of some persons affected to be brought to medical attention.

There has been one notable attempt to overcome this defect through a household survey in which a sample of homes were selected and their residents interrogated about the frequency of disease and disability. Although this technique overcomes the problems of under-reporting, it introduces other problems. Relying upon self-diagnosis is notoriously unreliable. Furthermore, memories are short. This technique then may lead to errors in ascertainment in either direction. Lastly, a household survey is a very expensive technique for collecting data.

Turning now to mortality, we find that the problem of ascertainment becomes considerably simplified. Essentially all countries require death notification and reporting can be assumed to be almost complete. As always, there are problems, and these relate to cause of death. As one example, suicide is felt to be widely under-reported as a cause of death. Certain cancers are far more susceptible to misclassification than are others. Cancer of the pancreas is a good example, partly because diagnostic tools for visualizing and evaluating pancreas are poor. This can be remedied somewhat by restricting cases to those confirmed by autopsy but autopsies are not routinely used and there are selective factors here which could badly bias diagnoses based on autopsy series. Furthermore, even at autopsy the cause of death may be in doubt.

It may be well here to introduce another problem facing the risk assessor and that is the problem of how to represent the risk to health for those factors which cause the increase in mortality.

The classic and most widely used technique is to estimate number of cases, e.g., of cancer, black lung disease or whatever and such a measure already has its uses. Alone, however, numbers of "excess" cases fails to capture important elements that one needs to fully understand the impact of the technology under study. Since everyone dies, there are in fact no "excess" cases of death. Some measure of life shortening would appear to be more meaningful and would have many advantages; for example, life shortening from chronic disease might well be aggregated or compared with life shortening from accidental injuries. Yet, it is frequently difficult if not impossible to make such estimates. For one thing, an estimate of life shortening would require some judgement about competitive risks, a notion which deserves comment, no matter how brief. The paradigm which dominates medical thinking about longevity is strongly related to concepts of disease. Specifically, it is widely believed that death follows from the effects of a disease process affecting one or more vital organs. It is furthermore assumed that individual diseases act independently of each other. The consequence of this belief is to direct medical treatment and research to the understanding and ultimately to the prevention of disease in the hope of lengthening life. For example, this paradigm would suggest that if cancer were prevented, those now dying of cancer would then have the same life expectancy as other members of the population.

It may be that this paradigm or medical model is defective, or, at least, incomplete. For example, it may be that rather than people dying because they have disease, that people have disease because they are dying. This implies that some fundamental aging process controls longevity and that the "cause" of death would then only be incidental to death, but having little effect on the duration of life. To complete this highly speculative direction of thought, persons dying of occupational illness might be experiencing little, if any, life shortening, but only an alteration in the pattern of disease present at the time of death.

## 3.4  Reproductive Effects

The question here for the risk assessor is the extent to which he wises to assess risks, not to members of the exposed generation (somatic effects), but rather to their offspring. There can be no reasonable doubt that human exposure to both physical and chemical agents can alter the genetic material and/ or the unborn fetus and therefore affect the health of future generations. Evidence for this statement comes from a wealth of information derived both from the direct observation of molecular and cellular behavior and from studies of observable mutations in mammalian species as well. There is also growing evidence from studies of human populations to attest to these effects (Table 2).

Yet, it remains true that we still have only a primitive ability to predict, with any accuracy at all, the quantitative consequences of environmental exposure to successive generations. Although geneticists generally agree that the majority of mutations are deleterious, they are unable to give us any guidance with respect to the magnitude of the risk. In fact, it is difficult, if not impossible to distinguish which are mutagenic effects, if any, among those reproductive effects shown in Table 2. The reasons are as follows:

1. We are only able to measure directly a few genetically determined traits. The hemoglobinpathies are good examples of genetically determined traits that can be observed, both in their heterozygous as well as homozygous condition.

2. Although much human disease (about 10%) is thought to be influenced by genetic characteristics, the extent to

Table 2
## SPECTRUM OF REPRODUCTIVE OUTCOMES NOTED IN ASSOCIATION WITH PARENTAL EXPOSURES TO ENVIRONMENTAL AGENTS

Maternal Exposure Before Conception

Oral Contraceptives
Irradiation (low doses)

Maternal Exposure at Conception and or
During Pregnancy

Ovulation Stimulants
Sex Hormones
DES
Anticonvulsants
Antimetabolites
Tranquilizers
Oral Anticoagulants
Diabetes
Alcohol
Irradiation (high doses)
Irradiation (low doses)
Anesthetic Gases
Mercury
Lead
Smoking

Paternal Exposure

Anticonvulsants
Antineoplastic agents
Irradiation (high doses)
Irradiation (low doses)
Anesthetic Gases
Vinyl Chloride
Hydrocarbons
Chloroprene
Kepone
Dibromochloropropane
Dioxin

Effects noted include: infertility, early fetal death, altered sex ratio, late fetal death, neonatal death, low birth weight, birth defects, developmental disabilities, childhood malignancies, childhood mortalities.

From Ref. 5

which genetic traits determine the expression of specific diseases is often unknown.  The point is that
recognition of most genetic disease is by no means
simple nor is the role of genetic factors in influencing disease at all well understood.  Therefore
the frequency of genetic disease in a population cannot as yet be used as a measure of the frequency of
mutation rates in the population.

3.  Many if not most mutations produce impacts on health
    which do not seriously disturb normal health.  Each
    of us harbors within us millions of mutations.  In
    other words, it is the rare mutation which exacts a
    significant toll on health.

4.  The power of most agents to produce mutations in animals
    is weak.  For this reason, mouse experiments designed to
    detect mutagenicity require enormous numbers of animals
    exposed to high doses of the suspect agent.  Here, too,
    as with somatic effects, dose rate is an important fac-
    tor, reducing the acute mutagenic effectiveness of ion-
    izing radiation by a factor of six when delivered
    chronically.

5.  In-vitro or laboratory means of testing materials for
    their ability to produce mutations, although being
    widely applied, have poorly understood predictive power for
    humans.  These tests are often carried out in single
    celled organisms where metabolism of the agent in question
    may be radically different from that of the intact animal.
    For example, the Ames test which utilizes a bacteria, the
    salmonella, as a test organism, attempts to overcome
    the metabolic differences between bacteria and humans
    by adding a homogenate of liver to the test medium.  Yet,
    the preparation of this material, or the animal source,
    can profoundly affect the outcome of the test.  One just-
    ifiable conclusion from these observed species differences
    is that a specific agent may be mutagenic in one species
    of animal and not in another.

To summarize the experience with physico-chemical mutagens,
knowledge is rarely available with which to predict effects on
subsequent generations from exposures that are experienced by
the present generation.

In addition to mutagenesis, environmental agents may affect
other measures of reproduction such as fertility or the health of
the child exposed in-utero to such agents.  For example, some
female children exposed in-utero to the synthetic estrogen diethylstilbestrol (DES), given to their mothers during pregnancy

have experienced many years later an increase in vaginal cancer. One can only speculate whether such an effect would have been recognized if the consequence had not been such a rare tumor as to excite the curiosity of medical investigators. How much other disease in humans is the result of exposure of the sensitive fetus to agents in the maternal circulation is not known.

Teratogenic effects may be produced by exposure of the father as well as the mother, and may be considerably more subtle than that illustrated by the example above. Adams has recently reported that male rats following recovery from exposure to certain chemicals sire offspring whose learning ability is defective[6]. This newly emerging field of investigation is likely to shed some important new light on human health and behavior in the next decade, but is still too new to be useful to the risk assessor.

## 4. ESTIMATING THE RISKS OF CANCER

We now turn our attention from a survey of general measures of health to a specific illness, cancer, and to our available techniques for generating risk estimates of cancer following exposure to specific agents or mixtures of agents.

Why special attention to cancer? There are several parts to the answer to this question.

1. Cancer is a relatively frequent cause of death (about 16% in the U.S.) and afflicts some 25% of all persons during their lifetime.

2. It is an illness which is probably more dreaded than any other and therefore an assessment of cancer risk is of paramount important to those contemplating a new technology.

3. The relationship of cancer to environmental agents has now been clearly demonstrated. The list of proven chemicals is rapidly increasing but is small compared to the number of suspect carcinogens. For no other chronic disease is there an equal interest in risk assessment, at least in a developed society. This is not to say that many other examples of environmentally induced chronic disease do not exist. The reader need only remind himself of the known relationships between such chemical agents as alcohol (liver disease), heavy metals (central nervous system dysfunction), coal dust (chronic pulmonary disease), and many biological agents with a variety of infectious diseases. None of these agents, however, are as closely linked with technology

as are the chemical emissions associated with industry.

Although, as noted above, evidence for human carcinogenicity exists for a number of chemical and physical agents, there are literally thousands of agents for which such evidence would be desirable. Why does it not exist? In the following paragraphs, I will examine briefly the four fundamental methodologies available for predicting cancer in humans: chemical structure-function analysis, in-vitro systems, animal studies, and human studies. For those interested in examining the extensive literature in greater detail, there is an extensive guide to the literature[7].

## 4.1  Structure-Function Relationships

It would clearly be of extreme value if the chemist, through structural analysis of a new compound, could provide some guidance with respect to the potential carcinogenicity of a compound. In order for him to do that, there would necessarily be important structural characteristics of carcinogens distinguishing these agents from noncarcinogenic agents. Although some patterns are suggestive, we are still a long way from such a goal. Recently, computer-assisted pattern recognition techniques have been applied to a large number of compounds representing a wide range of chemical groups, and the results of the studies indicate that these methods have considerable potential in the accurate prediction of carcinogenicity. Jurs et al.[8] worked on a set of 109 heterogeneous compounds of known carcinogenic activity. These compounds were drawn from more than 12 structural classes and included aromatic amines, alkyl halides, N-nitroso compounds, polycyclic aromatics, azo dyes and naturally occuring compounds such as simple sugars, amino acids and fungal toxins. The set was made up of 130 carcinogens and 79 noncarcinogens.

The structures of the chemicals were entered into the computer by sketching them on the screen of a graphics display terminal. The next step was to use the computer to generate molecular structure descriptors. The following types of descriptors were used: fragment descriptors (number of atoms of each type, number of basic rings, number of ring atoms), substructure descriptors (number of occurrences of a particular substructure of interest), environment descriptors (giving information about the interconnection of fragments and substructures), molecular-connectivity descriptors (a measure of the branching of the structure) and geometric descriptors (representing the shape of the molecule). A large number of descriptors was developed and tested. Pattern recognition techniques were then used to discriminate between carcinogens and noncarcinogens. No set of descriptors was found that would support a discriminant that could separate all the carcinogens from all the noncarcinogens. The best sets of descriptors produced recognition percentages of 90-95%.

Would computer-assisted studies identify nonmutagenic carcinogens? Could they be used to fill the gaps left by mutagenicity screening tests, such as the Ames test? How well would the computer-based SAR methods do when tested with genuinely "unknown" compounds, i.e., those for which no toxicity data exists? These are some of the questions that will need to be answered before the full value of these apparently promising new methods of carcinogenicity screening can be determined. Clearly they have considerable value in furthering our understanding of the structural features of molecules that can lead to carcinogenesis.

## 4.2  In-Vitro Systems

A major breakthrough in our understanding of the mechanism of action of carcinogens came in 1960 when Cramer et al.[9] reported the N-hydroxylation of the carcinogen N-2-fluorenylacetamide in the rat. This observation constituted the first direct evidence of the metabolic activation of a carcinogen to a more reactive form and gave rise to the current distinction between procarcinogens and carcinogens, and promutagens and mutagens. The metabolic activation of carcinogens in the presence of genetically sensitive organisms in vitro was the next major development in this area[10]. More recent developments are well reviewed in ref.[11] (To examine the usefulness of a decision tree for estimating toxicity, see ref. 12).

During recent years, the conviction has grown that mutagenesis plays a prominent, if not predominant, role in carcinogenesis. If mutagenesis is an important step in carcinogenesis, then simple systems in which mutagenesis can be detected might be considered predictive of human cancer. Based on this premise, a number of such systems involving viruses, bacteria, fungi and protozoa have been developed as screening techniques for chemical mutagens which by implication are presumably carcinogens. Since chemical mutagens may be quite specific in producing mutations in some test systems but not others, it is highly desirable that a battery of such tests be used in evaluating a single agent or suspect chemical. In fact, many agents proven to be human carcinogens are mutagenic in these in-vitro systems,[13] however, there are important exceptions (Table 3). Dioxin, for instance, which is a highly potent carcinogen in rodent studies, is AMES negative. Furthermore, it is difficult on the basis of in-vitro results to predict the potency of agents as carcinogens.

Clearly, it would be preferable if we could detect mutagenesis in humans. As pointed out above, our ability to detect new human mutations is still rudimentary.

TABLE 3
CORRELATION BETWEEN CARCINOGENICITY AND
MUTAGENICITY IN THE SALMONELLA/MICROSOME TEST
McCann and Ames[11]

|          |     | Carcinogenic |      |     |
|----------|-----|--------------|------|-----|
|          |     | Yes          | No   |     |
| Mutagenic | Yes | 157          | 14   | 171 |
|          | No  | 18           | 94   | 112 |
|          |     | 175          | 108  | 283 |

From Katz[14]

Changes in chromosomes which contain the genes which trans-
mit encoded information to offspring can be examined.  Furthermore, such analyses can be carried out on the lymphocytes of
peripheral blood, requiring only a simple venipuncture.  There
are three disadvantages, however:

a.  The technique is expensive.  Since cytogenetic changes
    occur in low frequency and since exposures produce a
    relatively small increase in the normal frequency of
    such changes, many cells must be examined, usually 200,
    before statistical inferences can be drawn.  A technician
    can count about 400 cells per week.

b.  There is a certain subjective element in counting cells
    so that the same technician should always be asked to
    simultaneously count cells taken from a normal or unex-
    posed population, preferably through "blind" technique,
    i.e., without knowing the source of the cells being
    examined.

c.  The significance of an increased frequency of cyto-
    genetic changes to human health is not known.  The
    procedure is based on the assumption that, if a given
    exposure has produced changes in the genetic material of
    circulating lymphocytes, changes may also have been in-
    duced in cells of other tissues, particularly gonadal
    tissues but also in cells which may ultimately become neo-
    plastic.  The technique has been successfully used to assess
    radiation dose following accidental exposures where the

extent of the exposure or the geometry of the exposure is unknown[15].

Finally, lest the reader be left with the conclusion that mutagenesis is a necessary and sufficient predecessor of carcinogenesis, he should be reminded that other mechanisms, involving endocrine manipulation, physical trauma (partial hepatectomy, chronic mechanical and thermal stimulation) are also known. There is strong evidence that immune and even psychologic factors may be involved.

In addition to the use of in-vitro systems for the detection of mutagenesis, these systems, utilizing both bacterial and mammalian cells, can be used to detect abnormalities in growth as measures of exposure to suspect carcinogens. For example, cells growing on an agar medium normally display a regular growth pattern in the presence of a carcinogenic agent adjacent cells will show irregular growth. This phenomenon has received some use as a measure of potential carcinogenicity[11].

## 4.3 Animal Models

Animals have been invaluable for the study of carcinogenesis and for the elucidation of certain principles or mechanisms involved. In the absence of human data, animal data provide our clearest indication of the hazard of an agent. Perhaps the most attractive feature of animal research for the cancer biologist is the opportunity to control those variables which so often plague the human epidemiologist: genetic variability, diet, smoking, age, climate, etc. Still, even with all of these parameters controlled, variability still occurs, so that results are rarely reproducible even under apparently identical conditions. Many agents (such as ionizing radiation) are known to be carcinogenic in both animals and man. Unfortunately, the model often fails, as when conflicting evidence appears among different species, or when a known human carcinogen fails to produce cancer in animals. For example, it is by no means uncommon for an agent to produce tumors in rats but not mice, or in mice but not rats. Sometimes an agent will be carcinogenic by one route of exposure and not by another. Even caloric level of the diet can be shown to affect tumor incidence (Table 4). Problems in animal research have been recently reviewed[17].

One interesting attempt to resolve some of these inconsistencies is the use of pharmacokinetics or the study of the comparative metabolism of chemical carcinogens. The hope is that by understanding differences in the way different species metabolize chemicals that some understanding of the apparent inconsistencies will emerge. Still, it is only in the past decade or so that these studies have been undertaken and it is still too early to evaluate this hypothesis.

TABLE 4
EFFECTS OF CALORIC RESTRICTIONS DURING
THE TWO STAGES OF CARCINOGENESIS*

| Group | Diet in Period of Carcinogen Application (10 weeks) | Diet in Period of Tumor Formation (52 weeks) | Tumor Incidence |
|-------|-----------------------------------------------------|----------------------------------------------|-----------------|
| HH | High Calorie | High Calorie | 69 |
| HL | High Calorie | Low Calorie | 34 |
| LH | Low Calorie | High Calorie | 55 |
| LL | Low Calorie | Low Calorie | 24 |

*Benzo (A) pyrene in skin carcinogenesis in mice

From:  Tannerbaum, A.[16]

Regulators, required by law to identify carcinogenic agents in the environment frequently must depend upon animal studies.  How does one assess animal data which is often conflicting and inconsistent and yet avoid arbitrary regulation which may have serious economic consequences?  Furthermore, there is certain knowledge that the world is not neatly divided into carcinogenic and noncarcinogenic agents, but rather that many agents are likely to fall on some spectrum of carcinogenicity.  Given a demonstrated effect in animal studies, it is not clear either which measure of exposure is most appropriate for extrapolating to human dose (Table 5).

Beyond the difficulties alluded to above, animal research suffers from the enormous economic costs of carrying out an adequate study.  Costs of such a study may well run to half a million dollars.  Risks are also great.  An unexpected epidemic in the animal colony can ruin such an experiment.  With hundreds of new chemicals appearing each year, clearly all cannot be thoroughly evaluated in animal studies.  Therein lies the great attraction of the in-vitro studies described above, namely, to screen such agents and select those which appear to be most likely to produce a threat to human health.

TABLE 5

RATIOS OF MOUSE OR RAT CARCINOGENIC POTENCY TO HUMAN POTENCY

| | A) POTENCY IN MG/KG BODY WT/DAY | | B) POTENCY IN PPM IN DIET | | C) POTENCY IN MG/M² SURFACE AREA/DAY | | D) POTENCY IN MG/KG Body WT/LIFETIME | |
|---|---|---|---|---|---|---|---|---|
| | MOUSE | RAT | MOUSE | RAT | MOUSE | RAT | MOUSE | RAT |
| Aflatoxin B1 | | 1.0 | | 2.8 | | 7.15 | | 35.0 |
| Diethelstibestrol | <9.2 | | <86.0 | | <113.0 | | <367.0 | |
| Vinyl Chloride | 13.71 | 4.04 | 128.0 | 11.3 | 169.0 | 28.7 | 547.7 | 141.0 |
| Chlornaphasine | 0.4 | | 3.5 | | 4.63 | | 15.0 | |
| Benzidine | 0.0007 | .11 | 0.077 | 0.31 | 0.009 | 0.785 | 0.030 | 3.9 |
| Smoking | 0.1 | | 0.94 | | 1.23 | | 4.0 | |

Source: NAS-National Academy of Sciences (1975)

## 4.4  Human Studies

No evidence is so convincing as that which results from study of the species in which we have greatest interest--man himself.  As noted above, disease which appears in humans as a result of environmental agents is not clinically distinctive from cases which appear among nonexposed populations, therefore evidence for causation comes always from comparisons of the frequency of disease among exposed populations with that among nonexposed populations, a technique known as epidemiology.  It is such studies from which we have learned of the cancer-causing qualities of abestos, vinyl chloride, and cigarette smoking, among others.

Perhaps the greatest challenge to the epidemiologist is the avoidance of confounding variables, i.e., through selecting comparison populations which are hopefully identical to the exposed population, the two different only in respect to exposure to the suspect agent.  It would also be highly desirable if among the exposed population, there are a number of subpopulations who have experienced a graded response to the agent.  Increasing frequency of response (i.e., disease) in parallel with increasing exposure, together with consideration of the degree of association, provide the most convincing evidence of a causal relationship.

Almost always, opportunities for study vary from the ideal in some significant way.  Commonly, the extent of exposure (dose) is unknown.  Even in an occupational environment, there are rarely adequate measures of exposure.  It is where such measures do exist that epidemiology has been most helpful in identifying carcinogens:  cigarette smoking (measured in pack-years) and ionizing radiation.  Otherwise, one must depend on such rough measures as years of occupational experience or estimates of dietary consumption.

Secondly, since there is almost always some element of self-selection in exposure, one can never be certain whether the agent under study is responsible for an observed effect or whether some other unidentified characteristic of the exposed population may be responsible.  Such factors as socioeconomic status, marital status, urban-rural differences as well as age and sex are all known to influence cancer incidence.

Very generally, there are many strategies available for epidemiological design.  In a prospective study, a population is chosen whose exposure to the agent under investigation is known.  They are compared with a population similar in all known respects and their exposure is known and confounding variables avoided through prior matching of controls.  The disadvantage is that such studies

are expensive and time consuming requiring often many years to
complete. If the health event being monitored is relatively
uncommon or rare, very large populations will have to be
followed if weak effects are to be detected between the two populations with statistical significance.

Much more expeditious studies can be undertaken by identifying populations whose exposure occured in the past, followed by examination of subsequent health experience. This is
relatively fast and inexpensive, but has the disadvantage that
exposures may not be known with any precision and often cannot
be excluded. Another alternative is the so-called case-control
method which examines the frequency of exposure to a suspect
agent within a group.

In spite of these difficulties, estimates of the effect
of environmental factors in producing human cancer have been
attempted. These estimates have widely varied. Doll and Peto[18]
have recently completed an analysis of cancer deaths in which
they have attempted to allocate the appropriate proportion to each
of several environmental exposures. These are shown in Table 6.

TABLE 6
PROPORTIONS OF CANCER DEATHS ATTRIBUTED
TO VARIOUS DIFFERENT FACTORS

|  | Percent of all Cancer Deaths | |
| --- | --- | --- |
|  | Best Estimate | Range of Acceptable Estimates |
| Tobacco | 30 | 25-40 |
| Alcohol | 3 | 2-4 |
| Diet | 35 | 10-70 |
| Food additives | <1 | -5[a]-2 |
| Reproductive[b] and sexual behavior | 7 | 1-13 |
| Occupation | 4 | 2-8 |
| Pollution | 2 | <1-5 |
| Industrial products | <1 | <1-2 |
| Medicines and medical procedures | 1 | 0.5-3 |
| Geophysical factors | 3 | 2-4 |
| Infection | 10? | 1-? |
| Unknown | ? | ? |

All of the above sections describing methodologies for assessing risk from environmental agents address only a single class of chronic diseases--cancer.  As noted above, cancer is associated with roughly 16% of deaths in the United States.  Of the quantitative causal relationship of the environment to other classes of human disease, we have only the crudest ability to produce meaningful dose-response relationships.  For an interesting discussion of the problems inherent in assessing risks to specific organ systems other than from malignant degeneration, the reader is referred to a recent symposium on health effects at chemical disposal sites.[19]

If, at this point, the reader is beginning to despair of establishing those conditions necessary to establish causal relationships between environmental agents and human disease, the writer may have too strongly emphasized the difficulties encountered in each of the techniques described above.  The wise and experienced analyst will examine <u>all</u> of the evidence related to an environmental agent and draw some shrewd judgements which should not be based on single experiments or observations.

## 5.  AN ILLUSTRATIVE EXAMPLE--THE RADIATION CASE

We now turn to a specific example to examine the commonly used risk estimate for ionizing radiation, its derivation, its weaknesses and strengths.

Research into the biologic effects of ionizing radiation began shortly after the discovery of x-rays by Professor William Roentgen in 1895.  Following World War II, the magnitude of such research escalated sharply to the extent that it now can be said that we know more about the biological effects of ionizing radiation than we do about any other environmental agent.

It is instructive to pause briefly and examine the reasons that have led to this:

1)  Dosimetry.  Unlike chemical exposures where absorption and metabolism make it difficult to estimate dose to target systems from known air, water, or food concentrations, the nature of penetrating radiation and availability of sensitive dosimeters make it rather easy to estimate dose, particularly in the experimental situation.

2)  Ample resources available for scientific research.  Because of the likely military and industrial applications of nuclear energy that could potentially expose human populations to radiation, governments made ample resources available for this work.  In the United States, several billion dollars have already been expended.

3) Exposed populations. A number of human populations had been exposed to radiation to a degree
sufficient to elicit toxic effects. These
included persons who had been subjected to radio-
therapy, occupationally exposed groups and also the
survivors of the Japanese atomic bombings[20] (extensive reviews of these experiences can be found
in references 21 and 22).

Together with this intense research effort, there was a
recognized need after World War II to assess risk and to provide guidelines for permissible exposures to the public. Other
organizations joined the International Commission on Radiation
Protection and the National Council on Radiation Protection in
monitoring the literature and in assessing risk: The National
Academy of Sciences formed the Committee on the Biological Effects
of Ionizing Radiation[20] and the United Nations formed the Scientific
Committee on the Effects of Atomic Radiation[21].

In recent decades these agencies have achieved a remarkable degree of unanimity in their estimates of radiation risk. This is not
surprising because these organizations have been reviewing the same
data and making the same assumptions. Yet today this consensus is
less clear. The validity of these assumptions has been challenged
by several recent studies and this dispute is central to the controversy over low-level radiation risk. Before moving to the
commonly accepted estimates of risk and some of the specifics of
the present controversy, it is important to review three of the
fundamental assumptions: dose rate, the so-called linear theory,
and total-body exposure.

Just as radiation effects are modified by dose, they are also
sensitive to the time over which the dose is delivered--the dose
rate. This is no different from essentially all other human
responses; for example, a given dose of alcohol, drugs, food, or
sunshine will be much less damaging if the exposure is experienced
slowly over a long period of time. Animal experimentation has shown
that protracting the exposure of some animals (low dose rates) produces a dramatic reduction in mortality compared with those animals
receiving the same exposure but at high dose rates. In fact, at a
dose less than 3 rem per week, no effects on longevity can be detected. This moderating effect of low dose rate can also be shown
in the development of cancer. Animal studies carried out in
several laboratories have shown that the development of tumors is
reduced if the dose of radiation is protracted.

The linear theory refers to a straight-line extrapolation from
high doses with demonstrable effects to low doses where there are
no demonstrable effects or, at least, questionable effects.
Studies of the radiation

effects in humans have mostly been carried out following exposures that were very intense (i.e., high dose and high dose rate). Noteable examples include the Japanese survivors of the atomic bombings and those persons treated with radiation for rheumatoid spondylitis, the two populations on whom greatest reliance is placed in developing risk estimates. But essentially no detectable effects exist in persons who have been exposed to low dose-rate radiation similar to that experienced in the operation of nuclear power plants. The nagging question then becomes how we should extrapolate from the high dose-rate exposure to the low dose, low dose-rate exposure. The assumption that has, in the past, nearly universal acceptance is that effects at low doses will be proportional to, or linear with, those that occur at high doses and high dose rates. Because among the Japanese A-bomb survivors approximately 200 cases of cancer occured after the entire group had been exposed to almost 2,000,000 person-rem, it has been assumed that each 10,000 person-rem of exposure will produce one future case of cancer, even if the average dose to each individual is small.

Many knowledgeable people believe this assumption of linearity probably leads to an exaggerated estimate of risk. In fact, the most recent report of the National Academy of Sciences shows a preference for a linear-quadratic model[20]. Two reviews of estimating cancer risk from low-level radiation exposure have recently been published[24,25].

The third assumption underlying the radiation risk estimates is that the entire body is exposed. If part of the body is shielded, particularly if the exposure is limited to an extremity, the effect of the exposure will be minimized or even absent. It is this phenomenon that permits radiotherapists to use doses to small target tissues that would be lethal if the whole body was exposed. Typically, several thousand rems are used for cancer therapy.

## 5.1  Risk Estimates:  Cancer and Genetic Effects

Having described the sources of data and the limiting assumptions, it may be useful to examine the magnitude of the risks and apply them to some occupational groups. Based on the reviews of the National Academy of Sciences and the United Nations, the risk estimate for cancer is about 100 cases per million person-rem. In other words, if a million persons are exposed to 1 rem above natural background during their lifetime, then the expected number of cancers in this group would be increased above the normally expected 200,000 cancer deaths by about 100 (i.e., from 200,000 to 200,100). Such a small increase, should it occur, could not be detected by statistical means, given the normal variability in cancer frequency.

Applying this risk estimate to the 60,000 people in the United

States who are involved with radiation in their work (excluding medical personnel) and assuming 40 years of continuous work at the average exposures experienced today (0.6 rad per year), one can calculate that the risk of cancer from occupational exposure would be increased by less than 3%. Since the frequency of cancer varies as much as 100% among areas of the United States, the risk of occupational radiation exposure, even over a lifetime, is small compared with other environmental causes of cancer.

In the early years following World War II, when little was known about the carcinogenic risks of radiation, genetic effects were considered among the more hazardous. The situation is now reversed. We now know that the cancer risks are greater and the genetic risks less than was previously thought.

Geneticists had at first thought that recessive mutations (those that require a damaged gene from both parents before becoming manifest in the offspring) might accumulate in each successive generation following several generations of exposure. The modern evidence is that such a phenomenon does not occur. Animal studies carried out over many generations, with exposures of 200 rem per generation, show no apparent change in fertility or evidence of poor health.

Furthermore, studies of the descendants of Japanese survivors of the atomic bombings show no evidence of genetic effects from the radiation exposure. Nor are genetic effects shown by any other studies. These statements, while reassuring, should not be taken to mean that radiation exposure is not mutagenic. Circumstantial evidence makes it almost a certainty that radiation can produce such an effect in humans. Our problem lies in the fact that we do not know exactly how genetic mutations will manifest themselves which makes it very difficult to adequately design experiments for their detection.

## 5.2 Fetal Effects

The human organism appears to be most fragile at the extremes of life. Whether it is air pollution, starvation, or infectious disease, the very young (particularly the in-utero fetus) are at greatest risk. The effects of radiation are no exception to this pattern. Early observations during the 1930's indicated that women who had been inadvertently treated with radiation during early pregnancy had an increased risk of bearing malformed children. Again, studies of the Japanese population who survived the atomic bombings have provided us with our best information on these effects. Of the women who were pregnant and heavily irradiated at the time of the bombings, many bore children who were mentally defective and/or had microcephaly (an underdeveloped head). Generally, the central nervous system seems to be the developing system most sensitive to radiation.

Data are too few and their variability too high to allow
any firm conclusion about the relative radiation susceptibility
of different embryologic structures within each species, but in a
few cases malformations have been ascribed to exposures as low
as 10 rem. In the vast majority of cases, the dose-response relationship is not linear, implying less effect per unit dose at
lower doses. Except for microcephaly, there are no accepted
estimates of risk because there have been no consistent findings
of defects at doses between 1 and 20 rem. For microcephaly, the
risk per rem of exposure is thought to be 1 per 1000 if the individual exposure is greater than 10 rem.

To place radiation risk in context, it is important to point out
that the normal risk to the fetus during its nine-month sojourn
in the uterus is high. Approximately 50 out of every 1000 children born alive have some developmental defect apparent at birth.
Therefore, if 100 pregnant women were exposed to 10 rem each, the
risk of a congenital malformation would be increased by only 2%
(i.e., from 50 to 51 per 1000). Airline stewardesses, whose annual occupational exposures are about 0.5 rem, would experience an
increased risk to the fetus of much less than 1%, even if they
continued to fly throughout their pregnancy.

In addition to congenital malformations, there is also evidence that the developing fetus is more sensitive than other age
groups to the carcinogenic effect of radiation. Studies in both
England and the united States have produced evidence of an increase in cancer among children exposed in utero.

6.  CONCLUSION

Assessing health risks of technology remains an inexact
science, just as our understanding of health is inexact. It is
also an art, requiring judgement and wisdom. The risk assessor
is urged to be explicit in his assumptions and value judgements
and to present ranges for his estimates where possible. If he
does so, he can make an important contribution to society and
its decision makers in their deliberations on technology
acceptance.

# REFERENCES

1.  Harris, J. E., Potential Risk of Lung Cancer from Diesel
    Engine Emissions. Report to the Diesel Impacts Study
    Committee, Assembly of Engineering, National Research
    Council, National Academy Press, Washington, D.C., 1981.

2.  Committee on Toxicology. Evaluation of the Health Risks
    of Ordinance Disposal Waste in Drinking Water. National
    Academy Press, Washington, D.C., 1982.

3.  Fishbein, L., Overview of Some Aspects of Quantitative Risk
    Assessment. Toxicol. Environ. Health, 6 1275-1296, (1980).

4.  Starr, C., Social Benefit Versus Technological Risk, Science
    165, 1232-1238, (1969).

5.  Buffler, P. A. and Aase, J. M., Genetic Risks and Environmental
    Surveillance, Journal Occup. Med. 24:305-314, (1982).

6.  Adams, P.M., Fabricant, J. D. and Legator, M. S., Cyclophos-
    phamide-Induced Spermatogenic Effects Detected in the $F_1$
    Generation by Behavioral Testing. Science 211: 80-81,
    (1981).

7.  Krewski, D. and Brown, C., Carcinogenic Risk Assessment:
    A Guide to the Literature. Biometrics 37:353-366, (1981).

8.  Jurs, P. C., Chou, J. T. and Yuan, M., Computer-Assisted
    Structure-Activity Studies of Chemical Carcinogens--Hetero-
    genous Data Set. J. Mednl. Chem. 22 276-483, (1979).

9.  Cramer, J. W., Miller, J. A., Miller, E. C., N-Hydroxylation:
    A New Metabolic Reaction Observed in the Rat with the Carcino-
    gen 2-Acetylaminofluorene. J. Biol. Chem. 235 885-888, (1960).

10. Slater, E. E., Anderson, M. D., Rosenkranz, H. S., Rapid
    Detection of Mutagens and Carcinogens. Cancer Res. 31 970-
    973, (1971).

11. Proposed System for Good Safety Assessment. Final Report of
    the Scientific Committee of the Food Safety Council, June,
    1980. Food Safety Council, 1725 K Street, N.W., Washington
    D.C. 20006.

12. Cramer, G. M. & Ford, R. A., Estimation of Toxic Hazard--A
    Decision Tree Approach. Food & Cosmetic Toxicology 16 255-
    276, 1978.

13. McCann, J. and Ames, B. N., Detection of Carcinogens as Mutagens in the Salmonella/Microsome Test: Assay of 300 Chemicals: Discussion, Proc. Natl. Acad. Sci. (U.S.A.) 73 950-954, (1976).

14. Katz, A. J., Letter to the Editor. Mutation Research 72 173-176, (1980).

15. Littlefield, L. G., Joiner, E. E.., Dufrain, R. J. Hubner, K. F., and Beck, W. J., Cytogenic Dose Estimates from In Vivo Samples from Persons Involved in Real or Syspected Radiation Exposures. K. F. Hubner, & S. A. Fry, eds. The Medical Basis for Radiation Accident Preparedness. Pages 375-390. Elsevier, North-Holland, (1980).

16. Tannenbaum, A., The Dependency of the Genesis of Induced Skin Tumors on the Caloric Intake During Different Stages of Carcinogenesis. Cancer Res. 4 673-677, (1944).

17. Ross, R. H., et. al., Scientific Rationale for the Selection of Toxicity Testing Methods: Human Health Assessment. ORNL/EIS-151, Oak Ridge National Laboratory, Oak Ridge, Tenn. 37830.

18. Doll, R. and Peto, R., The Causes of Cancer: Quantitative Estimates of Avoidable Risks of Cancer in the United States Today. JNCI 66/6:1192-1308, (1981).

19. Assessment of Health Effects at Chemical Disposal Sites. Proceedings of a Symposium Held in New York City on June 1-2, 1981. William W. Lowrence, Editor, Rockefeller University, NY (1981).

20. Moriyama, I. M., Capsule Summary of Results of Radiation Studies on Hiroshima and Nagasaki Atomic Bomb Survivors, 1945-1975. Technical Report RERF TR 5-77 Radiation Effects Research Foundation, Hiroshima, Japan.

21. Committee on the Biological Effects of Ionizing Radiation. The Effects on Populations of Exposure to Low Levels of Ionizing Radiation. National Academy of Sciences, Washington, D.C., (1980).

22. United Nations Scientific Committee on the Effects of Atomic Radiation. Sources and Effects of Ionizing Radiation. United Nations Report to the General Assembly, New York, (1977).

23. Cole, P. and Goldman, M. B., <u>Occupation in Persons at High Risk of Cancer</u>, J. Fraumeni, Ed., pp. 167-184, Academic Press, (1975).

24. Land, C., <u>Estimating Cancer Risks from Low Doses of Ionizing</u> Radiation. Science <u>209</u>:1197-1203, (1980).

25. United Nations Scientific Committee on the Effects of Atomic Radiation Dose-Response Relationships for Radiation-Induced Cancer,Twenty-Ninth Session of Unscear,Vienna, Sept. 1-12, 1980, A/Ac.82/R.377.

ISSUES RELATED TO CARCINOGENIC RISK ASSESSMENT
FROM ANIMAL BIOASSAY DATA

Kenny S. Crump

Science Research Systems, Inc.
Ruston, Louisiana 71270

## 1.  RISK ASSESSMENT METHODS

This paper considers some methodologies for estimating human
carcinogenic risks using animal data.  It must be recognized at
the outset that the data base for estimating human carcinogenic
risks is rather weak.  The primary sources of data are *in vitro*
mutagenesis studies, animal carcinogenesis bioassays, and human
epidemiologic studies.  Because many mutagens are also carcinogens, mutagenesis tests such as the Ames Test are used for
screening chemicals for carcinogenesis.  Some initial efforts
have been made to use results from these tests to estimate carcinogenic potency as well.  For example, Harris (1981) used data
from experiments on enhancement of viral transformation in Syrian
hamster embryo cells, and mutagenesis with and without metabolic
activation in L51784 mouse lymphoma cells, to estimate the human
carcinogenic potency of particulates in diesel exhaust.  These
particular estimates agree fairly well with estimates made from
a human population exposed to diesel exhaust in London bus garages.
However, the use of such mutagenesis data for estimating carcinogenic potency has not been widely accepted.

Human epidemiological data are clearly the most desirable
data for estimating potency in humans.  Even these data, however,
are often accompanied by serious problems.  Exposure levels often
are not known very accurately.  Frequently the exposed population
has had a different temporal pattern of exposure from the target
population for which risk estimates are desired.  Exposure often
will not have occurred far enough into the past for its full
effects to be manifested.  The greatest disadvantage of epidemiologic data is that humans must be exposed to potentially carcino-

genic substances for many years and possibly suffer irreparable damage, before the data are available. A major goal of risk assessment is to prevent harm to humans. Therefore, we hope that availability of positive human data will decrease in the future.

Thus, it appears that the primary source of data for estimating carcinogenic potency will be from animal carcinogenesis bioassays. Typically these tests are conducted using animals of both sexes from two species, frequently mice and rats. Animals from each sex-species category are randomly divided into a control group and one or more treatment groups. The treatment groups are exposed to constant dose rates of the chemical from weaning until late life, frequently until death. At the end of the experiment, frequencies of various types of cancers are noted as well as possibly the times of their appearance. The immediate question asked is not about the risk to humans, but rather the more limited question: "Was the chemical carcinogenic under the conditions of the test?" Even this limited question is often difficult to answer. Statistical problems which must be faced include those related to multiple comparisons, competing risks, and the interpretation of rare tumors. Discussions of these problems may be found in Fears et al. (1977), Gart et al. (1979), Peto et al. (1980); and Brown and Fears (1981).

Once it has been determined from a bioassay test that a chemical is carcinogenic, for regulatory purposes it is often required that a human "safe dose" be estimated. A safe dose is defined as one for which the risk is no larger than some specified low amount such as $10^{-5}$. The first step is to estimate a safe dose for the experimental animals. This involves fitting a dose response curve to the animal carcinogenesis response data and using this curve to predict doses which correspond to various levels of risk. The choice one makes for a dose response curve is critical because different curves can fit the data about equally well, but predict vastly different risks at low doses. Table 1 shows the fits of 5 different dose response models to benzopyrene skin painting data of Lee and O'Neil (1971). The equations for these dose response models are:

$$P(d) = \begin{cases} 1-\exp[-2.3 \times 10^{-4}(d-6)^2 - 4.1 \times 10^{-7}(d-6)^3] & \text{for } d \geq 6 \\ 0 & \text{for } d < 6; \end{cases} \quad (1)$$

$$P(d) = N(-5.806 + 3.224 \mathrm{Log}_{10}d) \quad (2)$$

where $N(x)$ is the standard normal distribution function;

$$P(d) = 1 - \exp(-9.7 \times 10^{-5}d^2 - 1.7 \times 10^{-6}d^3); \qquad (3)$$

$$P(d) = 1 - \exp(-9.5 \times 10^{-4}d - 2.5 \times 10^{-5}d^2 - 2.8 \times 10^{-6}d^3); \qquad (4)$$

and

$$P(d) = 1 - \exp(-2 \times 10^{-3}d^{0.1} - 9.6 \times 10^{-5}d^2 - 1.7 \times 10^{-6}d^3) \qquad (5)$$

Model (1) includes a threshold at a dose of d = 6 below which the risk is zero. Model (4) is called a "low dose linear" model; that is, at low doses the extra risk above background increases approximately linearly with dose or, more precisely, at d = 0 the slope of the tangent line of the graph of dose versus response probability is positive (i.e., $P'(0) > 0$). Models (2) and (3) are intermediate between threshold and low dose linear models; although there is no threshold, the slope of the tangent line is zero at d = 0 (i.e., $P'(0) = 0$). Model (5) could be called "low dose supralinear" because $P'(0) = \infty$. Table 1 reveals that these curves all fit the data of Lee and O'Neil about equally well. It would take an extremely large experiment for there to be much chance of a statistical test discriminating among these curves.

TABLE 1

COMPARISON OF OBSERVED AND EXPECTED RESPONSES FROM
SKIN-PAINTING EXPERIMENT OF LEE AND O'NEIL (1971)

| Dose Rate ( g/week) | Animals Tested | Animals Responding | Expected Responses Using Equation | | | | |
|---|---|---|---|---|---|---|---|
| | | | (1) | (2) | (3) | (4) | (5) |
| 6 | 300 | 0 | 0.0 | 0.15 | 1.2 | 2.2 | 1.9 |
| 12 | 300 | 4 | 2.5 | 3.0 | 5.0 | 5.9 | 5.7 |
| 24 | 300 | 27 | 22.2 | 26.4 | 22.8 | 21.9 | 23.5 |
| 48 | 300 | 99 | 105.0 | 100.8 | 101.8 | 101.4 | 101.3 |

However, as Figure 1 illustrates, below a dose of about 10 ppm these curves diverge dramatically. The doses corresponding to a risk of $10^{-5}$ are, in order from (1) to (5), 6.2 ppm, 2.0 ppm, 0.32 ppm, $1.1 \times 10^{-2}$ ppm, and $9.8 \times 10^{-24}$ ppm.

Calculations such as these show that one can get an extremely wide range of results by selecting different dose response curves and that one can not discriminate among these curves on the basis of how well they describe experimental data. One way out of this impasse is to determine whether some models correspond to what is

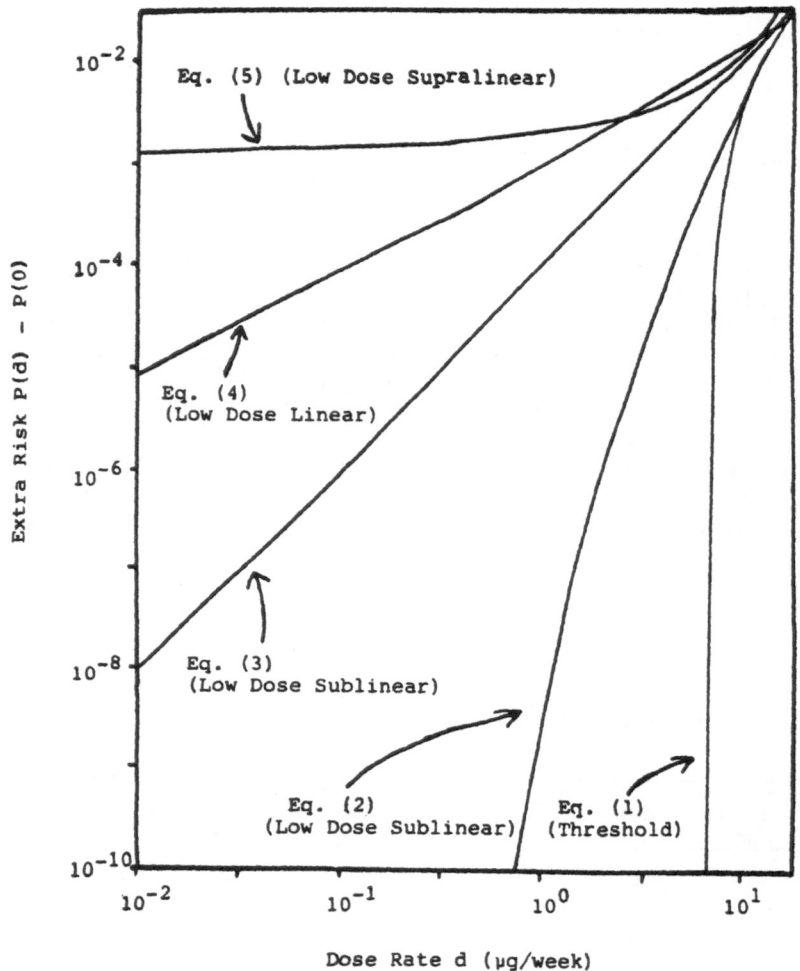

Fig. 1.  LOW DOSE PROPERTIES OF 5 DOSE RESPONSE FUNCTIONS

known about the biological mechanisms of carcinogenesis better
than others.  When this is done, it appears that low dose supralinear models such as exemplified by (5) can be disregarded; it
seems extremely unlikely that biological mechanisms could lead to
dose response curves having this shape at low doses.[1]  On the oth

---

[1] Please note that this argument does not apply to convex curves
which are linear at low doses; such models do not appear implausible.  In fact, the one-hit model of cancer, which has been use
extensively in risk assessment, is of this type.

hand, plausible and fairly general assumptions about the underlying biological mechanisms can lead to dose response curves such
as (4) which are linear at low dose (Crump and Masterman, 1979;
Peto, 1978; NAS, 1977, p. 38). Because of the possibility mutagenic origin of many cancers, a low dose linear curve shape is
also supported by the linearity of mutagenesis dose response data
(Ehrenburg, 1979).

Still, this evidence is far from conclusive, and many scientists feel that curves that predict much smaller risks at low
doses, including threshold curves, deserve consideration. For
example, in many carcinogenesis experiments, hepatocellular carcinomas appear only at very high doses associated with substantial
liver damage. In such situations, tumor induction may be a by-
product of the liver damage and might not occur at doses which
do not cause significant damage to the liver.

These considerations suggest that linear curves might be used
to set upper limits to low dose risks. Use of such limits in
setting allowable human exposure levels should tend to err on the
side of safety. In some cases these linear upper limits might
greatly overestimate low dose risks.

A linear approach was used recently by the United States Environmental Protection Agency (EPA) in setting water quality
criteria for carcinogenic substances (EPA, 1980). The particular
model used was the multistage model (Armitage and Doll, 1961;
Guess and Crump, 1976)

$$P(d) = 1 - \exp(-q_0 - q_1 d - \ldots - q_k d^k), \qquad q_i \geq 0. \qquad (6)$$

This model can be derived from plausible assumptions about the
biological processes, is flexible enough to fit a variety of curve
shapes, and is linear at low dose whenever $q_1 > 0$.

The statistical methodology applied to this model is described
in Crump (1981). The parameter k is set equal to 1 less than the
number of experimental groups, and the parameters $q_0, \ldots, q_k$ are
estimated by maximizing the likelihood of the quantal data consisting of the total numbers of animals and the total numbers of
these animals having the particular cancer under consideration.
A 95% upper limit $q_1^*$ for the parameter $q_1$ is calculated using the
asymptotic distribution of the likelihood ratio statistic.

**The** extra risk over background is defined as

$$[P(d) - P(0)]/[1-P(0)].$$

This quantity may be interpreted as the probability of a tumor
appearing after administering a dose d given that no tumor would
have appeared in the absence of any dose. For the multistage
model the extra risk is

$$1 - \exp(-q_1 d - \ldots - q_k d^k)$$

which is approximately equal to $q_1 d$ whenever d is small and $q_1 > 0$.
This implies that $q_1^* d$ is an approximate 95% upper limit on extra
risk at a dose d, and that $10^{-5}/q_1^*$ is an approximate 95% lower limit
for the dose corresponding to a risk of $10^{-5}$.

This method is a form of linear extrapolation because the
upper limit $q_1^*$ will always be positive. The procedure thus incorporates both model uncertainty and statistical uncertainty--the
former through the use of a low dose linear model and the latter
through the use of statistical confidence limits.

To illustrate the method, consider its application to the data
in Table 2 on the incidence of thyroid tumors in rats exposed to
ETU. (These data were not among those used by the EPA in setting
water quality criteria.) We note that the multistage model provides an excellent description of the data, as is frequently true
whenever the response does not decrease with increasing dose.
Table 2 also shows the maximum likelihood estimates of the multistage parameters. We note that both $q_1$ and $q_2$ are estimated as
zero and $q_3$ is estimated as being positive.

Table 3 gives both maximum likelihood estimates and 95%
upper confidence limits for extra risks corresponding to 4 dose
levels. There are several points with regard to these estimates
that are worth mentioning. The maximum likelihood estimates of
extra risk decrease very rapidly with decreasing dose--to be specific, they vary as the cube of dose. This is because, owing to
the fact that $q_1$ and $q_2$ are both estimated as zero, the maximum
likelihood estimate of extra risk is approximately $q_3 d^3$. The
upper limits on extra risk, on the other hand, vary linearly with
dose; a ten-fold reduction in dose corresponds to a ten-fold reduction in the upper limit. This results in the upper limits exceeding the maximum likelihood estimates by extremely large
amounts at low doses. Although the maximum likelihood estimate of
$q_1$ is zero, these data could have risen from a true multistage
model having a positive $q_1$. The methodology for calculating upper
confidence limits take this possibility into consideration because
they are based upon $q_1^*$, the largest value for the linear term $q_1$
which is consistent with these data (with the 95% figure measuring
the degree of consistency required). The large differences between
the upper limits and the maximum likelihood estimates reflect some
of the true uncertainty of these estimates.

TABLE 2
INCIDENCES OF THYROID CARCINOMAS
IN RATS EXPOSED TO ETU[a]

| Dietary Concentration (ppm) | Numbers of Animals | Numbers of Tumor-Bearing Animals | Expected numbers of Tumor-Bearing Animals[b] |
|---|---|---|---|
| 0 | 72 | 2 | 1.5 |
| 5 | 75 | 2 | 1.5 |
| 25 | 73 | 1 | 1.5 |
| 125 | 73 | 2 | 3.2 |
| 250 | 69 | 16 | 14.9 |
| 500 | 70 | 62 | 62.2 |

MAXIMUM LIKELIHOOD ESTIMATES OF MULTISTAGE MODEL PARAMETERS

$q_0 = .02077$      $q_3 = 1.101 \times 10^{-8}$

$q_1 = 0.0$      $q_4 = 1.276 \times 10^{-11}$

$q_2 = 0.0$      $q_5 = 0.0$

[a] Source: Graham et al. (1975).
[b] Fitting all six multistage parameters.

TABLE 3

ESTIMATES OF LOW-DOSE RISK FROM ETU
DERIVED FROM THE MULTISTAGE MODEL

| Dose Level (ppm) | Maximum Likelihood Estimates Of Additional Risk | 95% Upper Confidence Limits of Additional Risk |
|---|---|---|
| $10^{-1}$ | $1.0 \times 10^{-11}$ | $3.7 \times 10^{-5}$ |
| $10^{-2}$ | $1.0 \times 10^{-14}$ | $3.7 \times 10^{-6}$ |
| $10^{-3}$ | $1.0 \times 10^{-17}$ | $3.7 \times 10^{-7}$ |
| $10^{-4}$ | $1.0 \times 10^{-21}$ | $3.7 \times 10^{-8}$ |

## 2.  POSSIBILITIES FOR IMPROVING ESTIMATES AND REDUCING UNCERTAINTY

It is important that we make the best possible estimates of
low dose risks.  If safe doses are over-estimated, human health
may be adversely affected.  If they are estimated far lower than
what is required to protect human health, then unnecessarily stiff
economic penalties may result.  It should also be noted that
whereas the uncertainty as to the true safe dose may encompass
orders of magnitude, a change in an allowable exposure by a factor
as small as 2 or 3 can have far-reaching consequences to an affected
industry.  In this section we discuss briefly three ways the estimates discussed in the previous section might be improved.

### 2.1  Use of Time-to-Occurrence Data

The method discussed in the previous section used only the
quantal data on the total numbers of animals in the various treatment groups, and the number of these animals with tumors.  Time-
to-occurrence data such as the time to death and the time to
appearance of a tumor are also obtainable from an experiment.
Use of these data would permit the estimation of time-dependent
quantities, such as loss of life expectancy, which would in turn
allow one to account for the different consequences of late- and
early-occurring tumors.  These data might also enable one to
account more accurately for competing risks of death.

Methods for estimating safe doses from time-to-occurrence
data have been proposed by Hartley and Sielken (1977) and Daffer,
Crump and Masterman (1980).  Each of these uses maximum likelihood methods applied to the multistage model of cancer.  However,
they make different assumptions about the data and they use
different methods for computing confidence limits.  The Hartley-
Sielken model can be written as

$$P(t;d) = 1-\exp\left\{-\sum_{s=0}^{a} q_s d^s H(t)\right\}$$

where $q_i \geq 0$ and

$$H(t) = \sum_{r=1}^{b} b_r t^r, \ b_r \geq 0.$$

The Daffer et al. model is identical in form except that $H(t)$ is
not given a parametric form; $H(t)$ is an arbitrary nondecreasing
function and is estimated nonparametrically.  In the Hartley-

Sielken scheme P(t;d) represents the distribution of the time-to-tumor; in the Daffer et al. method P(t;d) represents the distribution of time to death from tumor.

An application of these methods to carcinogenicity data on p-cresidine (NCI, 1979) is given in Table 4 where maximum likelihood estimates of safe doses and 95% lower confidence limits for these safe doses are displayed. The data from which the calculations are made are given in Table 5. In the Daffer et al. analysis the underlined death times are assumed to represent deaths caused by cancer; i.e., none of the cancers are assumed to be incidental. In the Hartley-Sielken analysis it is assumed that time-to-tumor and time of death are stochastically independent; this implies that all of the cancers are assumed to be incidental. When time data are used, we must consider the age at which the estimates apply; the estimates in Table 4 all pertain to an age of 106 weeks, which corresponds to the termination of the experiment. For comparison purposes, the estimates obtained by applying the multistage methodology of Crump (1981) to the quantal data are also presented.

TABLE 4
"SAFE" DOSES OF P-CRESIDINE FROM DATA[a] ON MALE RATS, INCLUDING
TUMORS DISCOVERED AT TERMINAL SACRIFICE

| Incremental Risks | Hartley-Sielken (1977) Procedure 106 Weeks | Daffer et al. (1980) Procedure 106 Weeks | Crump (1981a) Procedure Using Quantal Data |
|---|---|---|---|
| MAXIMUM LIKELIHOOD ESTIMATES | | | |
| $10^{-4}$ | $.0^2 46$[b] | $.0^2 39$ | $.0^2 13$ |
| $10^{-5}$ | $.0^2 15$ | $.0^2 13$ | $.0^3 14$ |
| $10^{-6}$ | $.0^3 46$ | $.0^3 40$ | $.0^4 14$ |
| 95% LOWER CONFIDENCE LIMITS | | | |
| $10^{-4}$ | $.0^3 50$ | $.0^4 65$ | $.0^4 50$ |
| $10^{-5}$ | $.0^4 97$ | $.0^5 66$ | $.0^5 50$ |
| $10^{-6}$ | $.0^4 19$ | $.0^6 66$ | $.0^6 50$ |

[a] Primary tumors of the urinary bladder (NCI, 1979).

[b] $.0^2 46$ means .0046, $.0^3 15$ means .00015, etc.

TABLE 5
TIMES OF DEATH OF MALE RATS EXPOSED TO P-CRESIDINE

| DOSE (% of Diet) | INITIAL Group Size | TIMES OF DEATH IN WEEKS* | | | | | | | | | | | |
|---|---|---|---|---|---|---|---|---|---|---|---|---|---|
| 0 | 48 | 1 | 48 | 78 | 78 | 78 | 78 | 78 | 80 | 85 | 89 | 89 | 90 |
| | | 98 | 101 | 103 | 105 | 105 | 105 | 105 | 105 | 105 | 105 | 105 | 105 |
| | | 105 | 105 | 105 | 105 | 105 | 105 | 105 | 105 | 105 | 105 | 105 | 105 |
| | | 105 | 105 | 105 | 105 | 106 | 106 | 106 | 106 | 106 | 106 | 106 | 106 |
| 0.5 | 48 | 79 | 84 | 85 | 93 | 93 | 93 | 93 | 102 | 102 | 103 | 105 | 105 |
| | | 105 | 105 | 105 | 105 | 105 | 105 | 105 | 105 | 105 | 105 | 105 | 105 |
| | | 105 | 105 | 105 | 105 | 105 | 105 | 105 | 105 | 105 | 105 | 105 | 105 |
| | | 105 | 105 | 105 | 105 | 105 | 105 | 105 | 105 | 105 | 105 | 105 | 105 |
| 1.0 | 47 | 48 | 51 | 52 | 53 | 56 | 57 | 65 | 65 | 67 | 67 | 67 | 70 |
| | | 71 | 72 | 73 | 74 | 75 | 76 | 77 | 78 | 79 | 80 | 80 | 81 |
| | | 82 | 82 | 83 | 87 | 87 | 88 | 89 | 89 | 90 | 92 | 92 | 93 |
| | | 95 | 96 | 97 | 99 | 99 | 100 | 100 | 105 | 105 | 105 | 105 | |

Source: NCI (1979)

*Animals dying at underlined times had urinary bladder tumors.

In spite of the different interpretations of the data, the
maximum likelihood estimates of Hartley and Sielken and Daffer
et al. are quite similar. However, the lower confidence limits
differ considerably; whereas those of Daffer et al. vary linearly
with extra risk, those of Hartley and Sielken do not. The lower
limits of Daffer et al. are quite similar to those of Crump based
upon the quantal data. In view of the great uncertainty in quantifying human risks from animal data, this latter fact suggests
that use of time-to-tumor data will not yield substantial reduction in the uncertainty associated with low dose risk estimates.

## 2.2 Use of Pharmacokinetic Data

The "dose" used in the examples presented thus far has been
the dose to which the animal was exposed. It is possible that a
chemical carcinogen must be metabolized to an ultimate carcinogen
which combines with DNA to initiate a tumor. The chemical also
may be excreted by the body or may be metabolized into a non-
carcinogen. Once formed, the ultimate carcinogen might combine
with non-genetic material and pose no further carcinogenic threat.
These pharmacokinetic elements were incorporated by Cornfield
(1977) and by Gehring and Blau (1977) into mathematical models
relating the exposure dose to the metabolized dose reaching a
critical tissue. Both of these models predict under certain circumstances, a linear, but very slow, increase in metabolized dose
with increasing exposure dose up to an exposure at which the de-
toxification mechanisms are saturated. At exposure doses larger
than the saturation dose, the metabolized dose increases much more
rapidly with increasing exposure dose. If this situation actually
occurs in practice and if the metabolized dose could be accurately
measured as a function of the exposure dose, it is possible that
the uncertainty in low dose risk estimates could be greatly reduced.

To illustrate how this might happen, a hypothetical relationship is considered between exposure dose and metabolized dose for
ethylene thiourea (ETU), which incorporates a saturation dose such
as that predicted by the models of Cornfield and Gehring and Blau.
The results are illustrated in Figure 2. The graph in the upper
right shows the maximum likelihood fit of the multistage model to
these data using exposure dose. The graph in the lower right is
the hypothetical relationship between exposure dose and metabolized
dose curve, and indicates a saturation dose at about 200 ppm. The
graph in the lower right shows the maximum likelihood fit of the
multistage model to the data when metabolized dose is used in place
of exposure dose. The dotted line and arrows in Figure 2 illustrate how the data points (x's) are transformed when the % animals
with tumors versus exposure dose curve is transformed into the %
animals with tumors versus metabolized dose curve. The multistage
methodology for estimating safe doses can be applied using either

K. S. Crump

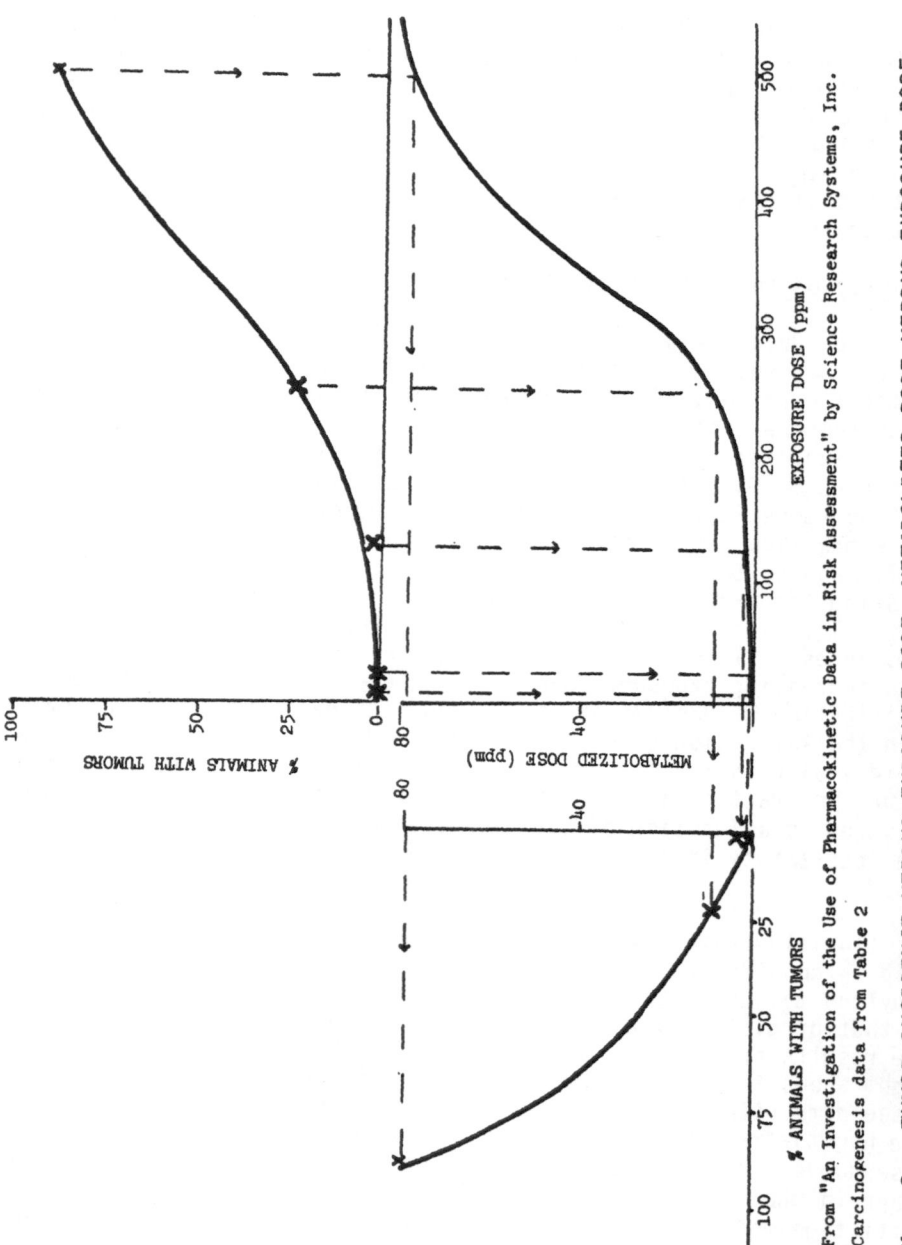

a From "An Investigation of the Use of Pharmacokinetic Data in Risk Assessment" by Science Research Systems, Inc.
b Carcinogenesis data from Table 2

Fig. 2. TUMOR INCIDENCE VERSUS EXPOSURE DOSE, METABOLIZED DOSE VERSUS EXPOSURE DOSE, AND TUMOR INCIDENCE VERSUS METABOLIZED DOSE[a,b]

the exposure doses or the metabolized doses. When exposure doses are used, the 95% lower confidence limit for the safe dose corresponding to a risk of $10^{-6}$ is $2.7 \times 10^{-3}$, as can be determined from Table 3. (From Table 3, $q_1^* = 3.7 \times 10^{-8}/10^{-4} = 3.7 \times 10^{-4}$ and therefore the 95% lower limit on dose corresponding to a risk of $10^{-6}$ is $10^{-6}/3.7 \times 10^{-4} = 2.7 \times 10^{-3}$.) When metabolized doses are used, the lower limit depends critically upon the slope of the graph of metabolized dose versus exposure dose below the saturation exposure dose of 200 ppm. When this slope is $10^{-2}$, the lower limit for metabolized dose is $3.4 \times 10^{-5}$ which translates into a lower limit for the exposure dose of $100 \times 3.4 \times 10^{-5} = 3.4 \times 10^{-3}$ ppm. This is larger than the safe dose obtained previously by 25%. However, if this slope is $10^{-4}$, the lower limit for the metabolized dose is still about $3.4 \times 10^{-5}$, but this now translates into an exposure dose of 0.34 which is over 100 times larger than that obtained from applying the multistage model to the exposure dose. Thus, if there is a saturation dose as predicted by the models of Cornfield and Gehring and Blau, application of the multistage methodology to these metabolized doses can in certain instances yield much larger, and probably much more accurate, safe doses than would be obtained from application of this methodology to the exposure doses.

This analysis assumed the metabolized dose versus exposure dose curve was known with certainty. In a real application an account would have to be made of the uncertainty in the estimate of this curve.

## 2.3 Choice of Experimental Design

A third way that uncertainty can possibly be reduced is through the use of an adequate experimental design. Here experimental design refers only to the selection of experimental doses and the proportioning of animals among these doses. Experimental design procedures for making these choices have been studied by Krewski and Kovar (1979); Krewski, et al. (1981); Wong (1979); Hoel and Jennrich (1979); and Crump (1982). These procedures are based upon various optimal design criteria, but they all require some prior estimate of the dose response function. This estimate might come from a pilot study or else experience with a similar substance. To illustrate the utility of optimal design methods, consider the problem of designing an experiment to maximize the expected 95% lower limit for the safe dose obtained from an application of the multistage methodology. This is a reasonable design criteria because the 95% lower limit for safe dose calculated from an experiment with this experimental design would still be a true 95% lower limit; however, it would avoid extreme conservatism that might result from a poor design.

Suppose 150 animals are available, the highest dose allowable
is 2500 ppm, and the prior estimate of the dose response function
is

$$P(d) = 1 - exp(-1.07 \times 10^{-7} d^2).$$

No analytic procedure for obtaining the optimal design has been
devised. However, an approximate solution can be obtained by
selecting various designs and calculating the expected safe dose
lower limit by simulation. Various designs and the corresponding
expected lower limits for the safe dose are given in Table 6.
Design I gives the largest expected lower limit. The expected
lower limit from Design I is larger than that of the more conventional Design II by a factor of almost 3. Design I was obtained
from the design methodology of Crump (1982) which uses a different
optimality criterion; intuitively, it should be nearly optimal for
this purpose as well.

TABLE 6
LOWER BOUNDS ON SAFE DOSE CORRESPONDING TO A RISK OF $10^{-6}$
FOR VARIOUS DESIGNS, ASSUMING A TWO-HIT DOSE
RESPONSE FUNCTION WITHOUT BACKGROUND[a,b]

|  |  |  |  | 95% LOWER BOUNDS | |
| --- | --- | --- | --- | --- | --- |
|  |  |  |  | Mean[c] | Variance |
| **DESIGN I** | | | | | |
| D:[d] |  | 474 | 2500 | $1.4 \times 10^{-2}$ | $6.4 \times 10^{-5}$ |
| N:[d] |  | 107 | 43 |  |  |
| **DESIGN II** | | | | | |
| D: | 0 | 1250 | 2500 | $5.8 \times 10^{-3}$ | $9.3 \times 10^{-6}$ |
| N: | 50 | 50 | 50 |  |  |
| **DESIGN III** | | | | | |
| D: | 1250 | 2500 |  | $7.4 \times 10^{-3}$ | $1.3 \times 10^{-5}$ |
| N: | 75 | 75 |  |  |  |
| **DESIGN IV** | | | | | |
| D: | 700 | 2500 |  | $1.2 \times 10^{-2}$ | $3.6 \times 10^{-5}$ |
| N: | 115 | 35 |  |  |  |
| **DESIGN V** | | | | | |
| D: | 1250 | 2500 |  | $7.8 \times 10^{-3}$ | $1.4 \times 10^{-5}$ |
| N: | 100 | 50 |  |  |  |

[a] From "Investigation of Theoretical Experimental Design Methodologies
for Carcinogenesis Bioassays" by Science Research Systems, Inc. (USEPA
Contract 68-01-5975).
[b] Specifically, $P(d) = 1 - exp(-1.07 \times 10^{-7} d^2)$.
[c] Each entry is calculated from 100 simulations.
[d] D = doses; N = number of animals.

Simulation studies such as this could be used to study other design criteria as well. They could also be used to study the important question of the robustness of the experimental design with respect to *a priori* specification of the dose response curve. These simulation studies are easy to carry out, relatively inexpensive, and I believe their wider use might significantly improve carcinogenesis bioassay designs.

# REFERENCES

Armitage, P., and Doll, R., 1961, Stochastic models for carcinogenesis, in: Proceedings of the Fourth Berkeley Symposium on Mathematical Statistics and Probability, Vol. 4, pp. 19-38, University of California Press, Berkeley, California.

Brown, C. C., and Fears, T. R., 1981, Exact significance levels for multiple binomial testing with application to carcinogenicity screens (sumbitted to Biometrics).

Cornfield, J., 1977, Carcinogenic risk assessment, Science, 193:693-699.

Crump, K. S., 1981, An improved procedure for low-dose carcinogenic risk assessment from animal data, Journal of Environmental Pathology and Toxicology, Vol. 5, No. 2, 675-684.

Crump, K. S., 1982, Designs for discriminating between binary dose response models with applications to animal carcinogenicity experiments, Communications in Statistics, 11(4), 375-393.

Crump, K. S., and Masterman, M. D., 1979, Review and Evaluation of Methods of Determining Risks from Chronic Low Level Carcinogenic Insult, Environmental Contaminants in Food: 154:165, Congress of the United States Office of Technology Assessment, Library of Congress Catalog Card No. 79-600207.

Daffer, P. Z., Crump, K. S., and Masterman, M. D., 1980, Asymptotic theory for analyzing dose response survival data with application to the low-dose extrapolation problem, Mathematical Biosciences, 50, 207-230.

Ehrenberg, L. S., 1979, Risk assessment of ethylene oxide and other compounds, in: The Banbury Report, V. K. McElheny and S. Abrahamson (eds), Cold Spring Harbor Laboratory, Lloyd Harbor, New York, 157-190.

EPA, 1980, Water quality criteria documents, Availability Federal Register 45, No. 231, Friday, Nov. 28, 79317-79379.

Fears, T. R., Tarone, R. E., and Chu, K. C., 1977, False positive and false negative rates for carcinogenicity screens, Cancer Research, 37, 1949-1945.

Gart, J. J., Chu, K. C., and Tarone, R. E., 1979, Statistical issues in interpretation of chronic bioassay tests for carcino-genicity, J. Nat. Cancer Institute, 62, 957-974.

Gehring, P. J., and Blau, G. E., 1977, Mechanisms of carcinogene-sis: dose response, J. of Environmental Pathology and Toxi-cology, 44, 581-591.

Graham, S. L., Davis, K. J., Hansen, W. G., and Graham, C. H., 1975, Effects of prolonged ethylene thiourea ingestion of the thyroid of the rat, Food and Cosmetics Toxicology, 13, 493-499.

Guess, H. A., and Crump, K. S., 1976, Low-dose extrapolation of data from animal carcinogenesis experiments--analysis of a new statistical technique, Mathematical Biosciences, 32, 15-36.

Harris, J. E., 1981, Potential Risk of Lung Cancer from Diesel Engine Emissions, Report to the Diesel Impacts Study Committee, Academy of Sciences, Washington, D. C., 62 pages.

Hartley, H. O., and Sielken, R. L., 1977, Estimation of "safe doses" in carcinogenic experiments, Biometrics, 33, 1-30.

Hoel, P. G., and Jennrich, R. I., 1979, Optimal design for certain cancer problems, Biometrika, 66, 307-316.

Krewski, D., Kover, J., and Arnold, D., 1981, Optimal experimental designs for low dose extrapolation (submitted).

Krewski, D., and Kovar, J., 1979, Low dose extrapolation under single parameter dose response models (submitted).

Lee, P. N., and O'Neill, J. A., 1971, The effect both of time and dose applied on tumour incidence rate of benzopyrene skin painting experiments, British J. of Cancer, 25, 759-770.

National Academy of Sciences, 1977, Drinking Water and Health, NAS Press, Washington, D. C., 939 pages.

NCI, 1979, Bioassay of P-Cresidine for Possible Carcinogenicity, National Cancer Institute Carcinogenesis Technical Report Series No. 142.

Peto, R., 1978, Carcinogenic effects of chronic exposure to very low levels of toxic substances, Environmental Health Perspectives, 22:155-159.

Peto, R., Pike, M. C., Day, N. E., Gray, R. G., Lee, P. N., Parish, S., Peto, J., Richards, S., and Wahrendorf, J., 1980, Guidelines For Simple, Sensitive Significance Tests for Carcinogenic Effects in Long-Term Animal Experiments, in: Long-Term and Short-Term Screening Assays for Carcinogens: A Critical Appraisal, IARC Monographs on the Evaluation of the Carcinogenic Risk to Humans, Supplement 2, International Agency for Research on Cancer, Lyon, France, Annex, 311-426.

Wong, S. C., 1979, Design for low dose extrapolation of carcinogenicity data, Technical Report No. 24, Department of Statistics, Stanford University, Stanford, California.

ENGINEERING RISK ANALYSIS


W. E. Vesely

Battelle Columbus Laboratories
Columbus, Ohio 43201


## 1. INTRODUCTION

"Risk" according to Webster's dictionary means a "danger" or
"hazard." Risk is always associated with an undesirable event
which can produce harmful consequences. Risk involves both the
frequency of the undesirable event and the severity of the consequences. When we say nuclear reactor risk is too large, we mean
the frequency of accidents is too large or the consequences of
the accidents are too severe - or both.

Engineering risk analysis is the art and science of
estimating the frequency and physical consequences of the undesirable event which produces the risk. By physical consequences,
we mean those physical outcomes of the event which can produce
harm to plant, animal, and human life. We do not include as part
of the physical consequences the actual health effects which can
result from the physical consequences. In this book, we consider
the evaluation of these health effects to be a separate type of
risk analysis which we term health risk analysis and which we
separate from engineering risk analysis.

Consider again nuclear reactor risk and consider the particular part of nuclear reactor risk which is associated with
reactor accidents. In performing an engineering risk analysis of
nuclear reactor accidents, we will be concerned with calculating
the frequencies and physical consequences of the accidents. The
frequency evaluations will involve constructing various accident
scenarios and quantifying the likelihood of these scenarios. The
physical consequence evaluations will involve estimating the
physical damage to the nuclear plant and estimating the amount of

radioactivity which escapes from the plant and is deposited on the land and is taken up by plant, animal, and human life. The evaluation of the health effects produced by the radioactivity will be the concern of health risk analysis which is treated elsewhere.

## 2. ACCIDENT FREQUENCY EVALUATIONS

The frequency, or rate, of an event is the probability that the event will occur in some unit time interval. The events of concern in engineering risk analyses are occurrences of fatalities, accident occurrences, and occurrences of severe natural events such as earthquakes, floods, and hurricanes. Fatality rates, accident frequencies, and natural event frequencies are usually expressed in units of per hour or per year. For example, according to the statistics given in The Great International Disaster Book,[1] the frequency of catastophic floods in the U.S. from 1938 to 1977 was 1 per year (a catastrophic flood was defined as one killing 10 or more people).

If we are defining the accident frequency for a single facility, then the frequency is expressed as per unit time per facility to explicitly denote the application to the single facility. For example, in the Nuclear Regulatory Commission's (NRC's) safety study of risks at nuclear power plants,[2] the accident frequency of having a core melt was calculated to be $5 \times 10^{-5}$ per reactor per year. Similarly, according to Accident Facts,[3] the frequency of a person dying from a fall, averaged over the U.S. population, is $6 \times 10^{-5}$ per person per year (the "facility" here is the person).*

The frequency of an event is one variable characterizing the risk; the other variable is the consequence. Associated with the event is a consequence or a range of possible consequences. If every event that is being analyzed has the same consequence, then the frequency of the event is really all that is necessary to characterize the hazard or risk of the event. An example of a single consequence event is a fatality, or a death, and the frequency of fatality is often used to characterize the risk of dying. If an event has different consequences, then the event frequency is not enough, and a characterization of the consequences is also necessary to describe the risk associated with the event. An example of a variable consequence event is a hurricane occurrence which can have different consequences with different likelihoods.

---

*The symbol $6 \times 10^{-5}$ is the exponent notation and is the same as 0.00006. The exponent of ten ignoring the negative sign (5 for our example) indicates how many places to the left the decimal place must be moved in the first number (6 for our example).

In engineering risk analyses, various techniques are used to calculate the frequencies and consequences of events. These techniques include:

1. Statistical analysis of past events having similar consequences,

2. Extrapolation techniques which extrapolate past occurrences of less severe events,

3. Event tree techniques which connect basic events to form accident sequences, and

4. Fault tree techniques which decompose the event into more basic causes.

We will now briefly discuss each of these approaches.

## 2.1 Statistical Analysis of Past Events

Statistical analysis of past events is performed when the risk-causing event has occurred before and has resulted in consequences at least as large as those of concern in the present risk assessment. Statistical analysis is what is used to estimate the various accident statistics and natural event statistics which are given in various reference documents. The type of analysis performed is often termed "actuarial risk analysis" or "historical risk analysis." The statistics which are usually estimated are the fatality rates (frequencies) for different population cross sections and for different causes. In addition, frequencies of natural catastrophies and frequencies of accidents versus consequences which have occurred in the past are tabulated. As source books for these statistics, the reader may wish to refer to 3,4, and 5 which tabulate the risks of deaths to individuals of different population cross sections. References 1 and 5 give frequencies of natural events and some man-made disasters causing various consequences. Table 1, for example, taken from references 3 and 5, gives the frequencies of death per person per year for various causes for various age groups in the U.S.

Various text books are available which describe the statistical methods which are used to estimate the frequencies and fatality rates given in the above source documents. References 6,7 and 8, for example, describe the theoretical statistical aspects and 9 and 10 describe the approaches from a demographic viewpoint. For all these applications, the basic estimation approach is rather straightforward. The frequency of a risk-causing event is estimated by taking the number of event occurrences in some time period and dividing by the appropriate "exposure time" in that time period.

TABLE 1. FREQUENCIES OF LEADING CAUSES OF DEATH FOR THE U.S. POPULATION IN 1977 (IN UNITS OF DEATHS PER YEAR PER 100,000 PEOPLE)

| | No. | Rate | | No. | Rate |
|---|---|---|---|---|---|
| All Ages | 1,899,597 | 878 | 25 to 44 Years | 103,042 | 182 |
| Heart disease | 718,850 | 332 | Accidents | 23,460 | 42 |
| Cancer | 386,686 | 179 | Motor vehicle | 13,031 | 23 |
| Stroke | 181,934 | 84 | Drowning | 1,690 | 3 |
| Accidents | 103,202 | 48 | Poison (solid, liq) | 1,349 | 2 |
| Motor vehicle | 49,510 | 23 | Fires, burns | 1,081 | 2 |
| Falls | 13,773 | 6 | Falls | 956 | 2 |
| Drowning | 7,126 | 3 | Other | 5,353 | 10 |
| Fires, burns | 6,357 | 3 | Cancer | 16,753 | 30 |
| Other | 26,436 | 13 | Heart Disease | 14,392 | 25 |
| Under 1 Year | 46,975 | 1,485 | 45 to 64 Years | 437,795 | 1,000 |
| Anoxia | 10,604 | 335 | Heart disease | 153,552 | 351 |
| Congenital anomolies | 8,420 | 266 | Cancer | 132,514 | 303 |
| Complications of preg- | | | Stroke | 22,925 | 52 |
| nancy and childbirth | 5,786 | 183 | Accidents | 19,167 | 44 |
| Immaturity | 3,714 | 117 | Motor vehicle | 8,000 | 18 |
| Pneumonia | 1,665 | 53 | Falls | 2,245 | 5 |
| Accidents | 1,173 | 37 | Fires, burns | 1,481 | 4 |
| Ingestion of food | 275 | 9 | Drowning | 940 | 2 |
| Motor vehicle | 253 | 8 | Surg. complications | 865 | 2 |
| Mech. suffocation | 206 | 6 | Other | 5,636 | 13 |
| Fires, burns | 159 | 5 | Cirrhosis of liver | 17,166 | 39 |
| Other | 280 | 9 | Suicide | 8,368 | 19 |
| 1 to 4 Years | 8,307 | 69 | 65 to 74 Years | 445,595 | 3,054 |
| Accidents | 3,297 | 27 | Heart Disease | 182,354 | 1,250 |
| Motor vehicle | 1,219 | 10 | Cancer | 115,587 | 792 |
| Drowning | 650 | 5 | Stroke | 37,896 | 260 |
| Fires, burns | 608 | 5 | Diabetes mellitus | 9,611 | 66 |
| Ingestion of food | 168 | 1 | Accidents | 9,006 | 62 |
| Falls | 121 | 1 | Motor vehicle | 3,060 | 24 |
| Other | 531 | 5 | Falls | 1,995 | 14 |
| Congenital anomalies | 1,066 | 9 | Fires, burns | 843 | 6 |
| Cancer | 631 | 5 | Surg. complications | 767 | 5 |
| 5 to 14 Years | 12,579 | 35 | Ingestion of food | 447 | 3 |
| Accidents | 6,305 | 17 | Other | 1,894 | 13 |
| Motor vehicle | 3,142 | 9 | Pneumonia | 8,335 | 57 |
| Drowning | 1,110 | 3 | Cirrhosis of liver | 6,208 | 43 |
| Fires, burns | 550 | 1 | 75 Years and Over | 797,318 | 8,941 |
| Firearms | 344 | 1 | Heart Disease | 366,141 | 4,106 |
| Other | 1,159 | 3 | Stroke | 116,753 | 1,309 |
| Cancer | 1,733 | 5 | Cancer | 116,675 | 1,308 |
| Congenital anomalies | 676 | 2 | Pneumonia | 30,487 | 342 |
| 15 to 24 Years | 47,986 | 117 | Arteriosclerosis | 23,683 | 266 |
| Accidents | 25,619 | 63 | Accidents | 15,175 | 170 |
| Motor vehicle | 18,092 | 44 | Falls | 7,762 | 87 |
| Drowning | 2,150 | 5 | Motor vehicle | 2,713 | 30 |
| Poison (solid, liq) | 709 | 2 | Surg. complications | 1,030 | 12 |
| Firearms | 665 | 2 | Fires, burns | 1,023 | 11 |
| Other | 4,003 | 10 | Ingestion of food | 725 | 8 |
| Suicide | 5,565 | 14 | Other | 1,924 | 22 |
| Homicide | 5,196 | 13 | Diabetes mellitus | 13,993 | 157 |
| Poison (solid, liq) | 709 | 2 | Emphysema | 6,190 | 69 |
| Firearms | 665 | 2 | | | |
| Other | 4,003 | 10 | | | |
| Suicide | 5,565 | 14 | | | |
| Homicide | 5,196 | 13 | | | |

If

F = the estimated event frequency, (1)

N = the number of events which have occurred
in some past time period, and (2)

D = the total "exposure time" in the observed
time period, (3)

then simply

$$F = \frac{N}{D}.$$ (4)

Equation (4) is the basic formula which is used in statistical risk analyses of past events.

The exposure time, D, in Equation (4) includes the number of units, or individuals, which were exposed to the possible event occurrence during the observed time period. If we are estimating the failure frequency for a particular type of component (such as a valve) at a chemical plant, then D is the total operating time for all the components of that type and N is the total number of failures of those components. If F is the frequency of death from a particular cause for a specific group of people, then D is the total lifelengths of the people in that time period and N is the number of these people who died from the cause.

We should interpret the frequency given by Equation (4) as being an "average" frequency since it involves an averaging of any detailed behavior of the units or individuals which are collected together to obtain the statistics N and D. For example, the frequence of dying from cancer for the whole U.S. population is an average of the individual frequencies for the different people combined together to form this average statistic. If the actual frequency significantly varies over the time period, then the estimate given by Equation (4) also will be a time average of the frequency over that time period. For example, the frequency of a person dying from a fall over his lifetime is an average of the frequencies for each year of his life.

Frequencies of risk-causing events are often assumed to be constant with time. The Poisson probability model for event occurrencies makes this assumption and is reasonably valid for many "rare" event occurrences such as catastrophic natural events. In various instances, however, frequencies are not constant. Figure 1, for example, gives the frequency of fatality versus a person's age which is sometimes referred to as the bathtub curve ("time" here is the time from birth). To model time behavior,

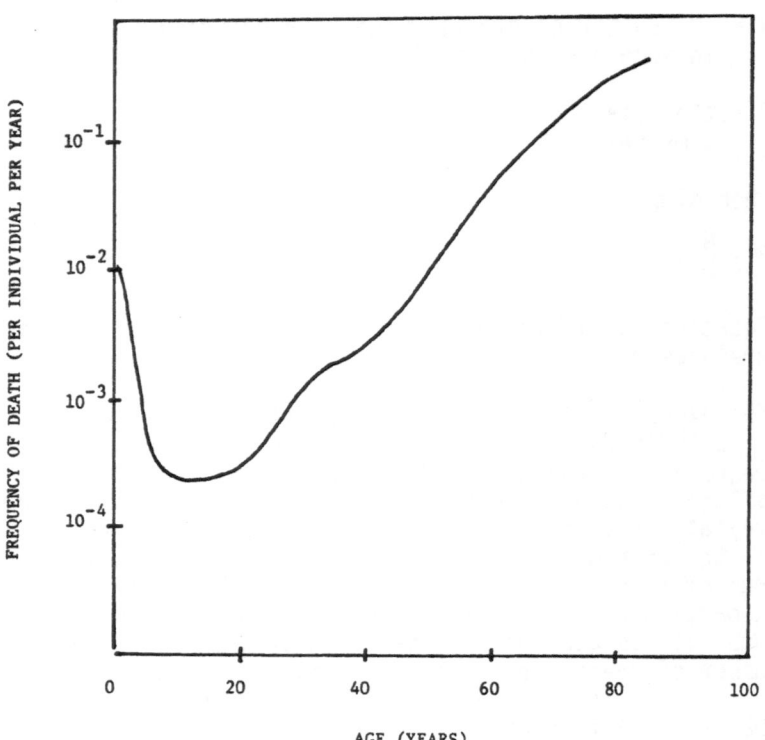

FIGURE 1.   FREQUENCY OF DEATH VERSUS AGE
(AVERAGED OVER U.S. POPULATION)

various approaches can be used, such as assuming a parametric
function and estimating the parameters from past data using
standard statistical techniques. References 6 and 7 describe
in more detail the Poisson probability model and the approaches
which are used when frequencies vary with time.

From the frequency we can calculate various other risk-
associated quantities. The expected number of risk-causing
events in some arbitrary time period is the frequency integrated
over that time period. If the frequency is assumed constant, as
is often done, then the expected number of events is simply the
frequency times the time period. If the expected number of
events over time period is less than one, then the expected
number can also be used as an approximation for the probability
of the event occurring in that time period. This approximation
is quite accurate (relative error less than 5%) if the numbers
are less than 0.1. If the expected number of events is greater
than one, then the appropriate probability model must be used to
obtain the probability of a given number of events occurring in
the time period.

In addition to applying to events with a single consequence,
Equation (4) is applicable to events of varying consequence. If
the events which have occurred have different consequences, then
the frequency as calculated by (4) pertains to all events which
can occur having the range of consequences which are associated
with the past events (i.e., which are associated with all the
past occurrences N used in calculating F). If we are interested
in the frequency of an event of a specified consequence such as
the frequency of fires killing more than 10 people, then we only
count the number of occurrences N with this consequence in
applying Equation (4) to calculate the frequency.

By choosing successively larger consequences values and
calculating frequencies of events having consequences greater
than these values, we can generate a plot of frequency versus
consequence. We can then draw a smooth curve through these
points to better show the behavior. This type of curve is often
constructed and is called in statistical terminology "the complementary cumulative distribution function" or simply "CCDF."
Of course, we can construct other curves, such as histograms,
giving frequencies in specified consequence intervals. Figure 2
taken from reference 5 shows the frequency of fires occurring in
the U.S. versus the number of people killed. The frequency shown
is the frequency of fires which killed <u>more</u> than the number of
people indicated on the X axis.

With the calculation of the frequency and frequency versus
consequence curves, we have completed our statistical data analysis tasks. Any additional tasks we have in our engineering

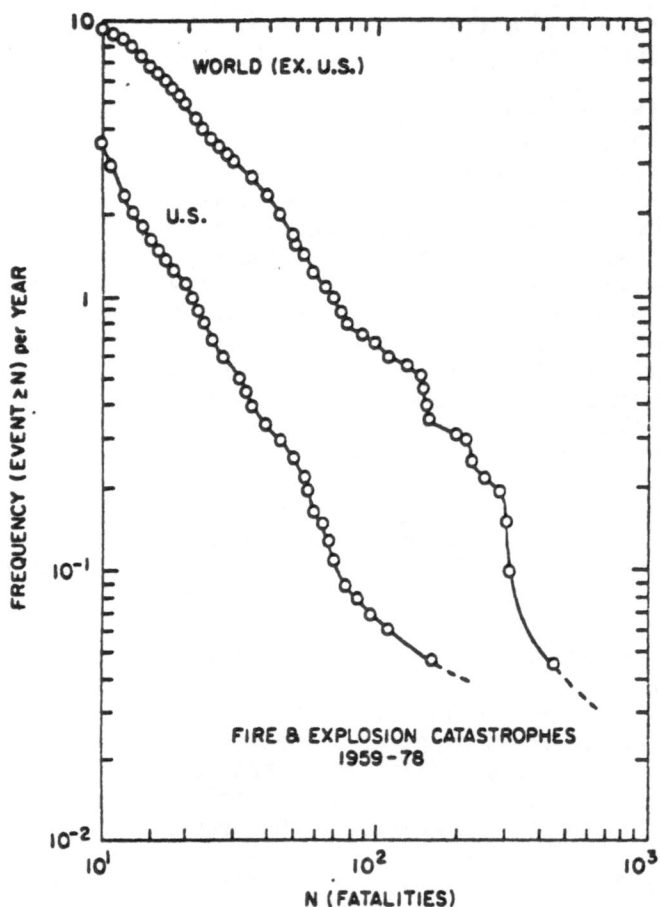

FIGURE 2.    FREQUENCY OF FIRES VERSUS
            THEIR CONSEQUENCES

risk analysis will be to obtain further numbers, or indices, which help to additionally characterize our results. As we have stated, for single consequence events, the frequency is the "risk." For example, from Table 1 in this section, the "risk" of dying from a fall for the U.S. population is $6 \times 10^{-5}$ per person per year. To depict the precision in our frequency estimate, it is important that we also calculate uncertainties or confidence intervals associated with our frequency estimates. This is done using standard statistical approaches as described in references 6-8.

If the event has varying consequences and we compute a frequency versus consequence curve, then we can obtain summary statistics (indices) from this curve. A commonly computed statistic is the frequency-averaged consequence which is sometimes called the "frequency times consequence." If $F_i$ is the frequency for events having consequences in a certain interval and if $C_i$ is a characteristic consequence of this interval (say the midpoint), then the frequency-averaged consequence $\bar{C}$ can be calculated as

$$\bar{C} = F_1 C_1 + F_2 C_2 + \ldots + F_N C_N . \qquad (5)$$

The sum on the right-hand side of Equation (5) is taken over all the consequence intervals into which the consequence range is divided.* The average consequence $\bar{C}$ is often termed the "risk" for consequence-varying events, however, we must remember it is only a characterization of the frequency versus consequence curve. Again, it is important to calculate uncertainties associated with our frequency-consequence curve or any summary statistics, and these uncertainty descriptions are obtained using standard statistical techniques.

## 2.2  Extrapolation Techniques

If we are concerned with events having extreme consequences, and if we've only observed events of lesser consequence, then we can't use the techniques of the previous section. If we want to use the information on the past occurrences to predict the risks of extreme consequence events, then we must extrapolate from the less severe to the more severe. Extrapolation techniques, or distribution techniques as they are sometimes called, are the body of techniques which estimate the frequency of severe events by smoothly extrapolating from less severe events which have occurred in the past.

---

*If we have a continuous frequency versus consequence curve (density function), then we can integrate the frequency times consequences to obtain the frequency-averaged consequence.

The word "smoothly" is very important in describing the concepts and approaches which are used. In applying extrapolation
techniques, the more severe events (catastrophies) are assumed
to be caused by the same physical mechanisms and processes which
caused the less severe events. The only difference is that the
catastrophic events are assumed to be more extreme realizations
of the same processes which caused the less severe events. In
using extrapolation techniques, we therefore assume a continuous
behavior in the processes and consequences from less severe to
more severe events.

Extrapolation techniques can be used to predict the frequencies of catastrophic floods from occurrences of minor flooding,
the frequencies of catastrophic earthquakes from minor earthquakes,
the frequencies of extremely high winds from occurrences of those
of lesser severity, and the frequencies of catastrophic structural
failures from occurrences of cracks and breaks. Figure 3 taken
from 2 shows the kinds of results which are obtained using extrapolation techniques. Figure 3 depicts the frequency of aircraft
crashes in the U.S. killing more than the number of people
indicated on the x-axis; the solid line is an extrapolation to
higher consequences based on an assumed extrapolation model.[*]

It is shown in references 11 and 12 that if the consequence
of the risk-causing event is related to the maximum value or
minimum value of some underlying process, and if certain
assumptions are valid, then a theory called "extreme value theory"
can be applied to extrapolate the event frequencies. It is shown
in extreme value theory that only six probability distributions
can describe the frequency of events which are the minimum or
maximum of infinitely many underlying independent processes. In
practice, extreme value theory is approximately valid if there
are a large number of processes whose maximum (or minimum) output
describes the consequence of the event and if there is not strong
dependency among the processes.

By viewing catastrophic events as being the maximum of a
sequence of recorded happenings, extreme value theory has been
applied to a variety of problems. For example, in predicting
frequencies of catastrophic floods on a given river, the maximum
river height is observed each year, and extreme value theory is
applied to predict the probability that the maximum flood height
in future years will exceed specified catastrophic flood levels.
This same reasoning is applied to predict maximum fire consequences,

---

[*]Only several of the historical data points are shown in the
figure; the crashes analyzed were those occurring from 1960-1973.
The figure is updated in reference 5 to include airline crashes
through 1978.

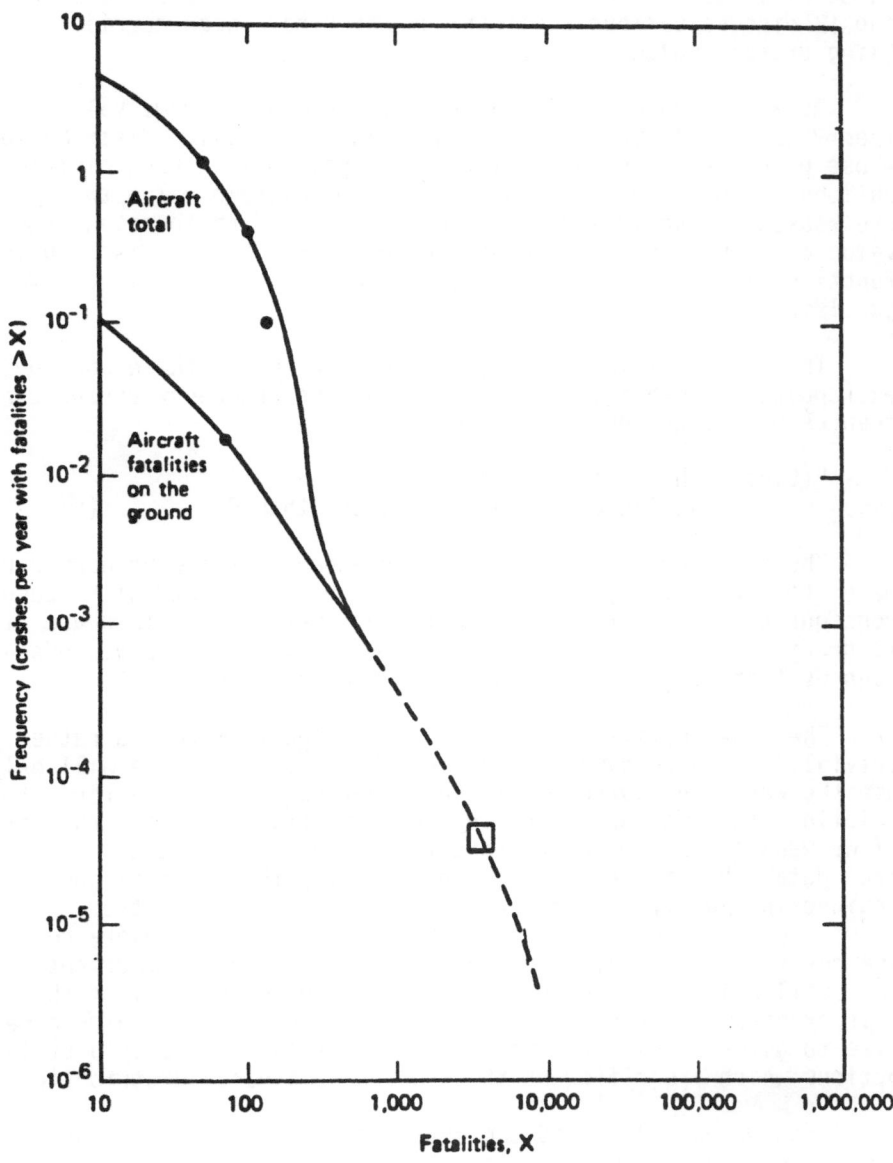

FIGURE 3. FREQUENCY OF AIRPLANE CRASH CONSEQUENCES

maximum tornado consequences, and maximum hurricane consequences which will occur in a given future time period. References 11 and 12 discuss various applications which have been carried out using extreme value theory.

As a consequence-extrapolation technique, extreme value theory can simply be viewed as supplying probability distributions whose parameters are to be fitted from past data; these distributions can be used if the assumptions of extreme value theory are reasonably satisfied by the data. Instead of these extreme value distributions, we can use any other continuous distribution functions to extrapolate to more severe consequences if we have justification for these functions.

The basic estimation process involved in all these continuous extrapolation techniques is to describe the frequency versus consequences by a function $F(C,\alpha)$ where

$F(C,\alpha)$ = the frequency of events
             having consequences greater than C.          (6)

The function $F(C,\alpha)$ is specified except for the unknown parameter (or parameters) $\alpha$. Having observed past events with measured consequences $C_i$, we can estimate the parameter $\alpha$ and then use it to predict the frequencies of events with extremely large consequences (catastrophies) which have not occurred yet.

The above estimation process is straightforward and rather trivial. If we're simply smoothing the data points and will only use the resulting curve of frequency versus consequence for interpolating, then we can choose any function that best fits the data. If we want to extrapolate beyond the largest consequences observed from data, then the critical points are (1) the physical and engineering justification for the form of $F(C,\alpha)$, (2) the assumption of smooth, continuous behavior from less severe consequences to catastrophic consequences, and (3), also important, the availability of measurements of the consequences $C_i$ of the past events. References 13,14 and 15 discuss models which give rise to alternative forms for the shape of the frequency distributions which are different than the extreme value distributions.

Once we have obtained the extrapolated frequency versus consequence curve, we have basically completed our engineering risk analyses. As for the frequencies calculated in the previous section, it is important to calculate the error spreads or confidence intervals for our frequency versus consequence curve which describe the uncertainties due to the data and due to extrapolation. The error spreads become larger and larger as extrapolation is made to larger and larger consequences and at some point the error spreads will become so large as to render the

curve meaningless. The statistical techniques used to calculate
the error spreads are given in 7 and 11. Finally, as in the
previous section, we can calculate summary risk statistics, such
as the average consequence, and the basic approaches given in the
previous section are also applicable here.

## 2.3  Event Tree Techniques

Event tree and fault tree techniques are used when direct
data does not exist on past occurrences of the event of interest
and extrapolation techniques are judged not to be applicable
(because the severe accident being analyzed is believed to involve different phenomena than those causing lesser accidents or
because specific extrapolation distribution functions are not
justifiable). The event tree and fault tree techniques are so
named because of the logic "tree" structures which each produce
to describe the basic events which must occur to cause the
accident or catastrophy to occur.

The event tree technique is an inductive logic technique
which can be applied when a chain of events must occur in order
for the accident to occur. A nuclear power plant, for example,
has various installed safety systems which are supposed to
protect against the release of radioactivity to the environment.
For an accident to occur which releases large amounts of radioactivity, an initiating event must first occur, such as a pipe
break or a fire, and in addition the safety system must fail to
operate as they are intended.

In event tree terminology, the initiating event and subsequent chain of system failures is termed an accident sequence.
More precisely, an accident sequence consists of a defined
initiating event and specific systems failing and other systems
succeeding which determines the severity of the accident. Consider
our hypothetical nuclear plant and assume it has two safety
systems, System A and System B, which are to protect against
reactor coolant pipe breaks. In reality, a nuclear power plant
has many more safety systems, however, we don't want to complicate
our discussion here. One possible accident sequence for our hypothetical nuclear plant would be

(Pipe break occurs) AND (System A fails) AND
(System B succeeds).

In addition to the above accident sequence, however, we can
define accident sequences for all combinations of System A and B
succeeding or failing. To facilitate the identification of all
the possible accident sequences, an event tree table is constructed.
At the top of the table we identify the initiating event, in our
case a pipe break, and then label the systems which can be called

upon to mitigate the accident when the initiating event occurs.
We then construct the event tree, which is simply a depiction of
all the possible accident sequences, by connecting the "forks"
which represent the possibilities for each system performance.
A particular accident sequence in the event tree is then a particular path from initiating event through each system's performance state to the final accident outcome.

Figure 4 shows the event tree for our hypothetical nuclear
plant. In Figure 4, "IE" denotes the initiating event. System
success is denoted by a step upward with the letter $A$ or $\bar{B}$ above
the line. For example, as shown in Figure 4, a particular accident
sequence is $IE \cdot A \cap \bar{B}$, which is a particular path in the event tree.
The event tree shown in Figure 4 also has a heading termed
"Accident Outcome" which describes the general consequence of
each accident sequence. The specific consequences associated
with each accident sequence are determined by a detailed consequence evaluation as we shall discuss in a later section. In
Figure 4, we've shown two possibilities for each system, "failure"
and "success," however, if each system had more than two performance states (to denote degraded operation, for example, but
not complete failure), then there would simply be more than two
branches for each system possibility.

In practice, event trees have been applied to analyze accidents at nuclear power plants, at chemical plants, at offshore
oil facilities, and at other industrial and military facilities.
Figure 5 shows an event tree which was constructed to actually
analyze pipe break accidents (called "loss of coolant accidents"
or "LOCA's") at a nuclear power plant; the event tree is taken
from the NRC Reactor Safety Study.[2] In application, event trees
can become very complex, consisting of hundreds or thousands of
accident sequences. To construct an event tree, the actual plant
or facility must be studied to identify the different kinds of
accident initiating events which can occur and the different
systems which can be called upon to mitigate the accident. For
example, for nuclear power plants more than 40 initiating events,
known as transients, can occur and can initiate different safety
system responses; when pipe breaks are considered, thousands of
different initiating events can be defined depending on the size
of the break and its location.[16]

Because event trees can be extremely large when many systems
can be called upon in an accident situation, the goal in such
problems is to reduce the number of accident sequences which must
be evaluated. The reduction techniques include eliminating
system forks which do not influence consequences or which cannot
exist (because of earlier outcomes) and subdividing large event
trees into products of smaller event trees which can be analyzed
more simply. When the times at which the systems fail influence

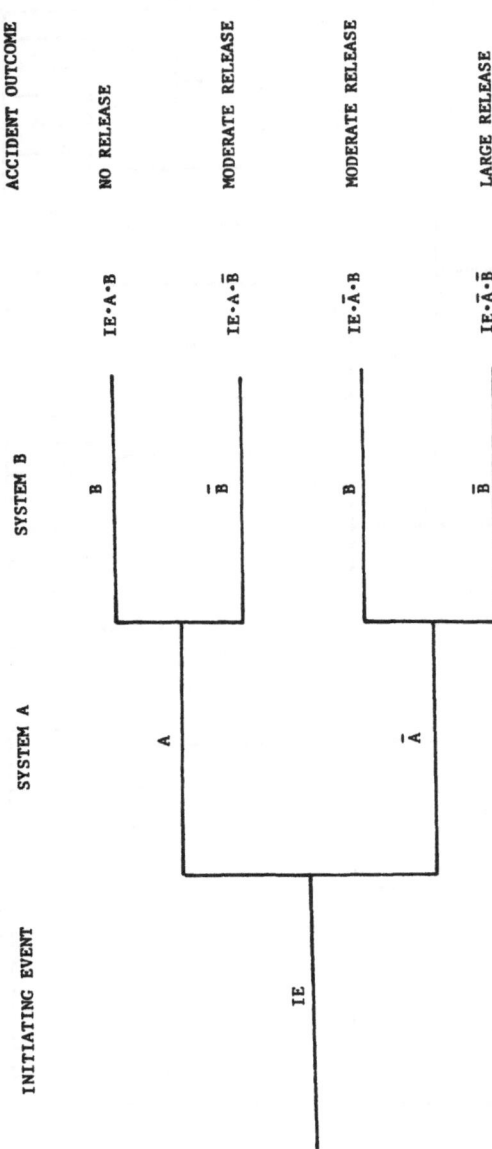

FIGURE 4.  EVENT TREE FOR THE HYPOTHETICAL NUCLEAR POWER PLANT

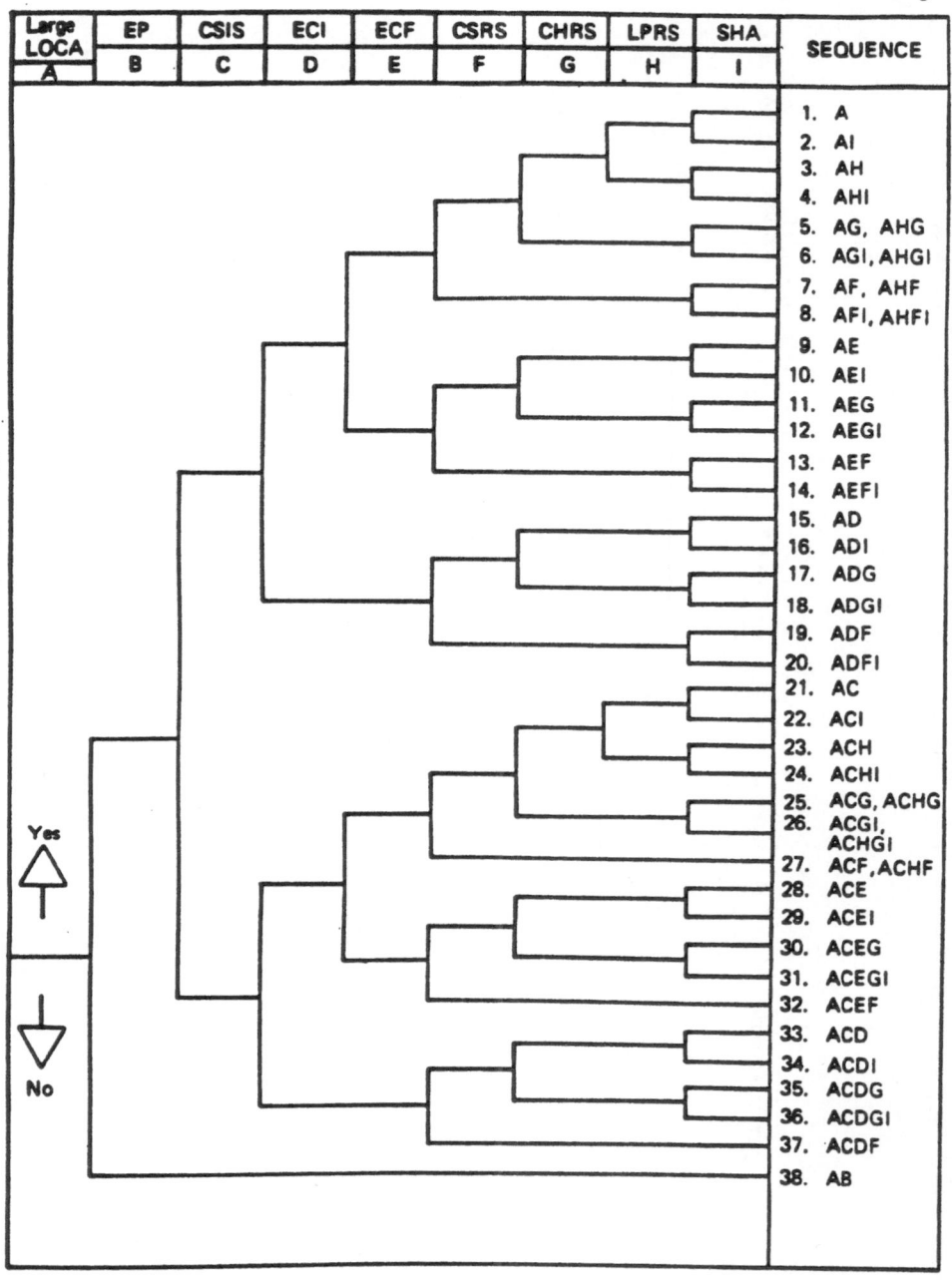

FIGURE 5.   EVENT TREE FOR LARGE LOSS OF COOLANT ACCIDENT IN A
            NUCLEAR REACTOR (THE HEADINGS DENOTE THE INITIATING
            EVENT OF A PIPE BREAK (A) AND SAFETY SYSTEMS WHICH
            CAN BE CALLED UPON)

the consequence outcomes, then different permutations of system failures and their times of occurrence must be considered. These more detailed considerations of event tree constructions are treated in 16. Even though this reference pertains to nuclear power plant analyses, the general principles are applicable to analyses of other accidents characterized by chains, or sequences, of events happening.

Once event trees have been constructed and have been reduced where possible, they can then be quantified to obtain the frequencies of the individual accident sequences. The quantification formulas are obtained using standard probability laws for chains (intersections) of events. For example, the probability of sequence IE·A·B̄ occurring in our simple event tree in Figure 4 is given by

$$P(IE\ A\ \bar{B}\ occurs) = P(IE\ occurs)P(A\ succeeds)\ *$$
$$P(B\ fails\ given\ A\ succeeds), \qquad (7)$$

where "P( )" denotes the probability of the event in parentheses.

The initiating event frequencies are usually calculated using the statistical analyses techniques which we have discussed in Section 2.1. The probabilities of system failure and system success which are required are in reliability nomenclature termed "system unavailabilities" and "system availabilities," respectively. When data on system failures and system successes exists, then traditional reliability techniques can be used to estimate these quantities.[7,17] (In essence, the system unavailability is the system failure frequency as estimated by the techniques in Section 2.1 multiplied by the average time the system is down for repair.)

When data are not available on past system successes and failures, then fault tree techniques are generally used to calculate the system unavailabilities and system availabilities from basic component reliability data which are assumed to be obtainable. The fault tree approach also allows the system interdependencies and the effects of the initiating event on system response to be taken into account.

---

*Note that the probabilities must be conditional probabilities because of the possible dependencies among the events. Furthermore, all probabilities of system failure and success are conditioned on the initiating event having occurred.

## 2.4 Fault Tree Techniques

The fault tree approach is a deductive logic approach which starts with an undesired event, such as system failure, and then systematically develops all the different possible basic causes which can lead to the event. The relationships between the undesired event and the basic causes are shown on a logic structure called the "fault tree." For a system comprised of hardware consisting of mechanical and electrical components, the fault tree expresses the system failure in terms of the basic mechanical and electrical component failures. For a more general event, the fault tree expresses the logic relationship between elemental events and the undesired event.

Figure 6 shows a simple fault tree from The Fault Tree Handbook.[18] The symbols in Figure 6 are the basic symbols used in a fault tree and are:

 : the rectangle signifies a fault, or event, which is described in the rectangle

 : the logical "OR," or union, relation

 : the logical "AND," or intersection, relationship

 : the circle signifies the designated primary events, such as basic component failures.

Stated in words, the fault tree in Figure 6 simply says that:

Event A occurs if Event X or Event B occurs, and
Event B occurs if Event Y and Event Z both occur.

In application, the symbols A, B, X, Y, and Z would be replaced by the event descriptors (such as "emergency core cooling system fails," "valve leaks," etc.) Event A is the undesired event of interest and is termed the "top event" (for obvious reasons). Events X, Y, and Z are designated basic causes, for example, for which data exists, and hence are depicted by circles.

In actual application, the fault tree is developed by determining the necessary, immediate causes of the top event. These

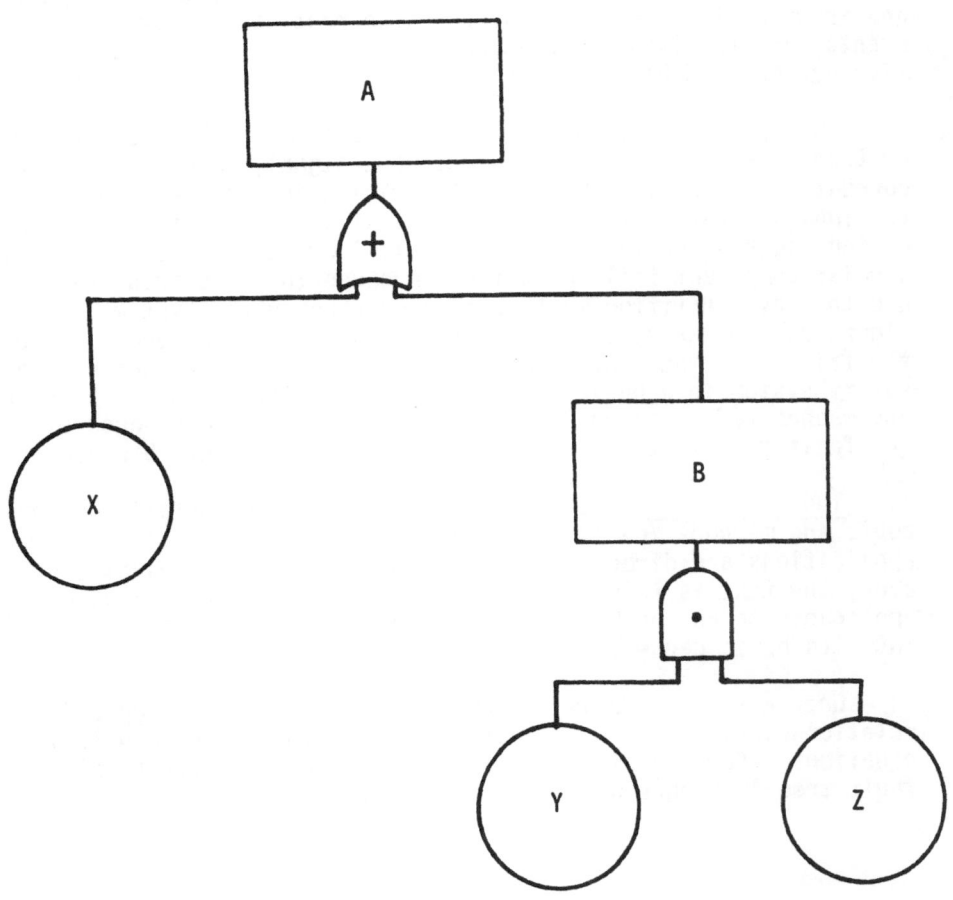

FIGURE 6.    A SIMPLE FAULT TREE STRUCTURE

causes are then expressed in terms of their immediate necessary causes. This process continues until the root causes are identified which is the limit of resolution of the fault tree analysis and is where the fault tree stops. The root causes, or primary events, are the events for which the analyst believes probabilities can be credibly estimated.

In practice, the fault tree for a system failure is developed by tracing the presence or absence of a signal, such as electrical current or fluid flow, back to its originating source. In this tracing, account is taken of the logical structure of the system design, such as whether it is a parallel or series system. Particular component failures which can cause the undesired event are thereby identified in this tracing through the system. Figure 7, for example, shows part of the fault tree developed for the failure of the containment spray injection system which is a safety system in a nuclear plant that basically cools the reactor and washes radionuclides out of the air if an accident occurs; the fault tree is taken from the NRC's Reactor Safety Study.[2]

The Fault Tree Handbook,[18] describes in more detail the concepts and methods involved in constructing fault trees: the applications are directed there toward nuclear power plants, however, the text is fairly readable, and the basic approaches are applicable to any problem where we want to decompose an event into its basic causes.

Once a fault tree is constructed, since it shows logical relationships, it can be translated into standard Boolean logic equations. For example, the Boolean equations for our simple fault tree in Figure 6 are

$$A = X + B \tag{8}$$
$$B = Y \cdot Z \tag{9}$$

where "+" and "·" are the "or" and "and" relationships, expressed now in Boolean symbols instead of fault tree symbols.

The advantage of expressing the fault tree in its equivalent Boolean logic equations is that Boolean algebra techniques can now be used to reduce the equations. The goal of this reduction process is to express the top or undesired event directly in terms of the primary events. We didn't do this directly since it is more methodical, and easier, to decompose events in a stage-wise fashion into their immediate causes, as constructed in the fault tree. In complex problems, it would also be virtually impossible to express the top event directly in terms of its basic causes.

The Boolean reduction process is quite straightforward and is done by successively substituting into the Boolean equations and

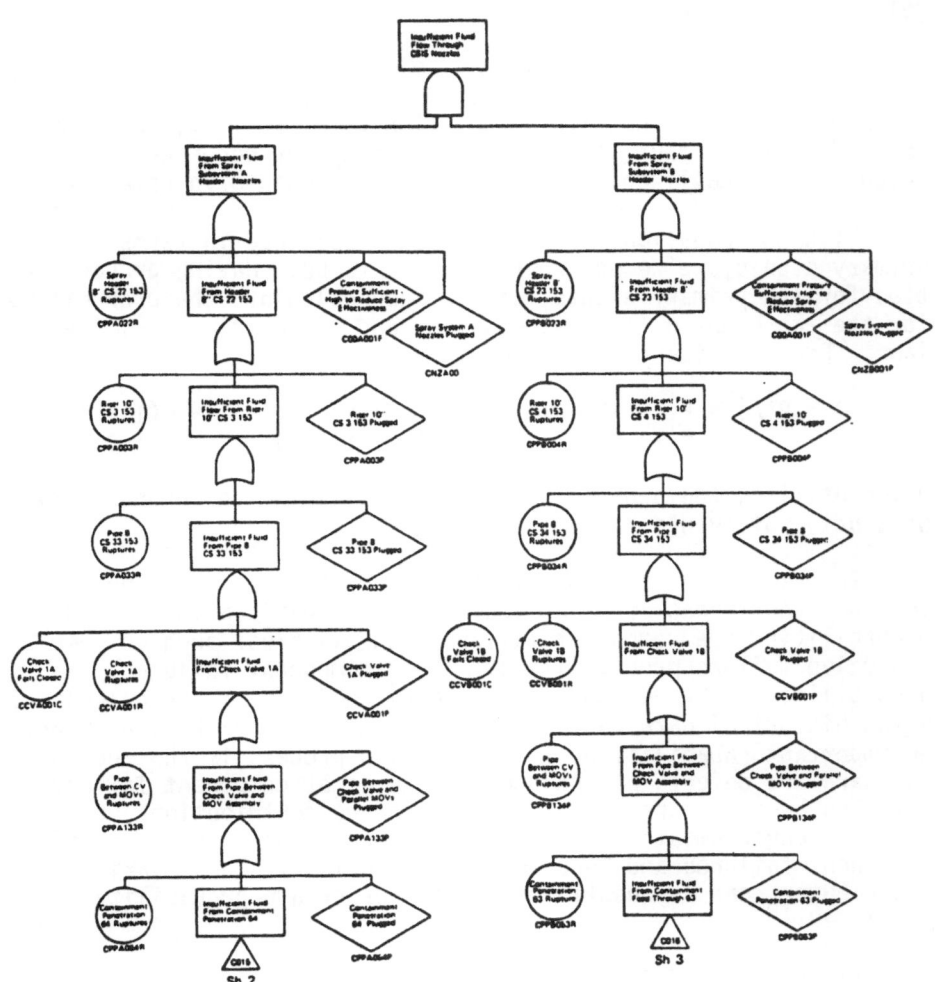

FIGURE 7.   CONTAINMENT SPRAY SYSTEM FAULT TREE FOR NUCLEAR
REACTOR RISK ANALYSIS (The diamonds on this tree
are types of primary events and the triangles
are continuation symbols)

simplifying using standard Boolean laws. For Equations (8) and
(9), the reduction process gives

$$A = X + Y \cdot Z \qquad (10)$$

where we've simply substituted Equation (9) for B into Equation
(8). For larger fault trees there are computer codes available
which automatically do this reduction as discussed in reference 18.

With the equation for the top event obtained in terms of the
primary events, we're now where we want to be. Having probabilities for the primary events, we can then calculate the probability
for the top event using this equation and standard probability
laws. For Equation (10), for example,

$$P(A) = P(X + Y \cdot Z) \qquad (11)$$
$$= P(X) + P(Y \cdot Z) - P(X \cdot Y \cdot Z) \qquad (12)$$

which utilizes the probability law for obtaining the probability
of a union of events.

In evaluating the probability of a system failure required
for the event trees of plant accidents, the basic primary event
probabilities are component failure probabilities, such as pump
and valve failure probabilities. These component failure
probabilities, termed "component unavailabilities," are calculated
using historical data records of component failures kept at plants.
An important capability of the evaluation process is that not only
can hardware defects be included as causes of component failures,
but also human errors, test and maintenance contributions, and
abnormal environmental effects can all be included as causes of
component failures and hence of system failure. These various
contributors can be quantified using appropriate probability
models.[16,18]

The ability of fault trees and event trees to include all
root causes of failure, such as human errors, is extremely
important since system unavailability, accident frequencies, and
finally risk is then tied to these root causes. These root causes
are the tangible ways in which risk can be controlled by design,
quality control, and management actions. The use of event trees
and fault trees to analyze complex accidents thus can be an
extremely powerful tool for decision-making.* Since they are a
key to the power of the techniques, the relationships among root

---

*It is interesting to note that in most risk analyses performed
to date on nuclear plants, the dominant contributors to risk were
found to be human errors, subtle component-to-component dependencies, and external events such as fires - and were not simple
hardware failures.

causes, fault trees, and event trees are depicted in Figure 8. (As we stated, we will take up the consequence evaluations and the utilization of the calculated risk results in the subsequent sections.) Reference 19 discusses the event tree-fault tree approach and is not as nuclear reactor specific (nor as detailed) as 16.

With the reduction of the fault tree and the calculation of the top event probability, we are in essence completed with the fault tree work. In actual applications, the fault tree can be very large and be complex, and hence there are computer codes available to automatically reduce the fault tree and calculate the probabilities.[18] We should note that in addition to the top event probability, a wide variety of information is obtainable from the fault trees and event trees, including various qualitative information about the risk contributors, quantitative rankings of the contributors, sensitivity analyses to show where deviations in values have greatest risk impact, and uncertainty analyses to show the impacts of our lack of knowledge on calculated risk measure.

Fault trees, either by themselves or in conjunction with event trees, can be applied to a wide variety of risk analysis problems. When the event trees define ways (accident sequences) in which the public can be endangered, then the event trees and fault trees can be used to determine the most effective ways of controlling and reducing public risk. They can also be used to determine where research dollars can be most effectively spent to reduce our uncertainties in risk.

When we are concerned with economic risk, then the event trees are constructed to define the ways in which the plant will be down, and the top events of the fault trees define failures of systems to provide product output. Using these event trees and fault trees, plant operations can be analyzed to increase output by identifying human errors which can be controlled, and components which need to be replaced with more reliable ones.

For all these different evaluations, the event trees give the frequencies of the accident (or plant shutdown) sequences occurring, and the fault trees give the probabilities of system failure. In applications where consequences are not of explicit concern, for example, because the consequences don't vary from sequence to sequence, these frequency and probability results are all that are computed and are taken as defining "risk." In utilizing these results, we are then concerned with controlling and minimizing accident frequencies and plant shutdown frequencies with regard to the root causes. When consequences can vary, then we must explicitly calculate the consequences for each sequence to adequately characterize the risk, and we now take up this topic.

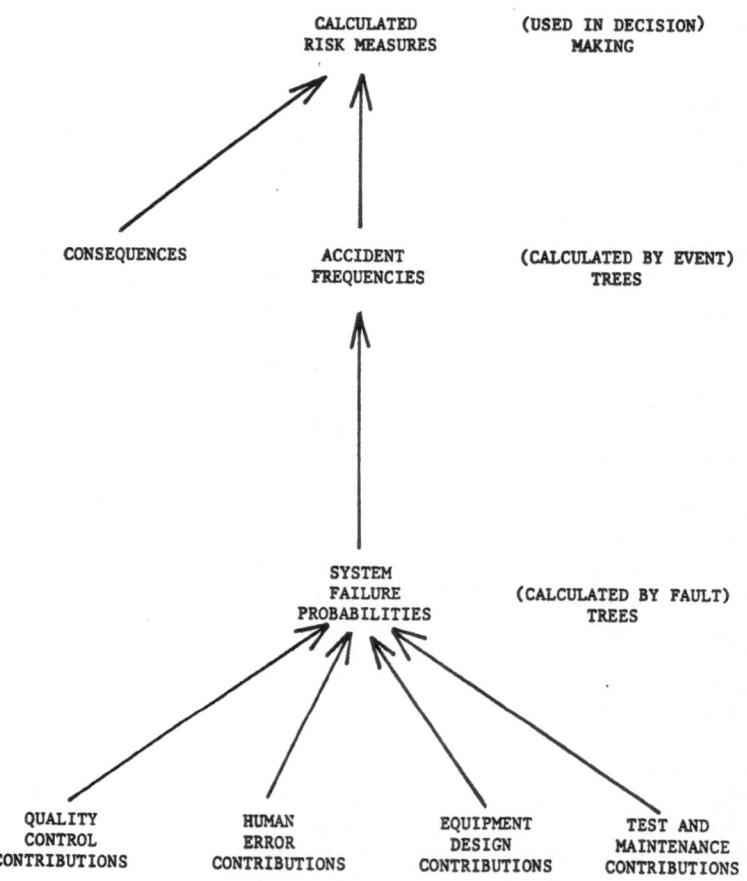

FIGURE 8.   EVENT TREE-FAULT TREE RISK ANALYSIS RELATIONSHIPS

## 3.  ACCIDENT CONSEQUENCE CALCULATIONS

When we use event trees and fault trees to model the accident occurrence, then we must separately calculate the consequences for each accident sequence when they differ from sequence to sequence. This consequence calculation is in essence done by determining the physical processes and phenomena which are associated with each accident sequence.  These physical processes and phenomena then determine, via appropriate accident models, the consequences which are generated by the accident sequence.

In an actual risk analysis of nuclear reactor accidents, each accident sequence in the event tree is analyzed for its consequences through an accident response evaluation.  The accident response evaluation consists of models which calculate the pressure, temperature, fluid flow, and neutron flux profiles which are generated by the accident sequence.  These profiles are then used to determine how much radioactivity is released from the reactor containment and in what form.  As a final step in the consequence evaluation, the released radioactivity is then transported to the surrounding environment using atmospheric dispersion models which account for the weather patterns and specific topography of the plant site.  The transport analysis includes evaluation of the various pathways by which the radionuclides can reach plant, animal and human life.*

The outputs of the nuclear accident consequence evaluations are the amounts of curies of various radionuclides which are deposited on the land and water and which are suspended in air for each accident sequence.  These curie amounts are given as a function of distance from the accident site and are also given as a function of the time from the initial occurrence of the accident. These curie amounts can then be translated into doses and health effects to the living population and economic damage to the plant and surrounding area.

The details of these nuclear consequence evaluations are described in references 16 and 19.  Reference 19 gives a good overview of the event tree-fault tree consequence evaluation process while reference 16 goes into more detail about technical considerations.  We will not go into these detailed calculations since they are peculiar to nuclear risk assessment.  However, the general steps which have been described above for nuclear

---

*In general, the consequence calculations use deterministic physical models to calculate the values of the physical variables and radioactivity released for each accident sequence.  The weather patterns are then modeled by probability distributions to give the distribution of radionuclides which results in the environment.

evaluations are also applicable to other types of risk analyses. For example, for chemical plant evaluations instead of radioactive curies being released, toxic chemicals would be the sources released and transported. For oil spills, the oil slick would be transported to determine consequences on the environment. Other examples will certainly come to the reader's mind. Figure 9 portrays the general steps which are involved in these consequence evaluations for a risk assessment.

As a final step in the consequence evaluations, these physical consequences, in terms of curies, toxic chemical concentrations, oil concentrations, etc., are translated into health effects. The health effects which are generally calculated for any risk assessment are numbers of early fatalities, latent fatalities (e.g., due to cancer), illnesses and genetic defects. In addition, economic consequences are also generally calculated including plant damage costs, property damage costs, and relocation costs of people who have to be evacuated.

## 4. RISK ANALYSIS RESULTS AND THEIR UTILIZATIONS

When direct data are available on accident occurrences and their consequences, then the statistical analysis methods described in Sections 2.1 and 2.2 of this chapter give the frequency versus consequence curves directly. When event trees and fault trees are used, as we have been most recently discussing, then the accident sequence frequencies and accompanying consequences must be separately combined to form frequency-consequence curves and other quantitative risk measures.

In the simplest type of evaluations, we obtain a frequency F and consequence C for each accident sequence, where the consequence can be a physical consequence or can be translated to health effects or can be expressed in units of economic loss. The set of accident sequences defined in the event tree then provides a set of frequency-consequence points (F, C) from which various risk curves and summary risk numbers can be obtained. The risk curve which is usually calculated is the complementary cumulative distribution function (CCDF) which is simply the sum of the accident frequencies having consequences greater than various values. The step function which is obtained is then smoothed using a continuous, fitting function. Figure 10, for example, shows a CCDF curve for early fatalities caused by reactor accidents as given in the NRC's Reactor Safety Study.[2] A point on the curve gives the probability per reactor year that an accident will occur and kill more than the number of people shown on the x-axis.

In addition to the CCDF, the other quantity usually calculated is the frequency-average consequence $\bar{C}$:

$$\bar{C} = F_1 C_1 + \ldots + F_N C_N \qquad (13)$$

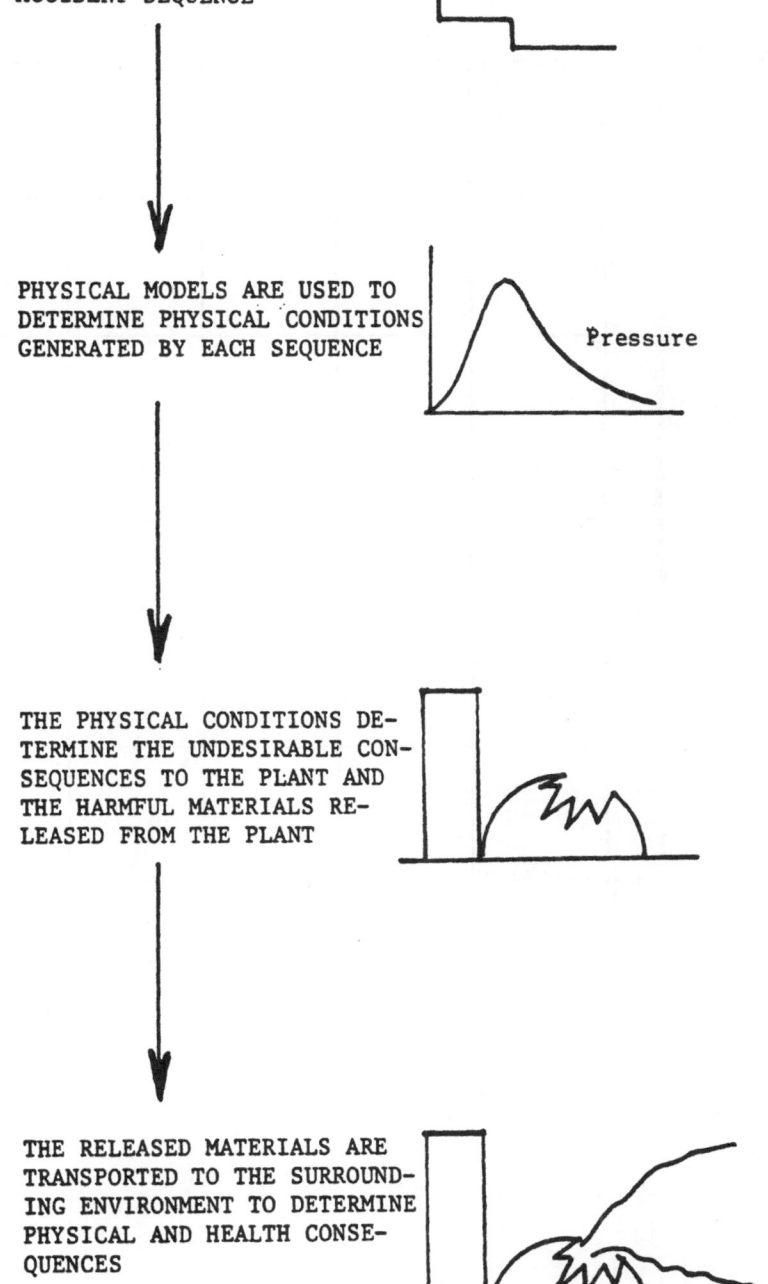

ACCIDENT SEQUENCE

PHYSICAL MODELS ARE USED TO
DETERMINE PHYSICAL CONDITIONS
GENERATED BY EACH SEQUENCE

Pressure

THE PHYSICAL CONDITIONS DE-
TERMINE THE UNDESIRABLE CON-
SEQUENCES TO THE PLANT AND
THE HARMFUL MATERIALS RE-
LEASED FROM THE PLANT

THE RELEASED MATERIALS ARE
TRANSPORTED TO THE SURROUND-
ING ENVIRONMENT TO DETERMINE
PHYSICAL AND HEALTH CONSE-
QUENCES

FIGURE 9. THE CALCULATION OF CONSEQUENCES
FOR AN ACCIDENT SEQUENCE

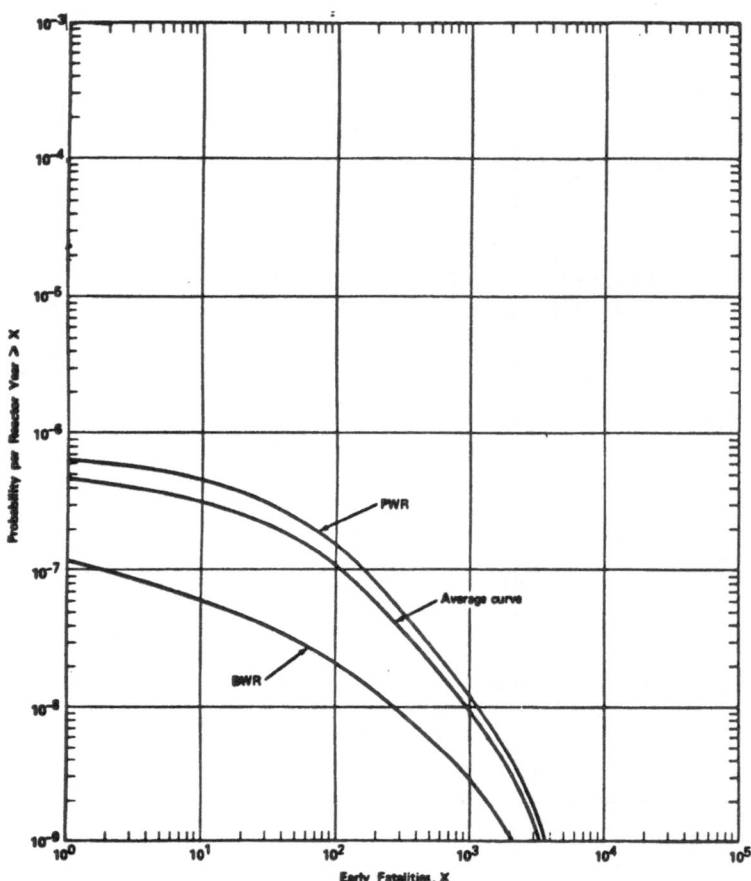

FIGURE 10.    EARLY FATALITY CCDF CURVE CALCULATED FOR REACTOR
              ACCIDENTS.  THE "PWR" CURVE IS FOR A PRESSURIZED
              WATER REACTOR, THE "BWR" CURVE IS FOR A BOILING
              WATER REACTOR, AND THE "AVERAGE CURVE" IS A
              WEIGHTED AVERAGE OF THE TWO (TO ACCOUNT FOR THE
              FACT THAT THERE ARE MORE PWR's THAN BWR's IN THE
              UNITED STATES).

where the sum is over the N accident sequences in the event tree. The quantity C is useful for giving the average consequence over many occurrences of the accident. However, in any given occurrence of the accident, the consequence can be quite different from C, and the likelihood of having a specific consequence is determined by the accident sequence frequencies.

In more complex evaluations where different consequences are possible from a given sequence, then instead of calculating a consequence C for each accident sequence, a consequence probability distribution is calculated for the sequence. The consequence probability distribution gives the likelihood of the sequence producing various consequences. The consequence probability distributions for each sequence are multiplied by their respective accident frequencies and are then combined to obtain the CCDF curve and associated summary statistics such as C. Reference 16 gives details of the models and calculations involved in this type of consequence treatment.

Because in general there are large uncertainties associated with event tree-fault tree risk assessments, it is important to describe the uncertainties which are associated with the frequency-consequence curves and summary statistics which are calculated. In general, there are three kinds of uncertainties which are associated with event tree-fault tree risk assessments:

1. Completeness uncertainties - due to our inability to conceive of all the possible accident sequences which might occur or all the possible ways a system can fail.

2. Modeling uncertainties - due to wrong assumptions or wrong equations being used for the complex accident frequency and consequence modeling.

3. Data uncertainties - due to uncertainties in basic input to the accident models because much of the accident data must be extrapolated from normal operating data.

The quantification of these uncertainties, particularly the completeness and modeling uncertainties, is difficult; however, sensitivity studies, classical statistical approaches, and Bayesian statistical approaches can be used to obtain some insights on the general sizes of the uncertainties in particular problems.[16] As ongoing research is conducted in this area, the methods of handling these uncertainties should become more systematic and more powerful.

        The uncertainty evaluations can be used to generate a band
of consequence-frequency curves to show the effects of the uncertainties in the final results. Figure 11 shows the band which
is produced for the CCDF curve. The percentage on each curve in
Figure 11 is the confidence we have in that curve.* For example,
for the 95% CCDF curve, we are 95% confident that the actual
frequency-consequence curve is below this curve; for the 50%
curve we are only 50% confident that the actual curve is below it.
The higher the confidence, the higher the curve, and the higher
the consequence for a given frequency.  The above CCDF confidence
band, though it is impressive looking, has a somewhat limited
utilization because of the restrictive assumptions which must be
made to calculate the different curves (e.g., that the uncertainties are perfectly quantifiable).

        In spite of the large uncertainties, event tree-fault tree
risk analysis still provides many meaningful results because, in
general, it is not the precise numbers that matter but the order
of magnitude of the risk measure (i.e. is it $10^{-2}$ or $10^{-6}$).  Also,
the relative contributors to the frequency-consequence curve or
the frequency-averaged consequence, expressed in terms of ratios
or rankings, often have less uncertainties than other risk
measures.  These risk contributors include the basic root causes
identified in the fault trees and are key results for decision-
making as we've discussed.

        Risk analysis using event trees and fault trees have been and
are being applied to a wide variety of problems.  In addition to
the applications we've already indicated, they have been applied
in some form or another to analysis of risks from nuclear fuel
reprocessing facilities,[20] from liquid chlorine transportation
accidents,[21] from gasoline transportation accidents,[22] from the
storage of liquid natural gas,[23] from sulfur emissions of coal-
fired plants,[24] and from tanker-ship collisions.[25]  As a final
addition, reference 26 discusses risk and reliability analysis,
and applications, from a general chemical plant perspective.  The
field of event tree-fault tree risk analysis is growing and pro-
gressing, and further strides in the methods and applications
should take place in the upcoming years.

5.  MORE SPECIALIZED RISK ANALYSIS METHODS

        As a final section in this chapter, we briefly discuss some
more specialized risk analysis approaches that are used for particular problems.  Specifically, we will touch on the approaches
that are used for probabilistic risk analysis of nuclear waste

---

*The confidence can either be a classical or Bayesian statistical
confidence.

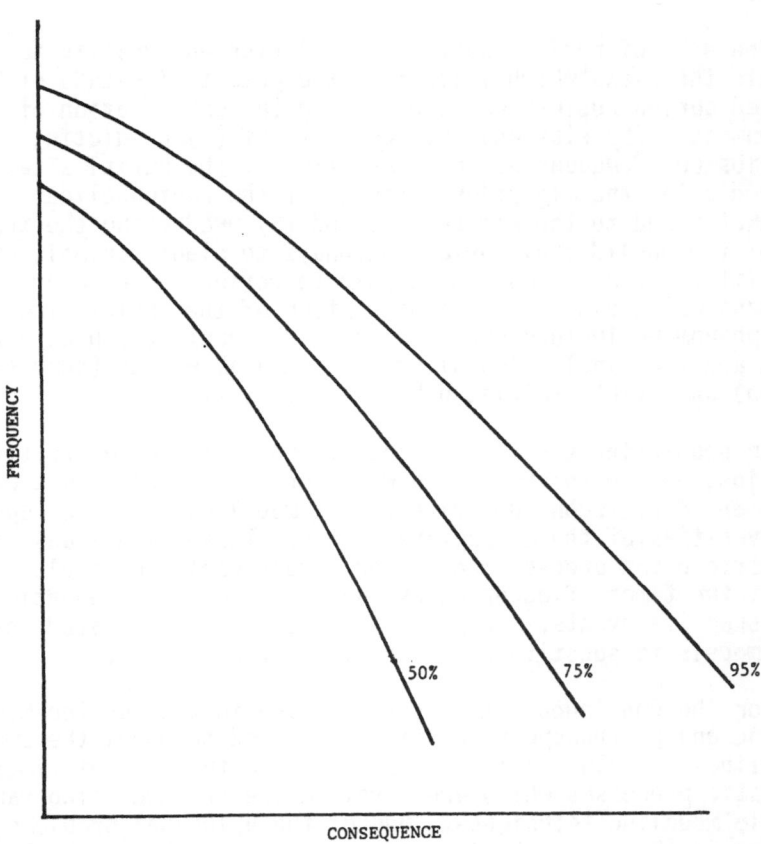

FIGURE 11.   CCDF FREQUENCY - CONSEQUENCE CURVES FOR DIFFERENT
             CONFIDENCE LEVELS

disposal risks, seismic risks, and fire risks. The outputs of
these risk analyses are the same as those we've discussed, e.g.,
frequency versus consequence curves and frequency averaged consequences.

The goal of nuclear waste disposal risk analysis is to
evaluate the risks which arise from the potential escape of hypo-
thesized buried nuclear waste resulting in contamination of the
environment. The risk analysis consists of (1) predicting the
scenarios and frequencies for disruption of the burial sites,
(2) predicting the migration patterns of the radionuclides to
groundwater and to the atmosphere, and (3) predicting the transport of the nuclides by various pathways to plant, animal, and
human life. Event trees can be used to define the categories of
phenomena which can result in disruption of the burial site.
These phenomena include sudden disruptive events (such as earthquakes and faulting), slow disruptive natural events (such as
erosion) and events initiated by man (e.g., mining).

In predicting the frequencies of the different disruptive
scenarios, the techniques which we've previously discussed are
supplemented by techniques directed at modeling the time dependence
and severities of the disruptive events. These techniques utilize
stochastic point process models and random variable models to
predict the times of occurrences and consequential properties of
the disruptive events.* Expert opinions are incorporated into
these models to substitute for the lack of hard data.

For the continuous processes involved in the nuclide migration,
mass and energy transport equations are used to model the amounts
of nuclides reaching various locations as a function of time; the
stochastic processes which are involved are treated using random
variable modeling techniques. One of the principal problems in the
modeling is the extremely long time periods which must be handled
in the predictions due to the long half lifes of some of the
radionuclides. Models and practical utilizations for nuclear waste
disposal risk analysis are still in the developmental stage. For
further details of the modeling techniques being developed, the
reader is referred to 27 and 28.

The goal of seismic risk analysis is to predict the risk from
earthquake occurrences. Of concern is the risk from building and
structural failures which can result in significant consequences

---

*Stochastic point process models produce probability distributions
for the time of occurrence of the event based on attributes of
the event. Random variable models assign a probability distribution to a variable to describe the likelihood of different
values being assumed by the variable.

or which can trigger further catastrophies. For example, an earthquake at a nuclear or chemical plant can cause a core melt to occur or can cause toxic chemicals to be released, thus compounding the consequences.

Seismic risk analysis consists of predicting the frequency, location, and size of the earthquake and the amount of damage produced by the earthquake. The damage evaluations consider the attenuation of the seismic forces by the soil, the response of structures to these forces, and the likelihood of the structures and components failing under the given forces.

The techniques which are used in the seismic risk analysis include:

1. Statistical analysis of past events for the earthquake frequencies.

2. Random variable modeling for the damage evaluations where probability distributions are assigned to the physical variables to account for the stochastic nature of the phenomena.

3. Event trees and fault trees to predict the probabilities of accidents occurring and systems failing.

The consequence evaluations are similar to those discussed in Section 3. As for the nuclear waste disposal risk analysis, many of the methods and the practical ways to utilize the seismic risk results are still in the developmental state. For further details on the approaches being developed, the reader is referred to 27 and 29.

Finally, the goal of fire risk analysis is to predict the risks from fires occurring in industrial and military complexes. The general fire risk analysis approach consists of predicting the frequency and location of fire initiation and the propagation possibilities for the fire, including their likelihood and associated damage. The analysis of the fire propagation possibilities incorporates the fire detection facilities and their efficiencies, automatic sprinkler system and manual extinguishing capabilities, and human error and success possibilities. The output from a fire risk analysis can potentially identify the critical fire locations, the impacts of fire detector and sprinkler system layouts, and cost effective ways of reducing the fire risk. The basic techniques used in fire risk analysis are similar to those listed for seismic risk analysis where "fires" are substituted for "earthquakes." This area again is still in the developmental stage. For details on fire risk models which have been studied, the reader is referred to 31, 32, and 33.

# REFERENCES

1. Cornell, J., The Great International Disaster Book, Simon and Schuster, New York (1979).

2. Reactor Safety Study: An Assessment of Accident Risks in U. S. Commercial Nuclear Power Plants, WASH-1400, NUREG/CR-75/014, U. S. Nuclear Regulatory Commission (October, 1975).

3. Accident Facts, 1979 Edition, National Safety Council, Washington, D.C.

4. Cohen, B. L., and Lee, I. S., "A Catalogue of Risks," Health Physics, Vol. VI, No. 36, pp 707-722 (June, 1979).

5. Coppola, A., and Hall, R. E., A Risk Comparison, NUREG/CR-1916, BNL-NUREG-51338, U. S. Nuclear Regulatory Commission (February, 1981).

6. Elandt-Johnson, R. C., and Johnson, N. L., Survival Models and Data Analysis, John Wiley and Sons, New York (1980).

7. Mann, N. R., Schafer, R. E., and Singpurwalla, N. D., Methods for Statistical Analysis of Reliability and Life Data, John Wiley and Sons, New York (1974).

8. Kalbfleisch, J. D., and Prentice, R. L., The Statistical Analysis of Failure Time Data, John Wiley and Sons, New York (1980).

9. Batten, R. W., Mortality Table Construction, Prentice-Hall, Englewood Cliffs, New Jersey (1978).

10. Benjamin, B., and Haycock, H. W., The Analysis of Mortality and Other Actuarial Statistics, Cambridge University Press, Cambridge, Massachusetts (1970).

11. Gumbel, E. J., Statistics of Extremes, Columbia University Press, New York, New York (1958).

12. Galambos, J., The Asymptotic Theory of Extreme Order Statistics, John Wiley and Sons, New York (1978).

13. Todorovic, P., "Stochastic Models of Floods," Water Resources Research, Vol. 14, No. 2, pp 345-356 (April, 1978).

14. Cramer, H., and Leadbetter, M. R., Stationary and Related Stochastic Processes, John Wiley and Sons, New York (1967).

15. Haugen, E. B., Probabilistic Mechanical Design, John Wiley and Sons, New York (1980).

16. PRA Procedures Guide, Review Draft, NUREG/CR-2300, U. S. Nuclear Regulatory Commission (September, 1981).

17. Green, A. E., and Bourne, A. J., Reliability Technology, Wiley Interscience, New York (1972).

18. Fault Tree Handbook, NUREG/CR-0492, U. S. Nuclear Regulatory Commission (January, 1981).

19. McCormick, N. J., Reliability and Risk Analysis, Academic Press, New York (1981).

20. Methodology Development for Risk Assessment of Fuel Processing, NUREG/CR-1604, DPST-NUREG-78-4, U. S. Nuclear Regulatory Commission (July, 1980).

21. An Assessment of the Risk of Transporting Liquid Chlorine by Rail, PNL-3376, TTC-0114, U. S. Department of Energy (March, 1980).

22. An Assessment of the Risk of Transporting Gasoline by Truck, PNL-2133, UC-71, U. S. Department of Energy (November, 1978).

23. Simmons, J. A., Risk Assessment of Storage and Transport of Liquefied Natural Gas and LP-Gas, Science Applications, Inc., Report for the Environmental Protection Agency (November, 1974).

24. Morgan, M. G., Morris, S. C., and Meir, A. K., "Sulfur Control in Coal Fired Power Plants: A Probabilistic Approach to Policy Analysis," Journal of the Air Pollution Control Association, Vol. 28, No. 10 (October, 1978).

25. Barlow, R. E., and Lambert, H. E., The Effect of U. S. Coast Guard Rules in Reducing the Probability of LNG Tanker-Ship Collision in Boston Harbor, Tera Report, Tera Corporation, 2150 Shattuck Avenue, Berkeley, CA 94704 (May, 1979).

26. Henley, E. J., and Kumamoto, H., Reliability Engineering and Risk Assessment, Prentice Hall, Englewood Cliffs, New Jersey (1981).

27. Risk Methodology for Geologic Disposal of Radioactive Waste: Interim Report, NUREG/CR-0458, U. S. Nuclear Regulatory Commission (October, 1978).

28. Scenario Development and Evaluation Related to the Risk Assessment of High Level Radioactive Waste Repositories, NUREG/CR-1608, U. S. Nuclear Regulatory Commission (August, 1980).

29.  Seismic Safety Margins Research Program (SSMRP), Phase I
     Final Report - Overview, NUREG/CR-2015, U. S. Nuclear
     Regulatory Commission (March, 1981).

30.  Proceedings of the the Fifth International Conference on
     Structural Mechanics in Reactor Technology, ISSN 0172-0465,
     Berlin, Germany (August, 1979).

31.  A Methodology for Risk Assessment of Major Fires and Its
     Application to an HTGR Plant, GA-A 15402, UC-77, prepared
     for the U.S. Department of Energy (July, 1979).

32.  Development and Testing of a Model for Fire Potential in
     Nuclear Power Plants, NUREG/CR-1819, U.S. Nuclear Energy
     Commission (November, 1980).

33.  Fire Risk Analysis for Nuclear Power Plants, NUREG/CR-2258,
     UCLA-ENG-8102, U.S. Nuclear Energy Commission (September,
     1981).

# QUANTITIES OF HAZARDOUS MATERIALS

## F. R. Farmer, O.B.E., F.R.S.

## 1. INTRODUCTION

Canvey Island is an industrial site on the north Thames shore, some 25 miles downstream from London. It is a petrochemical site measuring about 9 miles long and up to 2-1/2 miles deep. Ten separate companies have major operations on the site, the range of materials and quantities are listed in Table 1.

Two companies, Occidental Refineries Ltd. (ORL) and United Refineries Ltd. (URL) have planning permission for the construction for oil refineries and would extend the site on the West side.

The resident population, now 33,000, expressed their feelings that they were already at risk from the existing installation and this risk should not be increased. The Secretary for State for which the Environment ordered a public inquiry to be held as a result of which the Health and Safety Executive were asked to carry out a technical assessment. This assessment was carried out by members of the Safety and Reliability Directorate of the United Kingdom Atomic Energy Authority with links to and guidance from the Health and Safety Executive.

The terms of reference for the investigation were:
"In the light of the proposal by United Refineries Ltd to construct an additional refinery on Canvey Island, to investigate and determine the overall risks to health and safety arising from any possible major interactions between existing or proposed installations in the area,

TABLE 1

QUANTITIES OF HAZARDOUS MATERIALS

| Company | | Quantity or Capacity | |
|---|---|---|---|
| Shell | Ammonia | 14,000 tn | -33°C |
|  | Ammonia | 10,000 tn | Refrig. 3 ships in use |
| Fissons | Ammonia | 1,900 tn | -6°C (3.9 atmos) |
|  | Ammonia | ≈40 tn | In rail wagons |
| 3 Refineries | Hydrofluoric Acid | 20 tr | In process |
|  | Hydrofluoric Acid | 2 x 40 tn | In store |
| British Gas | LNG | 4x20,000tn | Refrig. - 161°C below ground tanks |
|  | LNG | 6x4,000tn | Regrig. - 161°C above ground |
|  |  | 12,000 tn | In ships |
|  | LPG | 2 x 5,000 tn | Butane refrig. below 0°C |
|  | LPG Butane | 1 x 10,000 tn | Butane refrig. below 0°C |
| Shell |  | 1 x 1,700 tn | Atmos temp. |
|  |  | +4 total 3,200 tn |  |
|  | Propane | 4 total 1,600 tn | Atmos temp. |
| Mobil | LPG | 4,000 tn (total) | Atmos temp. |
|  |  | 5,000 tn | Atmos temp. |
| Calor | LPG | 500 tn | Proposal 1 tank refrig. |
| Fisons | Ammonium Nitrate | 1 x 3,100 tn | 44,000 cyl. 14.7 kg |
|  |  | 1 x 6,200 tn | 92% Aqueous Solution |
|  |  |  | 92% Aqueous solution |

where significant quantities of dangerous substances are manufactured, stored, handled, processed and transported or used, including the loading and unloading of such substances to and from the vessels moored at jetties; to assess the risks; and to report to the Commission."

The expertise of over 30 engineers, chemists and other specialists took part in the study which took 2 years and cost about L400,000. The report is published as "Canvey: an investigation of potential hazards from operations in the Canvey Island/ Thurrock area". ISBM 0 11833200 Y.

1.    What was the nature of the risk?

Large quantities of flammable and toxic materials, in process, in store or transport could put substantial numbers of the public at risk from fire, explosion, missiles or from the spread of toxic gases. The study was to identify and estimate the severity of the risk.

2.    How to assess and describe the risk?

The assessment of risk in a complex situation is difficult. No method is perfect, all have limitations of one sort or another. We agreed with the investigating team that the quantitative approach was the most meaningful way of comparing different risks. The team should attempt to quantify the probability of various types of accidents that might occur, and then the probability of a whole range of possible consequences. This would enable the risks to be described not simply as "large," or "small," but in a quantified, numerical way. To describe the probability of killing 100 people as a result of one type of accident as 1 in 10,000 compared to 1 in 100 as a result of another type of accident, is much more meaningful than saying the chances of these types of accidents occurring are "very remote" or "quite high." To express risks in numerical terms provides one with a common denominator, a method of putting various risks in perspective and incorporating them with each other. The team have made the best estimates of particular events occurring the results of their estimates are expressed as chances in 10,000 a year in this part of the report.

Such a method of assessment involves reliance on historical data. Historical data may be suspect for two reasons. Firstly, there is the probability that new operations may have created new hazards which have not yet produced accidents and secondly, it is possible that th lessons of past accidents have been learned and they will not occur again. Where data are lacking one has to make judgement based on experience and speculation. Notwithstanding this weakness it was agreed that a quantitative approach was necessary, despite the inevitable difficulties and elements of controversy which such an approach would involve.

In presenting their assessment of the probability and consequence of various events, the team wished to convey their view of the merit to be attached to the quantified data; to do this they attached a suffix a, b, c or d to most numbers where

(a)  is assessed statistically from historical data
(b)  based on statistics as far as possible, some missing elements supplied by judgment, - could be improved with more effort
(c)  estimated by comparison with each other, assessed, operations thought to be reasonably similar
(d)  a judgement only and prospects are poor for improvement.

## 2.   WHAT COULD HAPPEN - INTRODUCTION

Ammonia ($NH_3$) and hydrofluoric acid are irritant (toxic) gases. If released to atmosphere they may drift over populated areas and, concievably, kill many people.

Liquefied Natural Gas (LNG) and Liquefied Petroleum Gases (LPG) are flammable and under some conditions may explode. A large release of gas could drift over populated areas and then meet ignition sources.

The behavor of gas clouds will be discussed and later a more detailed look at a range of possible accidents and consequences.

## 2.1  The Dispersion of Released Ammonia

Experiments in which liquid anhydrous ammonia was poured on to water have been reported and assessed by Raj et al. The report concludes that such releases are adequately described in terms of a buoyant plume rise model, in which it is assumed that ammonia is released as a pure undiluted vapor.

This was challenged by Griffiths noting in particular,

-    the maximum quantity released in one experiment was 130 kg; this was small compared with 100 to 1,000 tons.
-    Pasquill's dispersion model was used to obtain diffusion coefficients at distances of 20 to 100 ft; it was originally formulated for distances greater than 100 m.

- Pasquilli's diffusion coefficients seem valid for releases of reasonable duration, 10 minutes to 1 hour, but not for the experiment for which the release time was 10 sec.

A preferred interpretation is as follows: Some ammonia - perhaps one-half will dissolve in water. The balance will evaporate and the vapour will contain some droplet aerosol, sufficient to render the mixture non-buoyant. There may be some initial mixing with air which will be cold and this will assist in producing a non-buoyant mixture. The aerosol may develop into a mixed composition fog by interaction with moisture in the air and the high latent heat of vaporization will enhance the persistence of this fog.

This model may also describe the result of a spill on land under some conditions. In 1977 at Pensacola, Florida, about 50 tons was released from a damaged derailed tank car. The cloud of vapor was held near the ground by the canopy created by trees, the light rain and a light wind. The cloud traveled almost 15 miles and was seen on radar (nearby airfield) for about 1 hour. Trees and ground vegetation was discolored and birds and small wildlife were found dead within 1,000 ft.

Hence it is prudent to treat a spill of ammonia on to water as forming a heavy gas cloud. This will give much greater distances for given ground concentrations than would a buoyant plume model.

## 2.2 Heavy Gas Clouds

In most actual or hypothetical accidental releases of flammable or toxic vapors that have or could harm the public - the vapor cloud behaves as a heavy gas. Many gases such as chlorine or propane are heavy by virtue of a high molecular weight. Methane evolved from a spillage of LNG is heavy at $-161^{o}C$. As described above, ammonia may form a dense cool mixture with air and water. Even HF which has a nominal molecular weight of 20 seems to behave as a mixture of hexameric species with a monomer with an effective molecular weight similar to chlorine.

Experiments have been carried out to find the behavior of heavy gases under different conditions; some general patterns emerge. When released the gases slump under gravity and will move upwind as well as downwind. The gas will form a wider plume than Pasquill's model and would initially be less affected by different weather conditins than Pasquill's formulation. The heavy gas movement would be more affected by local topography and would give a greater area coverage for a given gas concentration than would be given by normal air diffusion.

However there is still considerable uncertainty in predicting the behavior of heavy gas clouds, it is thought that
the uncertainty in a hazard range could be a factor of 2 or 3
in current models but this should not detract from the assessment of possible accident consequences, particularly as the
uncertainty in the estimation of frequency will be greater (or
as great) as the uncertainty in the consequence.

2.3  To Derive the Toxic Range for a specified Release

There are several stages.  Chlorine and ammonia are irri-
tants and it is possible that the toxic effect is not directly
proportional to concentration - but to some power.  In the
Canvey study it was assumed proportional to $t \times c^n$ where n =
2.75.  (This may not be generally accepted.)

For chlorine, the data used gave 1 gm/$M^3$ as lethal in 50 secs
or 0.1 gm/$M^3$ as lethal in about 8,000 secs. - Chlorine was used
as an example and HF was assumed to have a comparable toxicity of
either 30 or 50 minutes - these two numbers were used to give
appropriate constants to derive D = Dose as the integral of
$C^{2.75}$ dt.(t in minutes) (C in gm/$M^3$).

There is a further step - the probability that a dose D
is fatal is taken to be P = a.b log D.  For the two cases
A = 30 minutes and B = 60 minutes the values derived were:

|   |   A   |   B   |
|---|-------|-------|
| a | 1.14  | -7.41 |
| b | 0.78  | 2.20  |

The accident assessed was the rapid airborne release of 1,000
tons of ammonia; initially slumping.  The contour for 90% LD
reaches between 5 and 6 KM for A and B with a maximum ½ width
of between 2 and 3 KM.  The 5% contour reaches 11KM for A and
9 KM for B.  The population at risk was displayed graphically;
the 5 to 90% contour could be rotated around its point of
origin, to allow for wind direction; and casualties obtained
by simple fractional addition.

Some allowance should be made for the shelter provided by
houses.  If the "length" of the cloud is 4 KM, at a wind speed
of 3 M/sec, it would take 20 minutes to pass - or less away
from the centreline.  A house may have an air change of 3
per hour so that staying indoors could reduce the concentration
by 2 and the dose D by 7.

2.4  Some Comments on Flammable Gas Clouds

The consequences of a large release of gas are assessed as follows:

A continuous discharge, as from a broken pipe, may ignite near the point of discharge and continue to burn, relatively safely.  If it does not ignite, the gas will form a cloud stretching downwind until it find ignition or becomes dilute below the lower flamability limit LFL.

The theory for dispersion of heavy gases is not good - as already mentioned for toxic gases and for large releases, say 10 tn/sec or a total of 1,000 tns - involve extrapolation by 2 orders of magnitude.  An indication of LFL is given as:

|  |  |
|---|---|
| 10 Kg/sec | 100 M approx. |
| 100 Kg/sec | 400 M |
| 1,000 Kg/sec | 1,600 M |

The distances may vary a factor of 2 up or down.  If a gas cloud is formed - by a fast, puff release, or continuous release not ignited at source, it is convenient to assume and assess the effects as though it were a hemispherical cloud of radius r. Ignition will then give a burnt cloud radius of 2r.  Some explosion or fast deflagration may cause damaging overpressures beyond 2r; it is assumed that 10% explodes, (or an explosion of 10% efficiency) and the distance to 0.2 atmos (3 psi) taken as causing some collapse of buildings; this turns out to be 4r.

When scaling from one quantity to another, the cube root law was used, so a cloud of 8 times the mass will have twice the radius and cover 4 times the ground area.

There is some feeling that a square root relation might be better - particularly for large masses.

For the quantity of:

|  | r (m) | To LFL |
|---|---|---|
| 100 tons | 80 | 2 KM approx. |
| 1,000 tons | 180 | 5 KM approx. |

Beneath the fireball it is assumed that all will die, as indeed at San Carlos (Spain) when a propylene tanker crashed and the fireball enveloped a camp-site killing 150 people. Beyond the fireball house collapse and flying fragments may injure about 10% of the population out to 0.2 atm overpressure.

If the gas cloud ignites at the edge of a populated area

the larger fireball covers a segment which includes N people.  If
the cloud has completely moved over the area before igniting,
the number of people affected - in a uniformly populated area -
would be 4N.  The probability of this second option must be lower
than the first, - there is likely to be sources of ignition
quite quickly in a populated area, but the gas cloud will have
considerable local variations in gas concentration.  It is convenient and somewhat logical to assume that an event having 4
times the severity will have 1/4 of the probability of occurrence.  This pattern of frequency/consequence will be discussed
later.

Taking this into account and the distribution of population in the affected areas, the various places from which a
gas cloud could originate, - the local wind pattern, it is
possible to estimate the consequence of a 1,000 ton gas escape,
not igniting at source, but travelling to touch or cover housing
at rates as follows:

The probability of exceeding the number of fatalities is

| Number | 10 | 750 | 1,500 | 2,500 |
|--------|-----|-----|-------|-------|
| Probability | 1 | 0.7 | 0.48 | 0.32 |

The next step is to discuss some of the accidents which might
occur and the probability.

2.5  Accident Frequency

I will consider a few cases - as examples.

The first is the carriage of ammonia, LNG, LPG by ship
alongside Canvey Island and berthing at one of several berths.
There is some chance that an accident might occur.  How to
assess this:  The study group made a search for relevant information from many sources - particularity the Port of London
Authoirty: Lloyds Register of Shipping, Norske Veritas and
World Data.

What is required is the frequency of accidents of moderate
severity and ships of a given size in coastal waters travelling
at given speeds or when berthing, taking into account the collision protection provided by the ships' structures.  The basis
for the assessment is presented in Appendices to the Canvey
report and the figure derived for LNG ship movements was:

$2 \times 10^{-6}$ per ship movement (of LNG) in the area

Additionally there is a risk of fire or explosion on or at
berthing.  From PLA data for 4 accidents to tankers at berth

attributed to fire in a 10 year period when over 37,000 tankers entered port. This gives $1 \times 10^{-4}$ per berthing or $5 \times 10^{-5}$ per harbour movement.

Norske Veritas data gives $3 \times 10^{-5}$ per harbour movement. In view of the known precautions by the two LNG carriers it was decided to reduce the estimated risk to:

$$3 \times 10^{-6} \text{ per harbour movement}$$

For ships carrying liquid ammonia or LPG, less was known about the protection given by the ships' structure and less known about the precautions taken in berthing so more conservative figures were used as:

$$2 \times 10^{-5} \text{ per ship movement - in transit}$$

and $\quad 2 \times 10^{-5}$ per ship movement - in berthing

from fire and explosion. These figures might be reduced.

Returning to the LNG traffic, a further risk contribution arises from pipe failure on ship to shore coupling, followed by a spill and fire. This was studied and assessed as a probability of $3 \times 10^{-5}$ per cargo transfer. Hence the LNG (from shipping) was summarized taking 50 berthings per year as:

Pipework failure $\quad 3 \times 10^{-5} \times 50 = 1.5 \times 10^{-3}$ pa

Collision $\qquad\quad 2 \times 10^{-6} \times 50 = 0.1 \times 10^{-3}$ pa

Fire/explosion $\quad\ 6 \times 10^{-6} \times 50 = \underline{0.3 \times 10^{-3}}$ pa

$$\qquad\qquad\qquad\qquad\qquad\qquad 1.9 \times 10^{-3} \text{ pa}$$

The total for ammonia ship traffic for 15 movements per year totalled $\qquad\qquad 7.8 \times 10^{-4}$ pa and for LPG traffic $\qquad\quad 5.5 \times 10^{-3}$ pa.

Further extension to this part of the study assessed changes which might come about as from extensions to the site or change in traffic density, - also from any ameliorating factors. One such is a speed restriction introduced by the PLA of 8 knots through the area - as compared with past practice of 11½ knots. A study of Lloyds Register critical impact speeds relating loaded displacement versus speed shows that the displacement of 4,000 tons at 11½ knots changes to 15,000 tons at 8 knots and could be higher if allowance is made for ship speed reduction prior to impact and impact angle to be less than 90°.

Not more than 13% of ships in this area have a displacement of over 15,000 tons.

This paper aims to give examples to illustrate the type of study, the uncertainties and assumptions, hence it is not proposed to discuss all accident initiators as from fires, flooding, missiles etc. These are presented in the full report.

It is timely now to consider the conclusions but first to indicate a range of possible actions - or investigations which may lead to a reduction of the assessed risk. These may be specifically related to one operator or site and are mostly subject to review as to their efficacy.

1.    An efficient water spray system could be installed on the jetty where liquid ammonia is unloaded - as a means of bring the gas into solution.

2.    Similarly water sprays might be useful where HF is processed or stored.

3.    A pipeline containing LPG could be emptied and taken out of service.

4.    Where large quantities of flammable liquids are stored additional retaining walls could limit the extent of fire spreading from a large spillage.

5.    Suggestions were made as to the use of a new road to serve a new refinery which could take traffic through less populated areas.

6.    The PLA should aim to ensure adherence to an 8 knot speed limit.

7.    Further assessment in specific cases might show the risks are less than assessed or that action might be identified to lessen the risk.

2.6   Results

Two types of risk were assessed. The first was the probability of an individual in the vicinity being killed as a result of a major accident and for this exercise 8 separate locations were identified as having different risks characteristics, - through their different proximity to any of the various sites having accident potential.

The second was the probability of events large enough to cause multiple casualities and for this exercise 4 areas were identified - as multiple casualities are influenced by population distribution and density and proximity is less important.

   - Individual  The risks are assessed for 25 events, -
(types of accident and location) and grouped for ammonia, HF, LNG,
LPG and flammable liquids, all for each of the 8 locations.  All
risks will be given as probabilities in units of $10^{-6}$ per year.

   The highest risk in one location was 2,500, but of that one
half arose from flammable liquids escaping from bunds.  The next
highest was 800 and others dropping below 500.

   In summary the results were as Table 2.

TABLE 2
ASSESSED INDIVIDUAL FATAL RISK AT CANVEY IS ($10^{-6}$/year)

| Plants | Initially | Improved |
|--------|-----------|----------|
| Existing | 530 | 130 |
| Existing & Proposed | 920 | 140 |

   - Societal  These are presented as present or proposed risks
of exceeding 1,500, 3,000, 4,500, etc. up to 18,000 casualties
from one event.  A breakdown in relation to "exceeding 1,500
casualties" is:

|  | New | | Improved | |
|--|------|-----|------|-----|
|  | Land | Sea | Land | Sea |
| Ammonia | 245 | 283 | 63 | 76 |
| LPG refineries | 170 | 17 | 99 | 165 |
| LPG British Gas |  | 320 |  |  |
| HF | 343 |  | 90 |  |
| LNG | 103 | 121 | 33 | 118 |

These are summarized in Table 3.

TABLE 3
ASSESSED RISK OF AN ACCIDENT AT CANVEY ISLAND
CAUSING MORE THAN 1,500 deaths ($10^{-6}$/YEAR)

| Plants | Initially | Improved |
|--------|-----------|----------|
| Existing | 1,700 | 300 |
| Existing and Proposed | 2,900 | 400 |

## 2.7 Comparison

I find it instructive to view these results against some background of natural and man-made disasters. A recent study has provided Figs. 1 and 2. Figure 1 shows the frequency with which casualities from natural causes exceed N in one year and is seen to reach $10^5$ and $10^6$ at $10^{-2}$ per year.

Figure 2 shows man-made only reaching N = $10^2$ to $10^3$ at a frequency of $10^{-2}$ to $10^{-3}$ per year.

The slopes of Figs. 1 and 2 are markedly different but this is exaggerated as on Fig. 1 there would be a low weighting for small accidents - (we do not hear of them) and on Fig. 2 the extrapolation even to $10^{-2}$ per year is very uncertain.

On Figure 3 is reproduced a well known curve from the Rasmussen report and on it is marked the upper to lower estimates for Canvey at N = 1,000 and 10,000.

As Canvey handles 24% of the U.K. petrochemical capacity, i.e. is a a fair proportion of the UK risk, the result so marked seems not out of place.

More work is required; most of the benefit is in the carrying out of the study with the cooperation of the companies concerned.

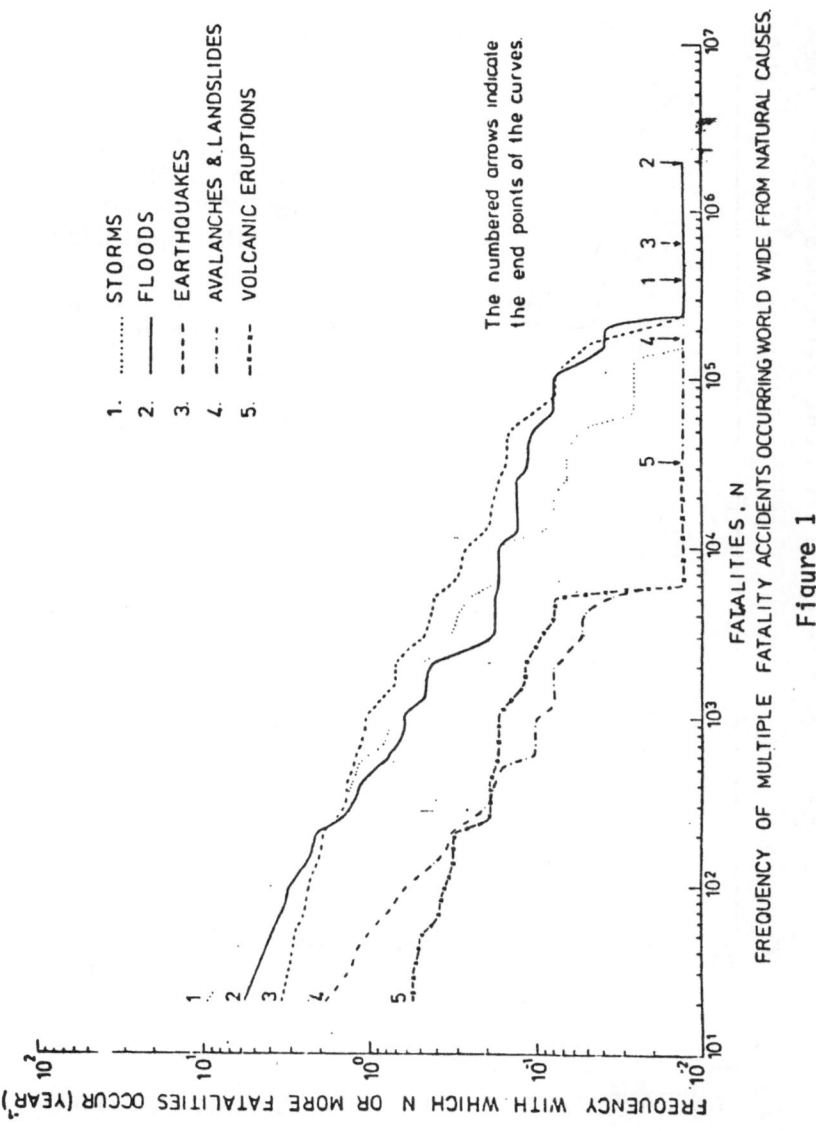

FREQUENCY OF MULTIPLE FATALITY ACCIDENTS OCCURRING WORLD WIDE FROM NATURAL CAUSES.

Figure 1

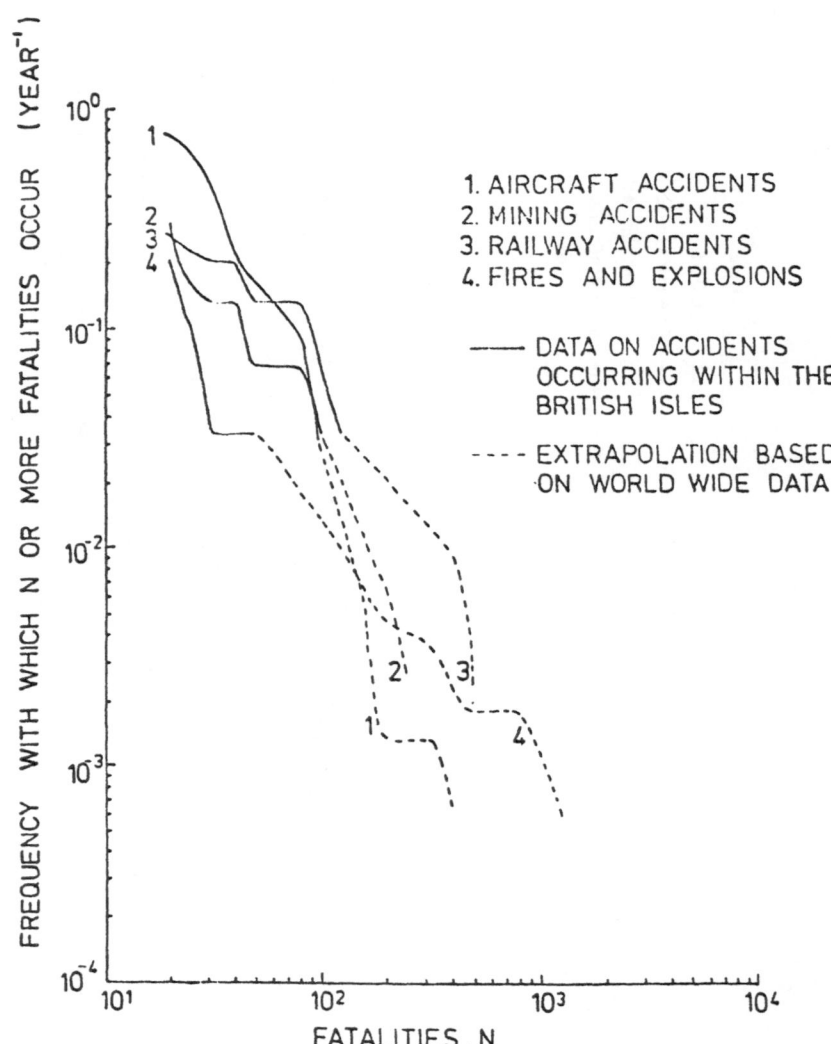

Figure 2
EXTRAPOLATION OF BRITISH ISLES ACCIDENT DATA

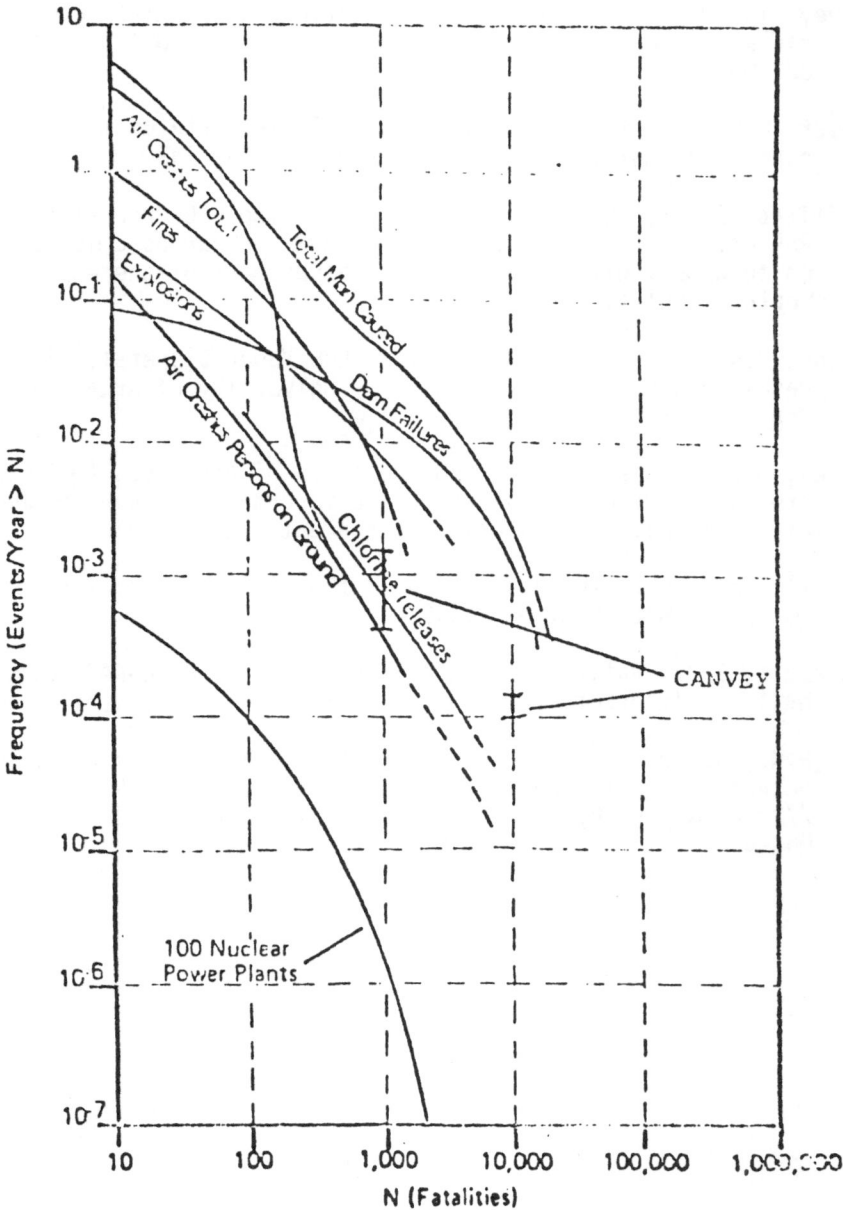

Figure 3
FREQUENCY OF MAN-CAUSED EVENTS WITH FATALITIES GREATER THAN N*

# REFERENCES

Canvey, 1978, an investigation of potential hazards from operations in the Canvey Island/Thurrock area. HMSO ISBN 0 11 883200 X.

Freyer, L.S., Griffiths, R.F., 1979, Worldwide data on the incidence of Multiple Fatality Accidents, SRD 149.

Griffiths, R. F., 1977, Critical review of the USCG report by RAH et. al. (1974) on spills of liquid anhydrous ammonia on to water, with an alternative assessment of the experimental results, SRD R67 HMSO.

Havens, J.A., 1977, Predictability of LNG Vapor Dispersion from catastrophic spills on to water. University of Arkansas, Report AD 525.

Kaïser, G.D., Walker, R.C., 1978, Releases of Anhydrous Ammonia from pressurized containers - The importance of denser than air mixtures. Atmospheric Environment, 12 2289-2300.

Pasquill, F., 1961, The estimation of the dispersion of windborne material, Met. Mag. 90 23449.

Pensacola, 1977, Railroad Accident report number NISB-RAR 78.4 National Transportation Safety Board, U.S. Government.

Raj, P.K., Haggrian, J. and Kalelkar, A. S., 1974, Prediction of hazards of spills anhydrous ammonia on to water, NTIS AD 779400, Report by Arthur D. Little, Inc. to U.S. Coast Guard.

THE WIDER IMPLICATIONS OF THE CANVEY ISLAND STUDY:
A DISCUSSION

A. V. Cohen - Deputy Head, Planning Branch, Health &
          Safety Executive Baynards House,
          1 Chepstow Place, London, W2, ENGLAND
B. G. Davies - Deputy Head, Major Hazards Assessment
          Unit, Health & Safety Executive
          25 Chapel St., London, NW1, ENGLAND

## 1. INTRODUCTION

We should say before we start that any views expressed will
be personal, and do not necessarily coincide with the views and
policy of the Health and Safety Executive (HSE), which can be
found in published documents such as the Canvey Study[1] itself.

The study is almost certainly the fullest one ever done on
a non-nuclear site, though it will surely stimulate others as
problems arise. Indeed, this effort has already begun. The
estimates of risk are, as Professor Farmer has shown, quantitative. Once you measure things, you have a better feel for them,
particularly if you put some reliability estimate on the figures.
But why was this particular study done?

Canvey Island is an extremely large, but by no means unique,
"chemical and refinery park" near a centre of population, closely
associated with the Port of London, so much of which is now to be
found near the mouth of the Thames. It was the proposal to build
two further refineries which lead to the Government requesting
HSE to undertake this substantial study. The study was not cheap.
Its direct was about £ 400,000 sterling, and the follow-up costs
probably doubled that figure.

The report is a good example of a full technical assessment
made on the best information available, and in the spirit of
British practice with assessments of this kind, it has been published to advance knowledge, to ensure that all interested parties
have access to technical data, and to stimulate criticisms of
techniques.

The immediate cause of the report was the need to make a specific input into the planning process. We imagine that our listeners will be much more interested in the wider implications of the report and the general lessons that can be drawn from it. It is a major technical assessment to be viewed as a study in its own right, and a source book of hazard assessments which are useful for other risk assessment purposes. We will deal with all these aspects in turn. But let us consider the planning aspects.

## 2. THE PLANNING ASPECTS

The most significant finding of the report was that existing installations possessed a quantifiable risk of killing up to 18,000 people. The process of analysis showed where these risks lay, and how the probabilities could be reduced by factors of about 10 by "reasonably practicable" means: better bunds, water sprinklers, and the like. These significant improvements could most effectively be found by such an analysis, and their implementation, which is already under way, costs operators many millions of pounds.

The term "reasonably practicable" has a specific meaning in English law, and in a significant judgment[2] was defined as follows:

"Reasonably practicable" is a narrower term than "physically possible", and implies that computation must be made in which the quantum of risk is placed in one scale and the sacrifice involved in the measures necessary for averting the risk (whether in money, time or trouble) is placed in the other, and that, if it be shown that there is a gross disproportion between them-the risk being insignificant in relation to the sacrifice-the defendants discharge the onus upon them. Moreover, this computation falls to be made by the owner at a point of time anterior to the accident.

Thus there is what might be termed an economic aspect to "reasonably practicable". We should note in passing that we are in an area where "reasonably practicable" is measured in millions, for capital installations costing hundreds of millions. The study arose because of the need to assess the risk of a proposed new installation, which would have been added to larger existing risks from neighbouring installations. The judgement of the Health and Safety Executive against the background of all the risks that people experience at any time, was that the new installation should go ahead, provided that "reasonably practicable" steps were taken in both existing and proposed plant. The latest phase of a planning inquiry has endorsed the idea that the new installation should go ahead, but has questioned whether the remaining risks in some aspects of existing installations can be tolerated without further safety measures.  This aspect in turn is still being examined.

The report must be regarded as a unique, pioneering exercise which will prove to be a major turning point in risk assessment work.  In Britain things will never be quite the same again for potentially hazardous installations.  Proposals[3] for UK regulations would require such studies in the case of plants with large inventories of hazardous materials to be made by the plant management, subject to selective review by the Major Hazards Assessment Unit of the Health and Safety Executive.  It is hoped that the Executive's reaction would enable the quality of the surveys to be progressively improved and that the main benefit from such surveys would be in the plant improvements achieved by the management themselves.

## 3.  EFFECTS OF THE REPORT ON THE PLANNING PROCESS

The companies involved at Canvey and several others, as well as consultants in the field of risk assessment have, since the report was published, adopted quantified risk techniques and the methodology is in greater use than it was prior to publication of the Canvey Report.

As far as the planning process is concerned the report has undoubtedly encouraged planning authorities, even in advance of any Regulations, to require companies to carry out hazard surveys on proposed developments, particularly of potential major hazards, to demonstrate to the satisfaction of HSE that their plants will be safely designed, operated and maintained before the plant becomes operational and before final planning approval is given for it.

## 4.  THE RISKS REASSESSED:  MAIN FACTORS

Since the report was published in June 1978 a great deal of follow-up work has been carried out.  The Health and Safety Executive intends to publish later this year[4] a further report dealing with progress with improvements and the subsequent reassessment of the risks, detailing the relevant analyses.  The main factors in the reassessment will be:

(i) Physical improvements, e.g. provision of additional deck protection to Liquefied Natural Gas (LNG) ships, improved bunding around storage tanks.

(ii) Changes in operations, e.g. cessation of ammonia storage at Shell.

(iii)   Detailed studies by firms which have made a properly
        argued and documented case for reducing a risk pre-
        viously assessed on a provisional basis, e.g. the risk
        from a limited spill of ammonia at Fisons.

(iv)    Changes in assessment techniques, e.g. improved under-
        standing of the dispersion of heavy gases.

It is clear that there will almost always be scope for physical improvements to the plant that will lead to reductions in
the estimated risk levels.  It is also true that a number of detailed studies carried out by firms since the report was published
have provided the additional information and insights that have
enabled the HSE and the Safety and Reliability Directorate of the
UK Atomic Energy Authority (SRD) to reduce with confidence a risk
previously based on a conservative judgement figure.  However in
some cases the arguments presented by firms for reducing the risk
have not been sufficiently convincing, and the original basis for
the assessment has therefore not been changed.

## 5.  THE ACHIEVEMENT SO FAR

The average individual risk of death or serious injury from
an industrial accident to a resident on Canvey Island itself was
originally assessed as about 5.3 chances in 10,000 a year.  In
1981 this figure has been reduced to about 0.5 chances in 10,000 a
year, which represents a ten-fold improvement.  At Region A on
Canvey Island, which was originally assessed as being exposed to the
highest level of risk, the reassessment shows a thirty-fold improvement.  At Stanford-le-Hope work currently in hand is expected
to confirm the present five-fold improvement.

We do not see risk assessment as an end in itself, merely as
an effective means of identifying the right steps in achieving
greater safety.  All in all, the benefits to local safety resulting
from the Canvey investigation are already very substantial.  It
has also given a great impetus to the development of risk assessment methodology.  Let us now turn to these wider implications of
the Canvey Report.

## 6.  THE TECHNIQUE

The salient point is that the assessment is quantitative, and
that an attempt is made to put a reliability estimate on each
figure.  At least in the UK, and probably elsewhere, this is a
new approach in non-nuclear areas.  But the technique is not a
full fault- or event-tree analysis of all the equipment:  to do
this for a multiple site would be an impossibly big job.  Instead,
it is done on what we might call a "broad block" technique, described in the report as being to "identify major accidents, to

assess the likelihood for their occurrence, and then to assess
the likelihood that they would give rise to various degrees of
severity in their consequences". This has the advantage of forcing
one to analyse all the possible things that can happen from significant activities, of attempting to collect historical data worldwide (which will all its limitations will certainly show up annual
frequencies of order $10^{-3}$ to $10^{-4}$) and then to attempt a limited
form of fault analysis thereafter to trace through probabilities
the components of which are of order $10^{-6}$ per year.

The great advantages of this technique are:

1. It forces an analytical approach and an examination of the
   true safety of components normally disregarded.

2. It pins the data to historical experience, and appreciates
   the limitations of it; and

3. It makes suitably realistic assumptions of probabilities; while

4. Systematising the possible effects and our knowledge and ignorance of them.

5. It distinguishes between societal and individual risk.

So much for the technique. What sort of criticisms have been
made of this?

1. Understandably, operators have said that the estimates of risk
   are self-evidently too high; in effect that they have operated
   safe installations like this for many years, with an excellent
   record. A real strength of this kind of study is that it puts
   such remarks in perspective. A lifetime's experience without
   major events in several plants still means no more than safety
   to, say, $10^{-3}$ per year.

2. That the analysts have exaggerated the consequences. Now this
   is partly fair. The assessors were asked, when in doubt, to
   err on the side of caution; and their own estimate was that
   they might have exaggerated by a factor of 2 or 3. A comprehensive critique has been made by Messrs. Cremer and Warner,[5]
   who conclude that for the very low probability consequences
   the factors could conceivably be 10 or more. All this illustrates the kind of uncertainty inherent in calculations of this
   kind. Errors of 100 or more had been alleged by critics of
   the Rasmussen Report[6]. These are usually quoted as being in
   the other direction. While the technical critiques are usually
   careful to point out that the errors could be either way, later
   discussion at this Conference might consider why the general
   public often assumes that the errors are one way for non-nuclear

and the other way for nuclear reports.

7. LESSONS ON RISK ACCEPTABILITY AND PHILOSOPHY OF RISK

We can draw the following conclusions:

1. Risks are, and indeed in practice have to be, tolerated in existing non-nuclear plant, which would call for even stricter control in a nuclear installation. Yet the chemical industry is highly competent technically and has a good safety record.

2. A more critical attitude is adopted by the public towards a new installation than to an existing long tolerated risk. Some of this is psychological - the new is always salient - but there may be an underlying "hard headed" reason: the cost of moving an existing installation must be borne in mind.

3. A risk will be viewed differently by those who draw direct benefits from it. Most of the inhabitants of Canvey Island get no direct benefit from the plant which employs relatively few people. They are commuters who live in Canvey and work in London. The benefit is enjoyed by the nation as a whole since Canvey is a significant proportion of British refinery capacity.

4. In this and similar cases judgements have therefore to be made nationally rather than locally, but always with due regard to local feelings. What is called for is a political judgement based on a technical assessment. The two must not be confused.

8. ARE THERE OTHER CANVEYS AND WHAT SHOULD WE BE DOING ABOUT THEM?

In one sense Canvey is unique: in another sense something like the Canvey Island situation must arise around nearly every large chemical park or refinery, or in any major port. Even a green field site will in due course attract some housing. Each case would normally have to be judged on its merits, but in the broadest sense, there are lessons which can be applied from Canvey to any large installation.

Should assessments be done for every installation? HSE would immediately answer that it has not got the resources itself to assess the several hundred sites of one kind or another up and down England. The key lies, as we noted earlier, in requirements to notify large sites of this kind, and for the largest, for a further requirement that the operators themselves present hazard analyses. It is unlikely that hazard analyses will always be done in quite the depth of Canvey, but they will still be expensive operations. The UK is not alone in this. We understand that the whole of the European community (EC) sees the need for requirements of this general kind, consistent with the views emerging in a Draft EC Directive.

## 8. THE USEFULNESS OF THE CANVEY REPORT IN GENERAL RISK ANALYSIS

The main benefit seems to us to be the encouragement of the quantitative approach in areas which have not previously been quantified, and the way that the judgement on risk is inevitably sharpened by this. But there are other benefits which flow from systematic thought.

In thinking through "what might happen" one is forced to examine processes in a new light, and also not only to think in terms of the "worst possible accident" but to examine systematically the medium probability moderately high consequence risk, which we are gradually learning must be thought of seriously. Direct experience teaches us about the high probability low consequence risk. Political salience forces us to estimate the low probability high consequence risk. We must not ignore the intermediate risks.

Another general benefit that comes from the accumulated systematic detail of hazard mechanisms. Probably none of us, if asked how a thousand tonnes of ammonia would behave, could have predicted what has come out of this study. We have found this report extremely useful in other more informal hazard analyses which underlie safety judgements of various kinds. Information on ammonia, chlorine, or gas dispersal is all to hand, and readily accessible. The techniques now exist which allow them to be applied to specific situations in real detail if this were judged worthwhile to do.

The results can also be applied to wider studies, but must always be treated with caution. We found this in the Comparative Risks Study, which one of us did in collaboration with D. K. Pritchard [8]. The risk of an oil-based electricity system must be associated in some way with that of a refinery. A straight proportional calculation indicates a risk for major incident linked with a 1 GW output for a year which coincidentally is rather larger than that associated with nuclear power. This is a fair reflection of the average national risk associated with an oil burning power system. But it does not necessarily reflect the risk associated with building the next oil burning power station: the marginal effect of an addition may even be of opposite sign to the national average. Deciding not to build an oil-fired power station might lead to a decision not to build an extension to a refinery: conversely it might lead to extra cracking facilities, to use up the residual fuel oil. Thus risk could be decreased or increased depending on specific circumstances.

The Table (drawn from figures quoted in the Comparative Risks paper) shows the oil catastrophic risk compared with those of

*A. V. Cohen and B. G. Davies*

typical nuclear installations.  It also shows the chlorine risks
associated with all power stations, and some rail risks associated
with coal transport for coal power stations, again deduced in
much the same spirit as the Canvey Report.

TABLE 1
THE PROBABILITY OF SOME MAJOR INCIDENTS*[9]

| | Within Weeks | Deaths Delayed 20-40 Years | Annual Probability |
|---|---|---|---|
| PWR remote | 0 | 30 | $10^{-5}$ |
| PWR | | | |
|   remote | 100 | 1,000 - 10,000 | $10^{-7}$ to $10^{-8}$ |
|   semi-urban | some 1000's | some tens of 1,000's | $10^{-8}$ |
| Oil Refinery ** | 1,500 18,000 | | $11 - 18 \times 10^{-6}$ $0.4 - 1.6 \times 10^{-6}$ |
| Coal Derailment and Collision | | | |
| 1.  Passenger Train | 20+ | | $3 \times 10^{-5}$ |
| 2.  Train with Dangerous Goods | 16+ | | $2 \times 10^{-6}$ |
| 3.  Both in Tunnel | 600 | | $10^{-8}$ |
| Power Station Chlorine Store | 100 | | $10^{-4}$ |

*Annual Chance Per Unit of Output (1000 MW(e) at 75% Load Factor
**Strictly speaking, these figures refer to casualties, about
half of which are fatal:  probably rather more than half, for the
case of 18,000 casualties.

One of the founders of that Canvey spirit is with us, and has just expounded it memorably to us. We have tried to show how this spirit can be applied to lead to some extremely interesting and stimulating conclusions. We are sure we will all be rewarded from the further discussions. We suggest the structure of our discussions might consider the significance of this approach for:

1. The planning process, both at Canvey itself and more generally.

2. General risk assessment.

3. Philosophy of risk and risk acceptability.

4. Applicability to other similar sites.

5. The use of the figures as a wider risk assessment reference book.

REFERENCES

1. Health and Safety Executive. Canvey: An Investigation of Potential Hazards from Operations in the Canvey Island/Thurrock Area. London, HMSO (1978).

2. Fife, I. and Machin, E.A., Redgrave's Health and Safety in Factories. London, Butterworth (1976).

3. Health and Safety Commission, Consultative Document on Proposed Hazardous Installations (Notification and Survey) Regulations. London, HMSO (1978).

4. This document has since appeared. Health and Safety Executive. Canvey: A Second Report. London, HMSO, (1981).

5. Oyez Intelligence Report. An Analysis of the Canvey Report Prepared by Cremer & Warner. London, Oyez Publishing Ltd. (1980).

6. U. S. Nuclear Regulatory Commission, Reactor Safety Study Wash 1400, October 1975.

7. The EC Directive has since appeared. EC Council Directive of 25 June 1982 on the Major Accident Hazards of Certain Industrial Activities (82/501/EEC). Official Journal Reference: OJ/No L230/1-5/8/82.

8. Cohen, A. V. and Pritchard, D. K., Comparative Risks of Electricity Production Systems: A Critical Review of the Literature. Health and Safety Executive, Research Paper 11. London, HMSO (1980).

9. Cohen, A. V., Relative Risk Studies: Their Relevance to Decision Making. The Analytical Approach in Perspective. In proceedings of the Sixth Annual Symposium of the Uranium Institute, London, September 1981. London, Butterworths Scientific Ltd. (1982).

BENEFITS AND RISKS OF INDUSTRIAL DEVELOPMENT WITH SPECIAL
REGARD TO THE FIELD OF POWER PRODUCTION

Arnaldo M. Angelini

Member of the Italian National Academy of Lincei
and Honorary President of ENEL

1. INTRODUCTION

The subject matter of the program of lectures which I have
the honor of introducing concerns one inevitable unfavorable
consequence of the activities aimed at the progress of mankind,
and especially at the attainment of better living conditions.

In effect, risk has always been and is so closely and insepar-
ably connected with every stage in the achievements of technology--
which since the dawn of civilization have led to what we generally
define as progress--as to constitute a necessary negative component of those achievements.

I trust that this consideration will fully justify some preliminary remarks about the positive aspects which play a decisive
role in the evolution of technology. In fact, I am convinced--
and I hope that this will clearly emerge from what I will be
saying--that a systematical approach to the risks of technology
must necessarily start from a review of those decisive factors.

If this were not the case, there would be no justification,
among other things, for the subject of one of the lecture of the
cycle, which concerns the application to our field of discussion
of the methods of cost/benefit analysis or more in general between
the positive and negative consequences of all human activities.

2. THE EVOLUTION OF WELL-BEING AND TECHNOLOGICAL PROGRESS

Having adopted this line of thought, it will not appear out
of place to make a digression concerning mankind's primordial and

basic need for progess, meaning changes from a given condition to another, regarded as preferable.

To give substance to the concept of progress, we shall indicate its components, that is those factors which involve or bring about an improvement in human conditions, both individuals and society:

a)  the uplifting of moral consciousness, which involves the attribution of ever increasing importance to moral values.

b)  The acquisition of broader and deeper knowledge, which mankind has always been seeking and which we regard as essential to progress.  In particular, while there is no question about the more immediate "usefulness" of scientific knowledge for the purposes of material progress, we cannot overlook its educational value on the philosophical and on the moral planes, and

c)  the improvement, as fast as possible, of material living conditions in the broadest sense of the term; this is an aim regarded as valid in the civilization in which we live* and which we must take into account, since the definition obviously depends on the general concepts which prevail in the world of today.

The attainment of these goals involves the need for a plan meaning the indication, for a community, of general and specific guidelines designed to promote and expedite the improvement of living conditions; the identification of the elements which constitute the essential presuppositions of the plan as goals to be attained; the time scheduling forecasts and the coordination of the stages of implementation of those goals; the indication of the resources and capabilities needed to attain them; finally, the prediction of the results which are expected could be achieved in the various sectors.

It should be borne in mind that man is not responsive only to impulses of a rational nature, and that therefore we cannot

---

*    It was so in many other civilizations, but not in all.  Where certain forms of mysticism have been and are prevalent, the renouncing of all material comforts was and is regarded as a factor of human progress.  Indeed, even in our own civilization certain material comforts it offers us are not universally accepted as factors of progress.  We do not wish to get into such delicate and vast issues, for we would stray from our subject.

count, even on a probabilistic basis, on a reliable prior knowledge of the responses that certain causes will elicit in society.

One typical example of this unpredictability is provided by
the reaction of public opinion to the peaceful use of nuclear
energy. It was favorably and even enthusiastically received in
the years that followed the first Geneva Conference organized by
the UN in 1955, while in the last decade, and especially after
1973 oil crisis, when the new source has become indispensable,
an unprecedented campaign was launched against the most effective
substitute for petroleum.

Our way of living and operating evolve in time. Time is the
universal and totally independent reference variable in the evolving
of any activity. The attainment of any end occupies a time interval, the achievement of a goal is marked by the moment, i.e. by the
time, in which it occurs, etc. It ensues that, for the attainment
of its goals, planning must include a time schedule for its implementation. It is customary to distinguish, in economic life,
between: the short term, as the time during which takes place
and is concluded that evolution of phenomena which does not have
as a necessary precondition some change in the structure of the
community's economy and does not necessarily derive from it; the
medium term, in which phenomena evolve not only because of facts
deriving from the existing structure but because of the changes
in the latter; the long term, in which the effect of changes in
structure prevails.

In the medium and long term the effects of those factors
which change the structures manifest themselves, those factors
are largely determined by scientific and technical progress. If
we could apply to this matter the language of mathematics, we
could say that the rate of progress, far from being uniform, follows an exponential law. This, of course, is merely a generic
consideration, since the "progress" of technology is not a
measurable quantity and cannot be expressed in figures.

3. THE EVOLUTION OF THE RISK CONNECTED WITH TECHNOLOGICAL
   PROGRESS

All human activities, from the simples to the most complex,
do involve risk, meaning the possibility of damages to persons
and things, depending on many different circumstances, some of
which are not predictable even in terms of probability.

Any technological innovation, in its application, involves
on the one hand risks, and therefore additional damages, and on
the other a decrease in risks, and therefore benefits. In effect
the balance of costs/ or, better, damage/benefits undergoes
changes as a consequence of the use in practical applications of

every contribution of applied science. This change, however,
does not remain confined to the area of application of the innovation, but very often extends to a much broader area.

To illustrate this concept, I shall mention a dam across a
river which, as regards risk, causes, in the case of collapse (a
very rare event which, however, has sometimes occurred), severe
damage to persons and property. The same dam, however, offers
certain benefits, such as for instance those resulting from irrigation, energy generation, possibly river navitagion, etc. Besides, in some cases the dam will also provide for the containment
of floods which, in its absence, would cause catastrophic consequences. Consider, in this connection the tremendous benefits
obtained from the regulation, by means of a number of dams, of the
waters of the Tennessee River and its tributaries.

To move to an entirely different field, the introduction of new
drugs does certainly involve risks, of which we have had unmistak-
able evidence, but at the same time it is to them that we largely
owe the substantial increase in life expectancy in civilized
countries.

Many other examples could be mentioned to show the importance
of risk analysis and risk assessment with respect to all significant consequences, direct and indirect, of technological innovations. And here it seems appropriate to mention the fact that, in
the assessment of the cost/ or, better, damage/benefits of technological innovation, there has been for some time a tendency to
overstress damages and underestimate benefits, especially of some
major technological innovations.

This tendency is mainly due to a persistent distortion of information, which heavily affects public opinion because of the
prevailing power of penetration and dissemination of information
which induce concern and anxiousness.

The responsibility for the consequences of certain distortions
of the information which reaches public opinion lies with those who
have no hesitation in using the very insidious weapon of an insis-
tent attack on the credibility of experts. The insidiousness of
this weapon becomes much greater when experts who enjoy prestige
in a certain branch of science or technology express opinions about
matters which lie outside their areas of expertise, and which they
know little about.

This fact leads us to extend the damage/benefits analysis of
industrial projects to information, with regard to the progress
of information media, and especially those which reach public
opinion.

4. THE NEED FOR A SYSTEMATICAL APPROACH TO RISK ANALYSIS AND
   ASSESSMENT

Given the interconnection and interaction between different
sectors of human activity, and in particular as regards the damages and benefits of technological innovations, whose effects often
"spin off" far beyond their specific scope of application, it might
perhaps be in order to refer in the first place to the sectors
of activity connected with three essential factors of economic
and social progress, and namely:

- raw material and products obtained from them,

- energy in its many forms, conversions and uses, and

- information in the broadest meaning of the term, which
  comprises all its forms and activities concerned with
  information processing, transmission and distribution.

These are three pillars of economic and social development,
so indispensable that the absence of any of the three will make
the others inoperative.

Although this is not the place for drawing a systematical picture of the human activities which advance apace with technology
and determine the economic and social development of various countries, it seems in order to stress that on these pillars can be
centered that articulation of the major production sectors, evi-
dencing the interconnections and interdependence which are also
relevant to the risk analysis and evaluation, extended also to
activities--like transportation--which though unconnected with
each of the above mentioned pillars, draw nourishment from the
first two and use the third.

This is one way to follow an approach to our subject that
will facilitate the use of the techniques, methods and tools of
systems engineering, aimed above all at the assessment of the
incidence on all system components of the innovations, or even
only variations, in the structure of the whole system.

To my knowledge, we still do not have any study which, referring to a generalized model of an extensive industrial system,
shows, also with regard to the problems of the environment, the consequences of any significant change in a component on the damage/
benefits relationship of the whole.

In the following part of this paper, rather than dwelling on
considerations of a general nature, I believe it more useful to
draw the attention of my audience to the features of various facets
of two concrete problems of great significance on an international

level, and very particularly for Italy.

The first concerns a comparison between the various ultimate alternatives for the generation of electric power, with regard to the damages and benefits of the possible solutions. As we will see, this problem involves among other things the control of major risks, such as those interconnected with the increasingly disturbing uncertainties involved in the importation of fossil fuels, especially oil, from the producing countries.

The other problem, also of considerable magnitude, concerns the siting of industrial plants in general, and particularly those for energy generation, in the dual intent of obtaining the highest degree of compatibility with the environment, while at the same time minimizing risks.

The first problem is of basic importance to Italy, which depends almost entirely on imports for the supply of primary energy sources. The second is no less important, for environment problems are particularly relevant to the case of Italy, due to its orography, to the seismicity of most of its territory, to the high population density, to the very special requirements of tourism and of the conservation of artistic and cultural assets. The two problems besides, are closely interconnected.

This paper ends with a chapter concerning one specific risk which in certain countries, the first of which in Europe is Italy, is becoming a cause for great concern.

I am talking about the consequences--which without any exaggeration can be termed catastrophic--of the actual moratorium in the implementation of electric power plant construction programs.

It appears from a study of our agenda that the subjects mentioned above do fall within the scope of the symposium and do not overlap with those discussed by other participants.

5. A COMPARISON BETWEEN PRODUCTION CYCLES, WITH SPECIAL REGARD TO ELECTRICITY GENERATION

It often happens that, in assessing and comparing risk, attention is concentrated on one of the stages of the production process of cycle, segregating it from the others, while the "systems approach" I advocate involves the extension of the comparison to the whole industrial process or production cycle, ranging, for instance, from the extraction of the raw material to the delivery of the product of the user.

It is obvious that no uniform guidelines can be offered for

all problems of this kind, because of their wide diversity. I therefore believe it appropriate to refer to the concrete case mentioned above, which is particularly significant both from the methodological viewpoint and because of its importance and present implications.

I am referring to a necessarily summary comparison between two electricity generation cycles: that from fossil fuels and that from the nuclear source.

For the sake of concision, I will dwell on some aspects of this comparison which, although important, have been somewhat underestimated. This is why, for instance, I will touch only briefly on power plant safety problems.

To this end, I shall draw on a discussion of the subject which I made last year at the International Conference on Current Nuclear Power Plant Safety Issues organized by the International Atomic Energy Agency in Vienna at the initiative of the United Nations.

With certain exceptions, sometimes also important, it is generally believed that the consequences of the oil crisis have grown so serious that we must abandon the use of this raw material for the generation of the electricity needed to meet new requirements.

Furthermore, it is acknowledged that the situation demands in general that oil and natural gas reserves be used for the purposes in which they cannot be replaced, in particular in the chemical and petrolchemical industry and in transportation.

This leads in some ways to a simplification of the problem, limiting the choice to that between coal and nuclear fuels. It should be added that the simplification is in part apparent, for the use of coal in thermal power plants poses as many and probably more problems than the use of fuel oil and, in any case, of natural gas.

We thus get to the most controversial question: the choice between the coal and/or nuclear fuels in the generation of thermal power to meet new requirements.

The purpose of this report is not that of making an analytic discussion of this choice--which would take time and make us stray from our subject--but calling attention to the factors that have a significant influence, in view of the risks and of the safety problems--in the broadest sense of these terms--towards an overall evaluation embracing, as I said before, not only power plants but the entire generation cycle and the problems connected with this choice.

The generation cycle referred to comprises all activities, beginning from the exploration for primary sources and ending with the delivery of power to the user.

The generation of electricity in power plants, using either coal (as well as from other fossil fuels) or nuclear fuels, constitutes one phase in the cycle, undoubtedly important but which cannot alone determine a final choice.

We must therefore direct our attention to the phases of the generation cycle which, following a natural order, must be taken into consideration for the purposes of an overall comparison.

a.  <u>Mineral exploration</u>:  does not pose special problems with respect to the goal to be achieved. The risks are limited and safety problems practically non-existent for the two sources forming the object of the comparison.

b.  <u>The working of coal mines</u> presents risks that are well known from the past experience. The victims are many, as it appears from official statistics. One index of relative risk for coal mining could be expressed in number of victims per million tons of coal mined.

From the risk standpoint, the working of uranium mines does not pose any problems. Uranium mining does require certain measures for the protection of miners against radiation; these measures can be applied without difficulties in the present state of our knowledge.

c.  <u>The treatment of uranium ores</u> does involve protection requirements but practically no risks, while the treatment of mined coal does not involve any special requirement.

d.  <u>The access to these primary sources</u> by importing countries raises transportation problems which, on equal energy contents, are very limited, almost insignificant for uranium, but very substantial for coal, not only as regards land transport (in particular by rail and pipelines) already very costly, but also ocean transport, which among other things requires the construction of often very costly facilities. The problems of access to primary sources involve major <u>risks</u> in the case of petroleum or its products--in our case fuel oil.

We are not talking here about the safety of persons
and property, but the extremely serious risk of a cut-
off in supplies resulting from international political
complications or even wars.

Without going into details, it appears evident that,
expecially for countries like Italy which depend almost
entirely on imports, this risk is very high in the case
of oil and its products, as regards both the access
to primary sources (which to a significant extent are
concentrated in politically rather unstable countries)
and shipments (for example, the situation in the Persian
Gulf).  The situation concerning coal is less critical.

To the problems of transportation may also be associated
those of:

e)    The stocking of coal which, as compared to that of nuclear
      fuels, is very costly and subject to considerable limita-
      tions.  While the amount of coal making up the reserve of
      a thermal power plant cannot be any larger than a few
      months supply for the running of the plant, nuclear fuels
      can be stocked for some years of production at a compara-
      tively low cost, without creating environmental problems.
      It should be stressed, however, that for the purpose of
      continuity of production, fuel elements for nuclear power
      plant are fabricated mostly in the country that uses them--
      in any event, from this viewpoint Italy is self-suffi-
      cient--and therefore, for the assurance of production con-
      tinuity, it is unnecessary to stockpile fuel elements but
      the raw material needed for their fabrication, i.e.
      uranium oxide, whose value is considerably lower.

It emerges from this consideration that the cost of fuel
stockpiling for an uranium-importing country like Italy
has a very low impact on the cost of energy.

When the breeder cycle will become industrially established, the amounts of depleted uranium from en-
richment and irradiated fuel processing plants may constitute a strategic stockpile sufficient for dozens of
years of operations.

It is hardly necessary to stress the very great importance
of this fact in view of the reduction of the blackout
risks deriving from transport difficulties that may derive from international complications and wars.

f)  One necessary phase of the cycle for the supply of
    light water and graphite nuclear power plants is the
    enrichment of the uranium required for the fabrication
    of fuel elements.

    This phase has a cost--which, however, does not affect
    the competitiveness of the power generated--but does not
    involve any particular risk.

    Uranium enrichment involves for certain countries a
    dependence on imports which carries very low supply
    continuity risks, since enrichment plants are located
    in countries which, politically, are far stabler than oil
    exporting countries.

    In any event, Italy is a 25% partner in the Eurodif con-
    sortium plant located in France, from which it can ob-
    tain its entire requirements.

    For enriched uranium, no transportation problems exist.

    Uranium enrichment is not necessary for the supply of
    heavy-water moderated reactors.

g)  I can omit a discussion of the questions, which are well
    known, concerning the phases of the nuclear generation
    cycle that takes place in the plants.  In particular,
    as regards safety, it is a known fact that such plants
    have formed the object of efforts never spent on any
    other sector of technology, as I will also discuss
    further on.

h)  As regards fission by-products, certain problems do arise
    with nuclear power plants, concerning:

    -   the radioactive effluents,

    -   the processing of irradiated fuels,

    -   the storage of plutonium, and

    -   the disposal of radioactive wastes.

    These problems were already widely treated and there-
    fore we do not have to discuss them in any detail.

    On the other hand, the problems concerning coal-burning
    plants are:

    -   the gaseous effluents of combustion, and in

particular carbon dioxide, sulphur dioxide, nitrogen oxides.

- The solid wastes, i.e. ashes, which create problems of storage, disposal, and when possible, utilization.

These problems concern the immediate, and sometimes farther, environment of the thermal power plants.

k) Some experts have expressed concerns of the emission into the higher atmosphere of the carbon dioxide which, in the long run, could cause an increase in temperatures of the earth's surface, with very significant adverse consequences. According to some assessments, the contribution to this phenomenon of the thermal power plants burning conventional fuels, including coal, could be a major one. This is therefore a long-term risk, like that often evoked for nuclear plant wastes containing radioactive isotopes having a very long life.

i) The problems of dismantling or decommissioning arise only for nuclear power plants.

Our purpose is not to dwell on the analysis of these questions, but to urge an objective examination, in view of an overall assessment comprising all phases of the production cycles compared and considering the problems in their space and time scope, and above all with all the necessary objectivity.

It is unnecessary to stress that the foregoing considerations make no pretence at being complete. They are merely intended to exemplify some of the concepts which I had previously expressed.

6. THE COMPATIBILITY BETWEEN INDUSTRIES AND THE ENVIRONMENT, WITH SPECIAL REGARD TO POWER GENERATING PLANTS

A while ago I mentioned the problems of electric power generation in relation to the environment, with regards to the cost/benefit comparison between the use of coal and that of the nuclear source.

The issue is one of those to which recently public opinion has been growing increasingly sensitive. To these problems and to the limitation of the risks of damages to the environment a special effort has been devoted in recent years by many industrial enterprises and public utilities.

Again with reference to a concrete case, specifically that of electric power generation, with which I am more familiar, I would like to stress once again that systems engineering is the most appropriate methodology also for the solution of environment problems, in relation to their eminently interdisciplinary nature and since it makes it possible to appreciate the benefits and damages of technological evolution in every sector of human activity--in the sense I mentioned previously--taking into account the existing or predictable interactions and interdependences.

I have followed these methodological principles in evaluating the relations existing between the selection of sites and the functional and safety requirements of nuclear power plants in Italy.

Efforts were made to obtain, through a rational choice of a site for a nuclear power plant, a limitation of its effects on the environment, apart from the very low probability of a possible plant accident.

The project, initiated more than one decade ago, concerns a systematical analysis of the features, not only physical but also and above all socio-economic, of Italy's entire coastal strip, measuring 7,500 kilometers, since along it the cooling requirements of thermal power plants, particularly nuclear, can be met with seawater. This project was completed with the publication of an Atlas of Coasts, the only work of its kind in the world. The Atlas, which is now well known to all concerned, has been received with interest both in Italy and abroad, it bears witness to ENEL's (Italian Electricity Authority) substantial and systematical efforts in searching for plant sites that will best reconcile the often conflicting requirements deriving from technical, economic, social, and land and environment conservation considerations.

A work similar to that done for the coasts has already been completed by ENEL, for the main Italian River course, the Po, and is approaching realization for other major Italian rivers, to study the possibilities of siting nuclear plants not only along the coasts but also inland.

It would take too much time to discuss the contents of the Coast Atlas.

7.   A POPULATION ANALYSIS EXTENDED TO THE ENTIRE ITALIAN
     TERRITORY FOR INDUSTRY SITING PURPOSES

The data in the Atlas, albeit complex and from many aspects comprehensive, are but part of the data being collected and studied by ENEL. ENEL has been pursuing several activities to obtain a detailed and comprehensive picture of the population data to be used in evaluating the possible nuclear power plant sites. This investigation has involved a careful survey of the entire national territory.

As it will clearly appear from a summary description of this study, it provides particularly effective guidance in making a preliminary selection of the areas best suited for the siting of industries and in particular electric power plants.  This population analysis supplies the essential information for the limitation of the risk connected with the plants, as regards population safety.

It goes without saying that other elements of evaluation also have a decisive effect on the siting of such plants:  they concern transportation, possibilities for the disposal of heat, etc.

The analysis that I am going to describe requires massive data handling, and therefore the use of a computer.  A "data bank" stores to each inhabited center the topographic and demographic data according to an appropriate geometric schematization of the shape of the center.  The entire national territory is rperesented by a grid of 400m square mesh, the nodes of which are the calculation points.  At each node the average population density in a circle of predetermined radius is computed as the ratio of the population in the circle and the area of the circle.

The local population density maps, offer a view both detailed and general of the population situation in the entire area of territory considered and make it possible to spot the areas having the lowest population density values.

As is readily understandable, this type of charting is much more significant than a simple population density diagram when it comes to selecting the site for a nuclear or non-nuclear power plant, as well as for any industrial factory.  However, the population density thus calculated does not take into account the radial distribution of the population around each point of the national territory, while it is interesting to know or at least characterize this distribution with a single index.  In fact atmospheric agents gradually dilute the effluents emitted by a plant as they carry them away, with the result that the distant population is less exposed.  An evaluation was therefore made, for each node of the grid, of a "demographic quality index", attributing to each population group a "weight" inversely proportional to the power 1.5 of the distance from the node considered, and then averaging this weighted population density over circles of predetermined radiis.

I believe it unnecessary to stress the usefulness of such a survey, extended on a capillary basis to the whole of Italy and expecially designed as a source of objective and reliable elements for judgement on the siting of industrial plants, not only for plant manufacturers and operators, but also for the central, regional and local government authorities who are called upon to issue the required permits and licenses.

There is no need to elaborate on the obvious consideration
that an industrial plant site is not chosen solely on the basis
of population analyses.  It will suffice to think of the problems
of electric power transmission to realize the importance of a
barycentric location of generation plants with respect to consumption centers, and no less weight should be given to many of
the other requirements that I mentioned when dealing with ENEL's
Coast Atlas.

8.    THE SEISMIC ANALYSIS OF THE TERRITORY AND THE DEFENSE AGAINST
      EARTHQUAKE RISKS

Earthquakes constitute a risk which, to various degrees, is
much more serious in Italy than in many other countries.

It should be noted at this point that earthquakes--even the
most catastrophic, such as those of the Belice Valley in Sicily,
of Friuli in the Friuli-Venezia Giulia and of Irpinia in the South--
have caused no significant damages to industries, and in particular to power generating and transmission facilities.  In effect,
the choice of sites for the latter, especially in the last few
decades, has taken into account the seismicity of the selected
areas.

Nevertheless, building construction activities in Italy are very
much concerned with the defense against earthquake risks, which have
proved very heavy, in terms of both casualities and material damages.  This is not the place for a discussion of the defense of
the population against earthquake and of the measures adopted by
the Government to this end.  I will therefore concentrate on the
investigations which have made it possible to achieve the highest
degree of safety now obtainable in thermal power plants--especially
nuclear--in the case of earthquakes.

The data which ENEL has been collecting and processing for
more than ten years for the study of possible sites for nuclear
power plants are not limited to those shown in general terms
in the Atlas, but also extend to geotectonic studies.

As is widely known, the degree of seismicity is another
element of judgement about the suitability of a site, of major
importance to the safety of nuclear power plants.  Complex geo-
logical and seismoligical surveys have to be conducted in a large
area (in the order of 200 to 300 km of radius) around the
chosen site, to make sure that the plant can operate under safe
conditions.  This basically involves looking for seismogenetic
structures, evaluating their seismic potential and predicting the
effects that such potential, if suddenly released in an earthquake, might induce on the plant.

Considering the number and size of the areas studied, also these investigations virtually covered the entire national territory; ENEL therefore decided to combine and coordinate the various studies in a single work of synthesis. To this end, with the help of the most qualified Italian experts, a methodology of investigation and mapping of the recent-evolution or active tectonic structures was developed; this methodology is entirely novel at least in Italy. A further thorough study was conducted on seismicity in Italy, taking into consideration all the earthquakes that have occured in the country from the year 1000 to 1975. The result of this program is a "Map of the Italian Neotectonic Features" and a "Catalog of Italian Earthquakes from the year 1000 through 1975". For each seism, the Catalog contains some 60 data; an appropriate computer program allows the management of the data stored, depending on the information required, and their graphic representation through a plotter.

The map of geotectonic elements and the earthquake catalogue will be followed by a certain number of seismologic charts the preparation of which will be accomplished in the framework of the finalized programs of CNR (Italian National Research Council).

CNR has just published a "Proposition for seismic re-classification of the national territory". This re-classification was carried out--I quote integrally from the foreword of this document--" in order to single out the parts of the national territory the exclusion of which from the areas of application of the codes for the construction in seismic zones, appears to be unjustified in the light of the knowledge available". This because we sought to supercede a classification that considers only the areas affected by the earthquakes in this century.

The consideration of the seismic events in Italy since the year 1000, and the availability of a better knowledge of the seismogenetic structures allowed this new map to be prepared that, even if representing only the first stage of the entire work, supplies a better representation of the seismic situation in the Italian territory.

9.   INVESTIGATION AND CHECKING THE INTERACTIONS BETWEEN ENERGY
     GENERATING PLANTS AND THE ENVIRONMENT

Even though a considerable part of the following considerations apply to any industrial plant I shall refer to the investigations made and measures taken to ensure the best compatibility between energy generating plants and the environment.

This, as we shall see, is a very extensive working program, to which I will be able to refer only very succinctly to describe its chief features.

In effect,a substantial contribution aimed at the limitation
of the environmental consequences of the operation of conventional
and nuclear power plants, is the product of an intensive research
and development activity on which I shall report briefly.

I shall begin with the activities concerning the atmosphere,
mentioning in the first place the studies and experiments carried
out to determine the atmosphere's capacity for dispersion and
self-cleaning with respect to pollutants. In this area, ENEL has
carried out complex studies on calculation models, leading to the
choice of a mathematical model which is used for predictive calculations relating to all new ENEL thermal power plants. The
soundness of this model has been tested by experimental surveys
conducted through the fixed monitoring networks of La Spezia and
Vado Ligure and some forty extemporaneous programs of chemical
and weather surveys carried out at and around several others of
ENEL's thermal power plants.

Mention should be made of the studies and surveys made of
local micro-meteorological conditions, which are carried out at all
new-plant sites by mobile laboratories and fixed weather stations,
and of the extemporaneous programs designed to determine the degree
of background pollution existing prior to the entry into service
of the plants.

Of special interest are: (i) studies for the optimization,
in the design stage, of dispersion characteristics and for the control of emissions, comprising design studies aimed and ensuring,
based on locally obtained chemical and weather information, the
most effective dispersing action of stacks; (ii) theoretical studies
and experiments towards the optimization of combustion processes;
(iii) the experimental use in the furnaces of several plants of
pollutant-control additives; (iv) the optimization of dust abatement methods, with special regard to the use of electrostatic pre-
cipitators; (v) studies on the progress of the methods for the
desulphuration of fuels and combustion gases, etc.

In addition to these studies and experiments we have the development and perfecting, at ENEL's Piacenza Central Laboratory, of
methods for the measurement of the various pollutants and weather
conditions, the installation and operation of fixed measurement
networks at the Piacenza, Chivasso, Turbigo, La Spezia, Vado
Ligure and Ostiglia power plants, the installation at all new
plants of monitoring systems equipped with instruments for the
measurement and continuous recording of the ground-level
concentration of sulphur dioxide, the collection and computer
processing of data obtained from the measurement and monitoring
networks.

Special attention is devoted to the studies and research concerning the influence of nuclear power plant on the environment. Close attention is paid to local weather and climate conditions: one or more anemographs and standard weather huts are installed at all nuclear plants; the Trino Vercellese Plant is equipped with a 390-foot weather tower, with provision for temperature measurements at different levels. Radioactivity measurements (for which methods of high sensitivity and energy discrimination have been developed) include measurements on specimens of environmental materials (milk, fish, grass, water, etc.) before and after the first start-up of the plant; measurements of the natural gamma-radiation background; field surveys and measurements by monitoring networks and radio-ecologic surveys conducted every 3 to 5 years. The purpose of the latter is to determine the actual dispersion into the environment of the radioactive discharges, and differ from those conducted by the monitoring networks chiefly because of the greater sensitivity of the instruments used and the larger size of the area checked.

## 10. THE DISPERSAL OF WASTE HEAT FROM THERMAL AND NUCLEAR POWER PLANTS - RESEARCH, DEVELOPMENT AND SOLUTIONS ADOPTED

Mention should be made of the formulation and development of predictive calculation models to evaluate heat dissipation in a water body and the checking of the soundness of these models through surveys and measurements conducted in the vicinity of power plants.

Very major efforts are devoted to the model studies aimed at determining the various interactions--mechanical, hydraulic and thermal--between a power plant and the body of water from which the required cooling water is drawn.

Again as regards the dispersion of cooling-water heat in water bodies it may be of interest to mention the survey programs conducted by ENEL in cooperation with a specialized U.S. company. In such surveys the temperature distributions in watercourses near several ENEL plants have been measured by infrared aerophotogrammetry; the method used is very advanced and was originally developed in the U.S. for use with some artificial satellites. While the aerial photographs were being taken, local water temperatures were also measured at certain sites as a check on the results; new survey programs are now planned.

Before concluding this part, I wish to mention a program of major importance initiated by ENEL for the purpose of concretely determining, through in-depth ecological investigation, the effects of the release of the discharge heat from a nuclear power plant into an ecosystem representative of the Po River. The choice of this river is an obvious one, since it is Italy's only river having a flow rate large enough to ensure the cooling of high-capacity nuclear power plants without cooling towers.

The ecosystem representative of the Po River has been located in the river stretch facing the Caorso Nuclear Power Plant.

The purpose of the program is to determine <u>quantitatively</u> the plant's total impact on the river resources and to provide useful information towards the interpretation of the possible mechanism through which any effects were produced.

The river's resources considered in the program are the following:

a)   diversion of water for drinking and other direct uses;

b)   irrigation;

c)   recreational activities (hunting, fishing, water sports, etc.);

d)   commercial fishing;

e)   discharging of wastes of various kinds (municipal, industrial, agricultural);

f)   other minor uses.

An evaluation of the extent of these resources necessarily involves a determination, in both the pre-operational and operational stages, of the state of the following components of the aquatic ecosystem:

--<u>fish population</u> (for commercial fishing and recreational activities; it is also an indicator of other resources);

--<u>fowl population</u> (for hunting activity and recreational activities in general, it is also an indicator of other resources);

--<u>communities of microorganism living on the bottom (bentonic) and at the surface (planktonic) of the river</u> (for the assessment of biological resources; and "indicator parameters");

--<u>associations of hydrophytes</u> (for the assessment of biological resources);

and

--<u>general quality of the water</u> (for drinking and industrial use, irrigation, recreational activities; it is an indicator of other resources).

I shall not go into a description of the program but merely mention that it involves tens of thousands of analyses and tests and that requires the work of many persons.  In addition to some 30 ENEL technicians employed on sample collecting and on physical-chemical and biological analyses, expert taxonomists from universities and natural history museums are making a valuable contribution in identifying and classifying biological samples.

I shall also mention that the soundness and interest of the program have been recognized by the European Economic Community, as well as in national scientific circles.

A similar work will be carried out for Montalto di Castro Nuclear Power Station under construction.  A research program is being studied that, even if different from the program under way at Caorso, owing to the different conditions of the two sites tends as well at evaluating and controlling the overall impact of the station on the environment.

I am ending here this listing of ENEL's research activity in the environment field which, as I have said, makes no pretence at being complete.  It shows the efforts and significant resources with which ENEL faces and solves the environmental problems which can be created by its plants.

## 11.  DAMAGES AND RISKS RESULTING FROM THE SLOWDOWN IN THE CONSTRUCTION OF POWER GENERATING PLANTS

For a country's economy, particular importance attaches, for the purposes of the cost/benefit and damage/benefit ratios, to the consequences of delays in the implementation of programs for the construction of plants required to meet the country's energy requirements.

This matter, which has formed the objective of discussion for over one decade, was discussed by a paper of mine presented at the Plenary Session of the 11th World Energy Conference held in Munich in September 1980*.

---

*Arnaldo M. Angelini, The Cost of Delays in Building Large Electric Power Plants Due to Obstacles in their Construction.  The Italian Situation.

For a detailed examination of the disastrous consequences of those delays, I would like to refer those interested in this matter to that paper and to the discussion that followed it.

Subsequently, in mid-1981, this issue was taken up again in Italy by the Associazione Italiana Ricerca Industriale (AIRI), which I have the honor to preside over, whose membership includes all Italy's large industries, most of the medium industries and many smaller ones; it is AIRI's intent to call again the attention of the Government and political leaders to the extremely serious consequences created by the stagnation in the construction of new energy generating plants.

I believe that, in order to illustrate the situation and its predictable developments in concise terms, as appropriate here, I could do nothing better than quoting below the substantial part of the Memorandum I have drafted with the members of AIRI for the purpose mentioned above.

The capacity of the Italian Generating plants is at the moment quite insufficient to cover the domestic demand; a proof of this is not only in the repeated service interruptions, but also in the increasing amounts of imported electricity notwithstanding the limitations dictated by technical difficulties and the unreliability of the supply due to various reasons.

The situations, which is already cause for great concern, will become more and more serious with time. In fact, the construction of a new station takes at present from eight to ten years from the final site selection, whereas, when there is no recession, the domestic demand for electricity grows by at least 60% in ten years. Therefore, in the period 1980-1990 there will be an energy shortage that will lead, among other things, to a contraction of the industrial production against any logic of economic and social development. This contraction can be limited only if the construction of the new stations, both nuclear and coal-fired, provided for in the National Energy Plan, is started immediately.

It will also be necessary to make an all-out effort to find new sources, bearing in mind that in the near future the increased demand of electricity cannot be covered only by energy saving (the importance of which must, however, be acknowledged) and renewable sources (hydraulic, geothermal, solar, aeolic--some of which are already exploited nearly to the maximum, others will not be available for some time notwithstanding the effort mentioned above).

As concerns costs, the opposition to the immediate construction of nuclear stations is particularly laden with heavily negative consequences, as it will make it necessary to resort to huge

amounts of coal with a rather heavy impact on the economy though
not as serious as the impact of using fuel oil.  In fact, the
cost of generating at least 100 billion kWh to cover the increased
demand for electricity anticipated for 1990--which will be possible
only if the construction of the necessary generating plant is
started within 1981--will total, over the 25 years of plant lifetime, more than 95,000 billion lire, of these, more than 60,000
billion will consist of fuel expenses, almost entirely in foreign
currency, if the demand is covered with coal-fired stations versus about 60,000 billion (of which only 25,000 billion in foreign
currency) if produced by nuclear stations.  Actually, according
to reliable estimates, with nuclear power stations the total
generating cost will be 32,000 billion lire less than with coal-
fired stations.

The lower operating cost results in a kWh cost of 25 lire
versus the 61 and 38 lire computed respectively for electricity
produced with fuel oil and coal, notwithstanding the greater investment required for nuclear stations (approximately twice that
of oil-fired stations).

The figures given here are based on April 1981 prices.  But
the prices will go up and they widen the cost gap considerably.
In addition, this gap does not contain the important investment
(without return), for the infrastructures--especially harbour fac-
ilities--required for coal transport and handling.

The forecast that France has quite recently made for 1990
indicates twice the price of the nuclear kWh for the coal kWh
and a treble price for the fuel oil kWh.

Again in view of the fact that the construction of a new
station takes between eight and ten years, Italy, over all this
period, will have to use fuel oil at least in the present amounts
and coal in increasing amounts with a relatively small contribution from the nuclear source.  Instead in France, for instance,
in 1990 nuclear stations will be covering more than 70% of the
total electricity requirement.  This means that starting in 1990
more than 70% of the electricity available in France will cost
about half that produced in Italy.

Nor should we ignore the important fact that the production
of electricity from the nuclear source is the only one that can
provide full assurance--and at a very reasonable cost--against
the growing risk of interruption of the supply of fossil fuels by
means of sufficient uranium resources to ensure plant operation
for many years.

Thus, because of the delay in the construction of power sta-
tions--which is isolating Italy from the other major industrialized

countries with serious adverse effects--the domestic productive sector will be penalized both as a result of the shortage of electricity and of the higher cost of the electricity available. This will obviously heavily jeopardize the competitivity of our products, particularly those that are characterized by a high specific consumption of electricity, and bear negatively on the areas--such as Southern Italy--where the demand grows at a faster rate than elsewhere in the country because of the need for a more intensive economic development.

AIRI's appeal is a pressing invitation to face the catastrophic consequences of inaction and to urgently take all the steps required to avoid them.

From the substance of the Memorandum it clearly appears that, to meet new energy requirements, the nuclear option, is taken for granted, which does not contradict the Government's choices.

Concerning this option, I could not add any considerations to those developed in the chapter dealing with the comparison between energy generation cycles, but to conclude this paper I wish to quote verbatim a general conclusion* recently reached by a group of independent experts appointed by the European Communities to study the problem:

"No amount of care will totally eliminate the risks of this, or any other sort of energy, but we have concluded, from our work and from our more general experience, that public health and safety would not benefit and might well suffer significantly from the replacement of nuclear energy by other readily available sources of energy."

I trust that you will agree with my conclusion that, generally speaking the risk of running short of energy is so serious, especially for countries possessing few energy resources, as to make an absolute priority of the attainment of the greatest degree of energy independence, especially as regards electricity generation. The nuclear source is the safest and most economical way to the attainment of this basic objective.

---

* Dunster, Latzko, Smidt, and Villani: Nuclear Safety in the Context of the European Communities.

# PRINCIPLES FOR SAVING AND VALUING LIVES[*]

Richard Zeckhauser
Harvard University

Donald S. Shephard
Harvard School of Public Health, and
U.S. Veterans Administration

[*]From, The Benefits of Health
and Safety Regulation,
Ferguson, A. and LeVeen, E. P. (Eds.)
Ballinger, Cambridge, Mass., 1981
Used by Permission.

## 1. INTRODUCTION

Many of the most pressing decisions of society directly or indirectly involve the saving or expenditure of lives. Energy planning, national health insurance, and occupational health and safety regulation, as well as national defense policy, represent major issues that invariably bring us back to the question: Which lives should be saved? Or, to reflect the process of life-saving more accurately, the question might be rephrased: Where should we spend whose money to undertake what programs to save which lives with what probability? Fifteen years ago, merely asking this question explicitly would have seemed unethical or at least repugnant to many, though its central issues were addressed implicitly in a whole range of individual and collective decisions. Today variants of this question are studied by theologians and sociologists, as well as economists and policymakers. The question of how lives should be valued is now an acceptable one for intellectual inquiry, although it is true that for some the answer cannot come through academic discovery processes.

Though study of the issue has begun, the economists who write on the issue of valuing life do not appear to speak to the regulators and interest-groups who deal with the issue, at least in an implicit fashion, on a day-to-day basis. Economists are accused, sometimes with justification, of concluding too quickly that policy choice to promote the saving of lives is merely a question of setting an appropriate price. Regulators are charged, often with merit, of ignoring (sometimes deliberately) the contribution that can be made by analytic approaches to policies for

the preservation of life. The critical question is how to inject
a bit of the thinking and concerns of each group into the
approaches of the other.

This chapter integrates and substantially abridges two
previous papers: "Procedures for Valuing Lives," Zeckhauser
(1975) and "Where Now for Saving Lives," Zeckhauser and Shepard
(1976). These papers provide numerous examples and references;
they also address a number of technical issues not covered here.

Our discussion here is directed for the most part to the
realm of public decisionmaking. The decisions to which it is
relevant are those by which a public decisionmaker allocates resources to enhance the probabilities of survival of private citizens. The resources involved may be public, as with highway
safety railings; private, but those of an unaffected party, as
when radiation standards are established for industrial processes;
or private and specifically those of the individual whose survival is affected, as with seat belt legislation.

Given a limited amount of resources to be spent in saving
lives, it is clearly desirable to allocate these resources as
efficiently as possible. The efficient use of resources for
saving lives is particularly important during times of budgetary
stringency, since expenditures for life-saving, like all other
expenses, are subject to more probing budget reviews. Moreover,
the ability of government agencies to shift life-saving expenditures to either individuals or business has been curtailed by the
concern over inflation and the continuing fervor for regulatory
reform. This suggests that even agencies that feel that present
expenditures on life-saving are too low will benefit from
rationalizing their interventions. And, pragmatically, the resource costs involved with saving lives inevitably enter into
decisions. Agencies that attempt to insulate themselves from
such considerations may find themselves overridden.

An analytic approach using only simple tools can help make
the regulatory process more efficient. Decisionmaking could be
substantially improved for instance, if agencies were required
to generate information on the costs of their life-preserving
programs and the benefits that they convey. The generation of
such information, even if never introduced as a formal part of
the decision process, would tend to make decisions more rational.

The concern of this chapter is with the analytical aspects
of the life-saving decision process and approaches that can
provide useful inputs into the process. We divide the process
into four areas: (1) *Valuation*. What values do we attach to the
inputs to and outputs from our policies? (2) *Prediction*. What
levels of outputs can we expect alternative policies to generate;

what levels of inputs can we expect them to consume?
(3) *Accounting*.  How should we add up these values so that we
do not misinterpret a quantity, miss anything of value, or
double count?  (4) *Incentives and the locus of decisionmaking*.
Recognizing the interests of all affected parties and the likely
differential access to information, who should be making the
appropriate decisions?

We do not attach any hierarchy of importance to these
questions:  each is considered below.  However, we believe that
future progress in formulating effective policy regarding life-
saving activities will require significantly greater attention
to questions of prediction, accounting, and incentives.

## 2.  PROCEDURES FOR VALUING LIVES

The perplexing problem of how we should value lives that
might be saved, injured, or expended through public or private
decision has not yielded to the substantial efforts of economists
and others.  Why has it proved so difficult to frame a mere
question of value:  What is a life worth?  Some factors can be
identified speculatively.  First, unlike traditional economic
commodities, there is only the slightest degree of standardization
for lives.  Second, unlike most commodities we value, lives are
not bartered on markets.  Indeed, it is against the law to sell
them.  Third, and perhaps partially explaining the second, the
question of whose life should be saved at what cost involves many
of the most fundamental values of our society.  Fourth, there are
many different producers of the commodity "increased probability
of preserving a life."  Individuals can do it for themselves;
we can impose traffic laws and vaccination regulations to help
protect them from other individuals; or society can provide incentives to induce them to preserve their own lives.  Finally,
through a variety of societal programs their lives can be saved
for them.  There are numerous other factors awaiting cataloguing
in an eventual intellectual history of the life-saving discussion.

There is no unambiguous procedure for valuing a human life;
indeed, evidence suggests that life valuation should not be
approached as a search for an elusive number.  Lives are different
from other commodities that our society produces, expends, or
merchandises.  The valuation of lives involves and reflects many
of the most basic beliefs and institutions of our society.  With
lives, it is not just the outcome of the valuation process that
is important.  The legitimacy and acceptability of the process
itself may exert a significant influence on welfare.

### 2.1  The Potential for Analytic Approaches

It might seem, then, that economists would have little to
contribute to the life-valuation discussion.  This chapter argues

the contrary. The complexity of the problem enhances the potential contribution of the organizing concepts of economics. Insights culled from the examination of a number of other sticky issues can be applied with profit. This chapter attempts to provide some of these insights.

It is critical that policymakers realize that there is no un-ambiguous procedure for valuing a human life. Not only do we lack a general approach that will apply in all circumstances; there is rarely any circumstance for which a specific approach could re-ceive universal approval.

Because there is no possibility for a scientific discovery of the loss inherent in the cancer-induced death of a 40-year-old father of two, for example, a major purpose of a study such as this should be to foster agreement on methodology. The next stage would be to gather some empirical materials that could be fed into such a methodology. With the aid of these supporting materials, some significant narrowing may be achieved among the estimates of different assessors. With present knowledge and in the context of the existing debate, great advances can be made merely by securing agreement on ground rules. Indeed, even within the theoretical literature there are extraordinary areas of non-agreement. (The term disagreement is really not appropriate, for the conflicts are rarely explicitly addressed. With a few excep-tions, such as Mishan's "Evaluation of Life and Limb: A Theo-retical Approach" (1971), there has been little attempt to resolve the issues at debate.)

Failure to arrive at unambiguous estimates in the past re-flects neither slack efforts nor stunted imagination. The assess-ment procedure is extraordinarily difficult. This suggests that whatever estimates are derived, whatever procedures are developed to secure estimates, there should be a continuous review process to note their successes and their implications. The valuation process may be simply too complex to reason through from beginning to end. An apparently attractive procedure may lead to valuations that are totally out of line with what seems to be intuitively reasonable. If so, it would be worthwhile to retrace the steps of the logic to search for possible deviation from what was truly intended. It is possible, of course, that the valuation pro-cedure was not in error, and that our original intuition guiding its methodology was more refined than our expectation of its out-comes. In other words, it seems ill advised to make the valuation process merely a once-around proposition from agreed-upon pro-cedure to accepted result. This may be the way of logicians, but it does not lead to sensible policy analysis.

Too often when analysts approach the problem of valuing life, they concentrate on philosophical issues that are inherently unresolvable. Sometimes they begin by identifying the difficulties. Then, if they have been scrupulously honest with themselves, they will tend to give up when they discover the most basic problems. At the other extreme, the analyst grinds out some numbers, however questionable. Such calculations are unlikely to have a positive effect. They will be effectively challenged by politically oriented individuals who oppose the actions they recommend, and by methodologically oriented decisionmakers who recognize inadequacies in their derivation.

If recognition of the difficulties leads to a surrender, and if plowing ahead leads to a discarded output, what should be done? Fortunately, a great deal can be accomplished. Most significantly, analysts can provide some basic building blocks so that the ultimate decisionmakers - and decisions are made every day, though frequently by inadvertence - can have some inputs for what they are doing. Sometimes these analytic inputs will make their greatest contribution by bolstering confidence. They may show, for example, that the choice between two options will not be affected by whether a human life is valued at \$X or at \$100X.

## 2.2 Willingness-to-Pay and the Valuation of Lives

Ask an economist how much a commodity is worth to an individual, and the likely answer is: The amount of other resources that the individual will sacrifice to secure it. The validity of a willingness-to-pay valuation is obvious when individuals are choosing goods for themselves. In the public sphere, however, goods are chosen for others, and payment as such will rarely be secured from the beneficiaries of public decisions. Nevertheless, the willingness-to-pay approach to valuation retains some attractive features. Most particularly, if willingness-to-pay amounts are employed to value outputs, and if programs are sought that provide the maximum excess of benefit over cost, then an efficient outcome will be secured. Conversely, if some selected programs are at variance with the maximization of net benefits using willingness-to-pay valuations, an inefficient outcome will be the inevitable result. As should be expected, the willingness-to-pay approach has been employed by those designing public programs for a range of goods from recreation days to waiting time for medical appointments. But the application has not been widespread,

*Private Decisions.* When the decisionmaker is also the payer and the predominant beneficiary (the person whose life is at stake), there is fairly widespread agreement that the decisionmaker's valuation should be the determining one. The analogy is made to market decisionmaking. Whenever the consumer is the only one affected by his or her purchases, and if markets are functioning perfectly, a socially desirable outcome is achieved by allowing

each person to make his or her own choices. The consensus from
economists about such situations is an important one:  The government should not intervene.

In the private decision context, the willingness-to-pay
criterion is relatively unquestioned.  Thomas Schelling, in an
essay entitled "The Life You Save May be Your Own" (1968), argued
eloquently for this criterion.  He hoped to help individuals get
their thinking straight when allocating resources to their own
benefit and to the benefit of others who value their continued
survival.

If individuals are willing to pay some amount to increase
their probability of survival, or the survival of someone else,
then they should be allowed to pay that amount and reap the benefits.  If there are other interested parties, and if social
arrangements can be worked out so that these others contribute
as well, then the sum total that interested individuals are
willing to contribute should be spent to that purpose.  No
necessary connection is implied between willingness-to-pay and
social value or intrinsic worth.  Indeed, willingness-to-pay as it
is traditionally employed to gauge the value of the outputs of
policy choice is not even the subject of inquiry.  Rather, individuals are merely being asked how much they will pay to secure
something they value.  The recommendation is being made that they
be allowed to purchase it.

*Public Decisions.*  The context of the problem changes when
the decisionmakers are public, not private, and the lives at
stake are not those of the decisionmakers or of others close to
them.  In general, public mechanisms for allocating resources
do not allow for individual purchases by which a citizen who
values an output highly can pay more for it and be assured of
securing it.  Most public resources are generated through tax
mechanisms; taxes are rarely imposed on a benefits-received basis.

Still, a rich town may choose to spend more per capita than
a poor town on public health or highway safety, thereby offering
higher probabilities of survival for its wealthy citizens.  It is
frequently alleged that within cities with wide disparities in
income, health-promoting services such as garbage collection are
superior in rich areas.  Because the rich would probably pay more
for these services, such an outcome - whether the result of political influence, a desire to attract well-to-do citizens, or
whatever - is closer to the hypothetical market outcome than an
equal provision of services.

What is noteworthy is that many citizens, including some of
the rich, find this unequal provision of services inequitable and
undesirable.  When decisions are made in the public domain, the

normative significance of what would be produced by a private
market is diminished.  This lesson, coupled with observations
about the distinctive qualities of life-preservation as an output,
suggests that determinations of willingness-to-pay should not be
widely accepted as sufficient guides to public decision in the
life-preservation area.  Nevertheless, willingness-to-pay calculations can provide a useful input to the decisionmaking process,
and they represent the motivating philosophy for most analytic
approaches to life valuation.

  *Identifying the Affected Parties*.  The willingness-to-pay
approach suggests that to value lives appropriately, one should
merely inquire what individuals would pay in a variety of contexts
to save the particular lives at risk.  When this process is undertaken, it should be recognized that the "interested parties" may
be a diverse group with very different concerns.

  The logical starting place is the individual whose life is to
be saved.  The reason for starting here, quite simply, is the expectation that his or her valuation will probably be the greatest,
though this is not necessarily the case.

  The second class of individuals who are likely to be
interested is the family and friends of the individual.  If the
potential deceased is a breadwinner, then this will include the
primary beneficiaries from the deceased individual's estate.  If
the individual at risk is a child, it will include people who
would be required to support him or her.

  The third category of individuals is society at large, consisting mostly of individuals who have only indirect connections
with the potential deceased.  Some indication of the magnitude of
society's concern might be given by the amount that the individual would contribute to or drain from society.  By this standard,
a big taxpayer would be valued more highly than a welfare mother.
However, it would seem that in American society, given the ex-
pressions of political feeling observed in other circumstances,
net dollar contribution is not a good indicator of the valuation
of the general society.  In most circumstances, following the
argument just made, it would be a substantial underestimate.

  From an analytic standpoint, a life preserved bears many
aspects of any good that may offer significant externalities.  If
the preservation of a specific life were up for sale, and if those
who benefited from saving it could be charged in proportion to
their benefits, everyone would be better off if the life were purchased for a price less than the sum of the valuations of all
affected parties:  the individual, the individual's family and
friends, and the rest of society.

*Alternative Procedures to Assess Willingness-to-Pay.*
Identifying the affected parties is a useful start to get a total
willingness-to-pay figure.  Next, dollar valuations must be
secured.  A number of analysts have attempted to make these
assessments; their results are instructive, though few of them
at this juncture would expect their empirical observations to be
used in policy application as a well-justified and fair assessment of the value of a life.

Jan Acton (1973) prepared and disseminated a questionnaire
that attempted to determine how much individuals would pay for a
mobile cardiac unit that would decrease the probability of death
if they had a heart attack.  His results suggested that individuals had difficulty responding to the types of questions he
posed, though  they provided answers that were not obviously unreasonable.  In response to questions about willingness-to-pay to
avoid 1/1,000 and 1/500 risks of death, Acton concluded that
"large groups of people would be willing to pay $28,000 and
$43,000, respectively, for each life saved at the stated probabili-
ties" (1973: 109-110).

It should be noted that Acton's question assesses the value
of a post-heart attack life, indeed one for which the attack would
have been fatal.  The quality of such a life and its expected
length are likely to be reduced relative to an individual who had
not had a heart attack.  As a general principle in valuing lives,
it is important to identify their expected quality and duration
should they be preserved.  We shall address this issue at length
in the section on Prediction.

Thaler and Rosen (1974) have looked to the labor market to
see how it rewards occupations that involve varying risks of injury and death.  Inserting appropriate qualifications, they conclude that workers "estimate the value of a life to be in the
neighborhood of $200,000" (1974: 38).  Their interesting methodolo-
gy begins with consideration of prices revealed in the market,
rather than an interview technique.

## 2.3  Frequently Proposed Alternative Measures
##       of the Values of Lives

Although conceptually oriented economists have given most
attention to the willingness-to-pay criterion as a means to
approach the life valuation issue, a number of alternative measures
have been presented in both the economics and public policy literature.

*Discounted Consumption.*  One frequently employed approach is
to look at discounted consumption as the total gain that an individual receives for remaining alive.  There are a variety of ob-

jections to this indicator. First, it in no way assesses how
pleasurable the individual finds his or her existence, or indeed
whether additional funds would make much of a difference. Some
people commit suicide, after all. Others might be willing to
give up a substantial amount in terms of survival probability
for an increase in yearly income. The major difficulty with relying on discounted consumption is that it really has no connection with the quantity that is to be determined - willingness-
to-pay for reduced probability of death. Total consumption is
determined more by an accounting relationship with lifetime net
income plus net transfers than it is by any marginal optimization
procedure.

*Discounted Production.* Any equally popular, though no more
compelling, measure of the value of life is discounted production.
In theory discounted production takes into account the resources
that the society as a whole would lose if the individual ceased
to exist. Once again, however, the concept has no connection to
tastes or preferences.

Do we really want to prevent a risk-sharing pool of individuals from spending 1 percent of their net production to protect
themselves against a 1/10 of 1 percent loss? Admittedly, if sure
death were the consequence being prevented, and if no one else
cared, discounted production might be the upper bound on valuation.
But fortunately that is not the situation, at least where such
matters as the control of environmental risks are concerned. At
the time that decisions are made on what levels of control to
employ, no individual faces more than a small fraction as a probability of death. This implies that monetary amounts significantly
in excess of lost productivity could be extracted to eliminate the
radiation peril, at the same time increasing everyone's welfare
above what it otherwise would be.

The key to this apparently paradoxical result is the absence
of significant income effects. For large probabilities of loss,
the constraint on payment imposed by one's discounted lifetime
income may become a significant factor in limiting one's willingness to pay. There is an additional, more psychological phenomenon
at play as well. Individuals seem to be willing to pay substantial amounts to protect themselves against any identified
risk, however small its probability. Insurance companies capitalize on this tendency by offering actuarially unfair double
indemnity coverage for very unlikely ways of dying. Once the
salesman suggests, "And if your death should be by so-and-so,"
the customer may be willing to expend a few extra pennies to get
many extra dollars in case of so-and-so. Assuming away anxiety -
and that is a massive assuming away - this course would make sense
only if so-and-so were an exceptionally expensive way to die.
(It may thus be that radiation-induced deaths are particularly to

be avoided and insured at extra value, for the significant chance
of large accompanying medical care makes them likely to be expensive. More likely the purchaser has not thought through his
decision, at least not like an economist would.)

*Net Contribution to Society.* It is frequently proposed that
an individual's life or health be valued by the amount of goods
that the remainder of society would lose on net, for example
through foregone taxes, were that person to die or become ill.
There are two basic objections to this standard. First, it fails
to take into account the valuation of some very relevant members
of society: the individual and his or her family. Second, it
looks only at the value that society would attach to the individual's economic contribution, and leaves aside measures such
as compassion, bereavement, and the fact that society does not
like to see its members die.

It is easy to illustrate the incompleteness of this approach.
Would we not pay for some measures that were designed to improve
the health and well-being of our elder citizens, even though they
are a net drain on society's resources? The conventional answer
is: Yes, but that is why we restrict the application of the net
contribution measure to individuals in their productive years, or
to those whose productive years stretch before them. That answer
is unacceptable. For whatever values lead us to have concern for
the nonproductive elderly - call them the X term - would apply to
our productive individuals as well. The magnitude of this X term
should be added to the value of the lost productivity to secure
any estimate of the value to the society itself. Perhaps the
"net contribution to society" approach is based on some empirical
observation that among productive individuals the X term is
trivial relative to net contribution. Unfortunately, no evidence
is generally given for that claim.

This all implies that net contribution to society should be
taken primarily as an indication of a clear lower bound on the
value that society in general should place on an individual's
life. It should be added to the amount that individuals themselves
and their families would pay to avoid the risk. If this does not
clearly preclude taking the risk, then it would be worthwhile to
assess further society's noneconomic losses.

*Valuations of Lives and Risks of Lives in Other Areas of
Society.* It is frequently asserted that we should observe how
much individuals who have high-risk occupations, or who live in
high-risk areas, charge for assuming these risks. This "revealed
preference," it is expected, will give us some indication as to
how society should assess the value of removing such a risk for
society in general. The comparison is not fully apt, as long as
the first class of risks is voluntarily assumed and compensated.

The logical problem is that the people who are assuming the risks are those who value them the least in relation to the benefits they get for risking them. They may be the poor, people who most under-estimate risks, people who legitimately have the lowest probability of being injured, people who will die soon anyway, or people who value their own lives the least.

Differential benefits may also account for certain indi-viduals or groups accepting voluntary risks of death. Evel Knievel commands a small fortune for engaging in his dangerous activities. People who would otherwise die or suffer greatly are likely to be willing to undergo life-threatening survery. Individuals who attach great value to their time might take a plane when a boat would be safer. One conclusion is evident. People who choose to engage in activities that place their lives at greater than usual risk are not representative of society in general. If the benefits and costs of taking such risks are independently distributed across the population, then a valuation of lives based on the assessments of those who voluntarily assume risks would tend to be an underestimate.

Chauncy Starr (1972), operating from the quite different per-spective of a physical scientist, has looked at life-saving or life-expending undertakings across a spectrum of activities. Starr draws the important distinction between voluntary and in-voluntary activities, with the suggestion that free individual choice cannot be expected to yield an efficient outcome when risks are externally imposed, and when individuals cannot inex-pensively purchase protection. Starr surveys a potpourri of risks and provides some guidelines for the assessment of alternative categories of risk. The strength of the Starr analysis is that it provides us with useful numerical indicators of the magnitude of present risks and the way these risks may be valued. It stops short, however, of providing us with a coherent methodology that can be employed for setting regulatory standards, or more general-ly, in forging public policy.

The controversial study directed by Norman Rasmussen of the Massachusetts Institute of Technology (U.S. Atomic Energy Com-mission 1974), has employed a variant of the Starr approach as one way of conveying the meaning of its assessment of the risks associated with nuclear power generation. The report provides the reader with some comparisons with dangers associated with other sorts of hazards. The implicit assumption running through the analysis is that there is a collection of potential risks of death or illness, of which money and technology are sufficient to eliminate a limited number. By seeing just how much a risk under study contributes to the aggregate risk, we can find out whether that risk is worth accepting.

This brief review makes it clear that there are conceptual and philosophical difficulties inherent in any procedure that attempts to attach a value to life, though conducting assessments with the aid of such procedures may nevertheless be helpful. In many circumstances policy choices may not change substantially if estimates of the value of a life vary by a factor of 10. Getting a valuation that is accurate within a factor of 3 might be very useful.

Clearly, the analyst has a great number of suggestive techniques to provide. By gathering assessments of life valuation from two or three different approaches, the policy decisionmaker can attempt to "triangulate" on a final valuation. If the operation of the policy by itself generates information on the number of lives that are sacrificed, and at what price, then the political process may be well equipped to provide updated decisions about the way and the magnitude at which lives should be valued. A number of useful analyses and surveys of the burgeoning literature have been developed recently. See, in particular, Viscusi (1978) for an estimate, Bailey (1980) for a survey of the field, and Crouch and Wilson (1980) for promising work on comparative risk analysis.

*Special Problems in Valuing Lives.* Procedures for valuing lives run up against a host of special problems that are unlikely to be encountered with the valuation of traditional commodities. Three of these problems - difficulties in comprehending small risks, anxiety, and risk aversion with respect to total number of lives lost, are addressed here.

*Difficulty in Comprehending Small Risks.* Most of the health and safety issues with which regulatory agencies are involved concern low-probability risks, which most people have great difficulty evaluating. Often the probabilities are exceedingly low. A typical individual told that he or she had either an additional 1/10 of 1 percent or 1/100 of 1 percent chance of death over the next year might find it perplexing to distinguish between the two risks. (It would be more perplexing if the numbers were $10^{-10}$ and $10^{-11}$, as they often are in risk analyses.) Fortunately, decision and risk analysis can offer some guidelines. We might ask the individual how much he or she would pay to avoid a 10 percent increased chance of cancer over the same period. Then we might divide this amount by 100 to get an approximate idea of what 1/10 of 1 percent chance is worth.

*Anxiety.* It should be recognized that the procedure above can be misleading, since it fails to consider anxiety. (It also neglects income effects [Zeckhauser, 1975].) Once a new element of risk is announced, individuals have something new to think and worry about. It would seem rather unlikely that a 1/10 of 1 percent

risk would generate only 1/10 the anxiety of a 1 percent risk of
the same loss.  In other words, the amount that would be paid to
avoid the anxiety associated with a risk would be very nonlinear
with the probability of the risk.  For smaller risks it would be
proportionately greater.

Because the types of risks that arise from most regulated
risks are of the low-probability variety, we must expect the
anxiety cost to be a fairly substantial proportion of the amount
that an individual would pay to avoid the risk.  Mere extrapolation from more significant risks in other areas would not seem
to be valid.  The medical area may provide the best example of
the anxiety-eliminating expenditures that should be borne in mind
here.  A 40-year-old expectant mother whose future child is at a
risk from Down's syndrome, a genetic ailment that is detectable
by amniocentesis, might welcome that procedure even though she
would not have an abortion whatever it reveals.  There is a high
probability that the results of the test will be negative, and
will thereby relieve her anxiety during the course of her pregnancy.

*Risk-taking or Risk-averting Behavior on Number of Lives Lost.*
Should there be risk aversion related to total number of lives
lost?  The answer seems to depend in part on the scale of the
example considered.  If there were some threshold, say three lives,
below which the general public was not informed of a death, the
loss of a single life would not come to the attention of anyone
except those immediately involved.  Society at large might thus
prefer risk aversion on lives lost when a limited number of lives
was involved; that is, society should be more willing to take one
chance in 1,000 of losing one life rather than one chance in
10,000 of losing 10 lives.  However, if the numbers of this hypothetical problem are multiplied one hundredfold to 100 and 1,000,
the preference of the uninvolved public might reverse to risk-
taking.  The loss of either 100 or 1,000 lives would be regarded
as a major catastrophe, but the latter would be only 10 percent
as likely to occur.  The larger loss, using a utility function to
define intensities of preference, might not be considered ten
times as bad.

Consideration of the valuations of those immediately involved with those who die might strengthen any societal preference
for risk-taking.  Suppose a happily married childless couple were
deciding whether to fly on separate airplanes for a trip.  If
they thought the matter through, they should almost certainly
prefer to take a joint risk of death rather than each taking it
separately.  Their risks of death would be the same; but all the
unfortunate consequences of widowhood would be eliminated.

The spirit of this example can be carried over into the analysis of risks that are of consequence to the general society. If any explosion wipes out a community of 10,000 individuals, most of the people who would have placed a high value on the lives of those killed will have been killed themselves. By contrast, if an additional 10,000 people are killed in auto accidents, most of the major externality sufferers will still be alive. Other things equal, concentrating the lives lost on a geographic basis reduces the externality loss per death. The general lesson is that, at least for those closely connected to individuals with lives at risk, it may be beneficial for society to exhibit risk-preferring behavior.

## 2.4  The Importance of Process

For many societal decisions that affect life-threatening activities, the procedure by which the decision is made may be as important as the outcome. Many analysts dismiss too quickly the significance of having an equitable and widely accepted process. (The criminal justice field is an area in which our convictions about the integrity of process have made it difficult for traditional cost-benefit analysts to have much of an impact. It is no surprise that "process" is a watch word of the legal profession.) When process is important, analysts can take any of three tacks.

First, they can labor earnestly to provide the inputs that are required by the process. They might, for instance, suggest to the Occupational Safety and Health Administration (OSHA) administrator who sets standards for benzene exposure just how many lives the alternative standards will affect, what it will cost to save those lives (at the margin), and what is being spent to save lives in other environmentally affected areas. Second, analysts can investigate the outcomes of the process - has it been producing desirable results? Finally, they can examine the process itself in hope of improving its performance.

All three of these approaches have the virtue of complementing and informing presently accepted procedures for decision-making, which is likely to enhance the final outcome of the analysis. Individuals may far prefer an outcome that they believe has been justly derived, to one that rests on unimpeachable but nevertheless distasteful calculations.

This point about the importance of process may find support in elementary economics. (1) Most of the decisions that we are making in the area of environmental and workplace protection, though they may cost or save substantial numbers of absolute lives, can have at most a marginal impact on total life expectancy in the society. (2) The dollars that we will be spending because we

tighten standards on various pollutants can run into the billions
or even tens of billions of dollars. Still, this expenditure is
unlikely to be more than a small percentage of our GNP. (3) The
basis for decisions is generally not a continuous variable that
ranges between fully acceptable and unacceptable, but a dichoto-
mous choice. Either we are following some procedures that have
general adherence, or we are not. The use of an acceptable process for environmental decisionmaking has been described as an
"on-off" variable (Zeckhauser 1975). If we can keep this variable
"on" without drastically affecting the number of lives or dollars
saved, we may gain an extraordinary amount at small cost. (A key
question for society, addressed at least implicitly in the Supreme
Court's 1980 decision on OSHA's benzene standard, is when do these
other costs get too high.)

The more closely a process for making policy choices accords
with valued beliefs, the better accepted it will be. One important
valued belief is that society will not give up a life to save
dollars, even a great many dollars. Rarely is this belief, widely
held albeit mistaken, put to a clear test. When it is, it may be
desirable for society to spend an inordinate amount on each of a
few lives to preserve a comforting myth. Such a myth-preserving
action was taken when the federal government assumed the costs of
renal dialysis. The specific individuals who would have died in
the absence of the government program were known. The lives at
stake in this policy context were identified lives. They can be
contrasted with the nonpersonalized, statistical lives that are
saved, for example, by expenditures to construct highway safety
barriers. The valued belief that society will not sacrifice lives
for dollars is more strongly maintained when the specific identi-
ties of the victims are known. An effort to preserve this valued
belief may explain in part the frequently noted difference between
the resources expended to save statistical and identified lives.
Society, acting collectively, shows itself willing to pay much
more to save the latter.

Acceptance of the importance of process in life valuation can
have some discomforting implications. We may find that we are
spending $100,000 to save a life in one area, but sacrificing
lives in others that could be saved for an expenditure of $10,000.
The consequence is that with a reallocation of resources we could
have both more lives and more money. Yet, if the decisions in
the two areas were well accepted by the society, then it might be
preferable not to change. Lives and dollars are sacrificed in
return for more satisfaction with the ways these decisions have
been made. Agreement on process enables society to avoid what
might be difficult conflicts involving equity.

In our discussion below under the headings of Prediction,
Accounting, and Incentives, we shall have more and harsher words
about the problem of misordering (saving expensive years of life,
before implementing inexpensive approaches).

*Income Distribution.* One of the most troubling areas of the life and health evaluation issue is the need to take income distribution into account. Yet it seems that this issue is sometimes introduced as a red herring. One is often told that a certain procedure such as the use of willingness-to-pay as a measure of benefit when establishing safety regulations is unacceptable because it does not concern itself with distributional implications, or because it would value the lives of the poor less than the lives of the rich. These objections could be telling, but only if the objector would accept the particular procedures that would exist if there were a totally egalitarian income distribution. But such acceptance would frequently not be forthcoming, and the distribution-based objection is perhaps merely the most convenient one at hand.

What about the cases in which income distribution or lack of attention to it may be a decisive issue? First, as discussed above, process is important. If the lives or dollars involved, though consequential, are not significant in relation to the total magnitudes for society, then we may strongly object to a scheme that appears to disadvantage the poor, certain ethnic groups, or people who live in particular places, however efficient the scheme may appear. This issue relates to the whole question of efficient and inefficient transfers, whether compensation will be paid for losses incurred, and ex ante versus ex post welfare assessments.

Suppose, for instance, that a poor man would pay up to $5,000 to eliminate a particular risk on his life, and that a rich man would pay up to $50,000. The policy choice is whether to spend $10,000 to eliminate such a risk on either of the two men, or $20,000 to eliminate it on both of them. If income transfers could be made without cost, it would be preferable to leave the life of the poor man at risk while eliminating the risk for the rich man. With the $10,000 saved, we could give the poor man his indifference price for continued exposure to the risk plus a $5,000 bonus. The rich man could be charged part of his $40,000 surplus.

Now income transfer schemes are rarely undertaken to compensate for particular low-level risks that are imposed on individuals, so this sort of transfer cannot be expected to take place. Still there are societal policies that work to some extent in this direction. For example, tax assessments may be revised downward in areas that are newly polluted. If public policymakers really followed the individualistic ethic, they would monitor who gains and who loses on average and at the margin from each policy measure. Then by adjusting policies so that the relative weights placed on different income groups were consistent, a more equitable income distribution could be achieved in an efficient manner. The policies that redistribute the most for a

given efficiency loss would carry the heaviest redistributional
load. In contrast to the present situation, in which some poorly
chosen redistributional efforts incur tremendous efficiency losses,
the poor would be substantially better off, and so would the rest
of society.

Attention to problems of distribution affects the willingness
of many observers to allow the free market to operate. It may
well be argued, for instance, that it is unfair to allow individuals to engage in high-risk professions, since the poor will
be induced to pursue those professions. But are they not better
off to have the opportunity available to them? Provided that they
understand the risks they are taking, the answer is yes. What of
the objection that we should introduce social policies that reduce their poverty and that would therefore remove their incentive
to assume these risks? That objection may be valid, but until
those policies are introduced, we should not further reduce the
welfare of the poor by denying them occupations just because
middle-class individuals would not be willing to accept them. The
nineteenth century satirist Anatole France remarked on a variety
of prohibitions that discriminate against the poor in this manner:
"The law, in its majestic equality, forbids the rich as well as
the poor to sleep under bridges, to beg in the streets, and to
steal bread."

The whole issue of denying the poor risks that they view as
acceptable gets tied up with our perceptions of the income distribution. If we observe that poor people have to sell themselves
into potential physical infirmities, we are forced to recognize
that the income distribution is much more uneven than we had
previously perceived, or that the consequences of that unevenness
are more distressing. Prohibiting the poor from taking such risks
may be a way of salving the conscience of the middle class at the
expense of the welfare of the poor. To the extent that it clouds
perceptions about inequalities in the income distribution, it may
do the poor a double disservice.

*Judging Overall Results.* How people feel about the society
in which they are living matters a tremendous amount. Is the
government affording them adequate protection? Are large cor-
porations being allowed to foul portions of the environment that
by right should be common resources? What sorts of compensation
schemes are being carried out? The questions might continue, but
the basic point is clear. Because of quite separate factors, a
society may prefer a set of policies that produces fewer dollars
and fewer lives than would be available through another. That is,
depending on how we get there, we may prefer to arrive at point *B*
than point *A* in Figure 1. Note that the preference for the
apparently dominated point is not due to a variation in the lives
that are lost; rather, it may relate to the way in which decisions

were achieved, the way compensation was carried out, and the like. The omitted dimension is the attractiveness of the processes through which lives were saved and expended.

It is not possible to disentangle completely the attractiveness of a process and the quality of the outcomes that it generates. If a choice procedure usually (though not always) leads to the right results, protects against extremely serious mistakes, and has a logic and internal consistency that supports its integrity, then the procedure itself is likely to be well regarded. If the procedure occasionally selects an inefficient outcome, such as point *B* over point *A* in Figure 1, that may turn out to be a cheap price to pay for acceptable process. It would seem unwise to tamper with the procedure on the few occasions when it makes such selections. To do so might well undermine its legitimacy.

Procedures that regularly lead to inferior outcomes, however, are less likely to be respected and cherished, certainly over the long run. This is not to assert that when assessing the attractiveness and legitimacy of a process the members of society are likely to be guided predominantly by the outcomes that they observe or predict it will produce. They will probably also ask, does the procedure pay attention to all lives that are threatened? Does it appear that resource costs are appropriately recognized in its deliberations? Are its methods consistent across choice situations? Using criteria such as these will tend to favor procedures that generally perform effectively. This suggests that attractive procedures can be found that will generate efficient outcomes with a degree of regularity.

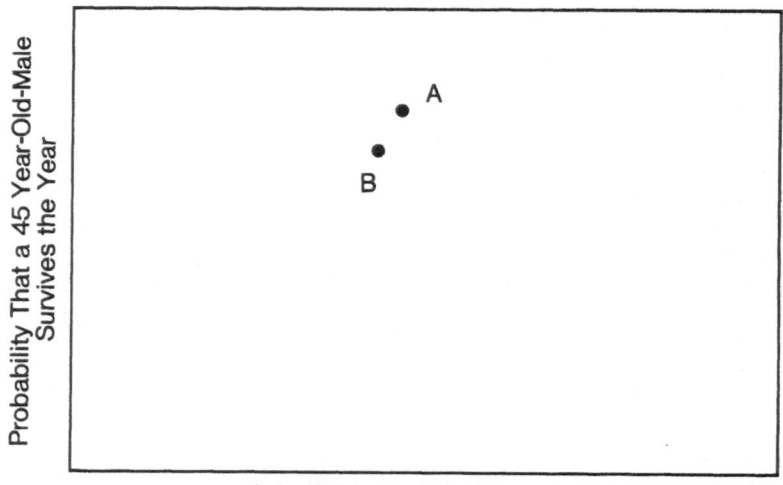

Figure 1.    The attractiveness of process in the evaluation of life.

## 2.5  Where Should we Go From Here?

We outlined four classes of issues in the beginning of the
chapter:  valuation, prediction, accounting, and incentives.  None
of these issues is more important than the others.  Like the four
legs of a stool, each contributes to a common endeavor.  The
valuation question, the one that was the most difficult to confront a decade or so ago, has received the greatest attention since
then from economists and analysts, certainly when judged in relation
to the definitiveness of the answers and the degree of refinement
of methods that can be expected.  It continues to offer an abundance
of intriguing intellectual issues (a number of which are addressed
in other chapters in this volume).  We believe, however, that
future progress in formulating effective policy regarding life-
saving activities will require significantly greater attention to
questions of prediction, accounting, and incentives.  Below, we
outline some issues in and approaches to these three areas.  The
reader is referred to Zeckhauser and Shepard (1976) for further
detail.

## 3.  PREDICTION, INTEGRATED WITH VALUATION

Our interest in improved prediction reflects a desire to pursue issues that can be resolved in preference to those that can
merely prolong debate.  It is our belief that many policy issues,
both broad and narrow, could be effectively resolved if our predictive capabilities were improved.

Many of the important policy issues that affect life-saving
are the subject of spirited policy debate.  What sort of health
system should we have in the United States?  How should we generate
electricity?  We suggest that the most significant disagreements
involved in these debates could be resolved if we had the ability
to make more accurate predictions about the health, dollar, and
other consequences of alternative policies.  It is not differences
in trade-off rates that lead the proponents and opponents of
nuclear power to their conflicting policy conclusions.  Rather,
those two parties provide quite different estimates of the potential
costs of nuclear power, measured in terms of both dollars and the
probabilities of loss of health and life.  Similarly, the advo-
cates of prepaid health plans (Health Maintenance Organizations)
differ from those who support a fee-for-service system in their
predictions about the ultimate consequences in terms of costs,
efficacy, and acceptability of the care delivered.  The arguments
of either camp would be refuted, if the predictions of the other
could be shown to be accurate.

Milton Friedman, whose views on economic policy frequently
diverge from the mainstream, made an equivalent point, "Differences
about economic policy among disinterested citizens derive predomi-

nantly from different predictions about the economic consequences
of taking action - differences that in principle can be eliminated
by the progress of positive economics - rather than from fundamental differences in basic values, differences about which men
can ultimately only fight," (1953: 5).

There is not space to outline prediction procedures here. The
reader is referred to Zeckhauser and Shepard (1976), which provides sample predictions of the benefits from mobile cardiac units
and diet control of cholesterol, both designed to reduce deaths
due to acute complications of atherosclerosis. It then examines
air bags for cars and lower speed limits, interventions designed
to reduce motor vehicle fatalities.

## 3.1 Quality-Adjusted Life Years

Significant gains in terms of the effective design of life
programs could come from a simple refinement of output measures
that would take into account the age of individuals saved, the
number of years saved, and the quality of those years. This unit
of output will be termed quality-adjusted life year and will be
referred to by the acronym QALY. It will be tallied on a year-by-
year basis, with QALYs received in year $i$ indicated by $q_i$ and the
stream of QALYs as $q_1$, $q_2$, . . . . A government policymaker
choosing among alternative health-promoting policies should look
to their consequences for individuals' QALY streams. We suggest
that the appropriate measure for the output of a health-promoting
program is the total gain in discounted QALYs it provides to all
members of the population. (Discounting imposes severe, but
generally not implausible, restrictions on preferences for health
at different times.)

Quality levels could be indexed on a variety of arbitrary
scales. In order to gain a number of useful properties, we propose they be calibrated using von Neumann-Morgenstern utility, in
the manner illustrated by the following example. Assign a year
at full function a utility of 1, and a year without life a utility
of 0. An individual has a choice between living the rest of his
or her life with a specific impairment or having an operation.
The operation has a probability $x$ of restoring full function
(this will not extend the lifespan, however), and a probability
$1-x$ of being immediately fatal. The value of $x$ that would leave
the patient indifferent between having and not having the operation
is the QALY level for the patient's particular level of impairment. When an alternative will affect different years in different
manners, the utility value for each year must be scaled separately.

Let us look at a hypothetical, costless medical procedure
using the QALY analysis. (Resource costs to the individual could
be included in the calculations by including consumption levels

as a determinant of the quality of life, hence the $q$ value, within a period.) An individual has a maximum lifespan of two years. There is a $A$ chance of death at the end of the first year whether or not the procedure is performed. The QALY level for death is scaled to be 0. The procedure, which may be conducted at the beginning of any year, entails a mortality rate of .2. If the procedure is a success, it will restore the individual to full function, so that $q_1 = 1$ and $q_2 = 1$. In the absence of the procedure, the individual will have a QALY level of .9 the first year and, if still alive, a QALY level of .7 in the second year. Consistent with von Neumann-Morgenstern utility, the QALY value in a period is computed as an expected value. That is, it is a weighted average of the $q$ values for the different outcomes, with the probabilities that the respective outcomes are achieved employed as weights. The individual's alternative lotteries are shown in Table 1. The procedure should be undertaken at the beginning of the second year. The QALY stream for that alternative dominates the other two streams.

| Outcomes | No Procedure | | Procedure First Year | | Procedure Second Year | |
|---|---|---|---|---|---|---|
| | Prob-ability | QALY Stream | Prob-ability | QALY Stream | Prob-ability | QALY Stream |
| Survive both years | .6 | [.9,.7] | .48 | [1,1] | .48 | [.9,1] |
| Die at end of year | .4 | [.9,0] | .32 | [1,0] | .4 | [.9,0] |
| Die from procedure first year | - | - | .2 | [0,0] | - | - |
| Die from procedure second year | - | - | - | - | .12 | [.9,0] |
| Overall QALY stream | | [.9,.42] | | [.8,.48] | | [.9,.48] |

Table 1    An individual's QALY stream for three hypothetical choices.

## 3.2  Life-Saving in the Context of Present Policy - A Model that Incorporates Prediction and Valuation

What should we do with our measure once we have it? A simple supply and demand diagram may prove useful in helping us keep our thinking straight on some of the issues to be considered below. The supply curve in Figure 2 represents alternative ways to secure one additional QALY. Following the cost-benefit approach, and leaving aside the possibly important question of who receives

the years and who pays for them, we would wish to start by purchasing the lives that are cheapest. We would continue purchasing
these quality-adjusted life years until the last unit purchased
cost us just the amount we were willing to pay for it. If that
socially optimal cost per QALY is $V_1$, then we would purchase up
to but not beyond point $A$.

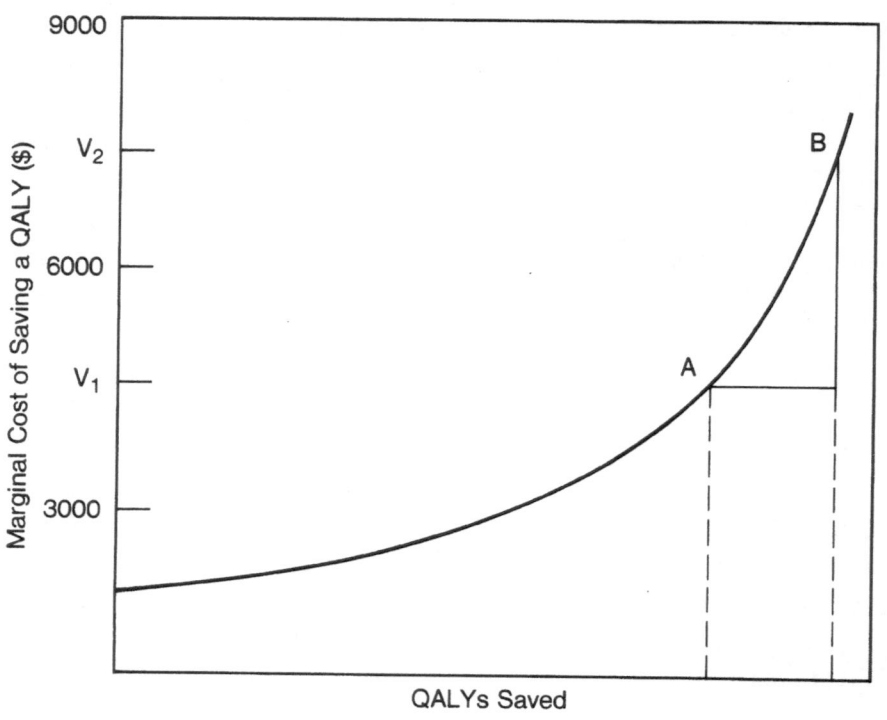

Figure 2    Supply curve for QALYs.

There has been a great deal of discussion, as we have mentioned,
about the way $V_1$ should be defined. However fascinating the discussion, the diagram shows that it may not be of great operational
importance. If the supply curve for lives available is inelastic
in the range of values under consideration, neither the total number
of life years nor the total amount spent on preserving life years
will vary dramatically if the value placed on a QALY is, say, increased by 70 percent to $V_2$. Therefore, even if we were to thrash
out disagreements on which measures of life valuation are appropriate, or on how they should be estimated, we would not substantially improve performance on our implicit objective function. For
example, if our true valuation were $V_1$, but we mistakenly employed
$V_2$, the shaded area in the diagram would be the cost of our error.

### 3.3  Saving Expensive Life Years Before Those
###      That are Less Costly:  Misordering

Even if the relevant portion of the supply curve is inelastic, we might still reap substantial gains through careful and thoughtful attention to decisions for saving lives.  The primary difficulty with present decisions, we would argue, is that we are not proceeding smoothly up the supply curve; that is, we are not saving the expected life years that become available to us most cheaply.

The issue of "incorrect" ordering for life-saving has been flagged and flogged in connection with the assertion that we spend much more to save "identified" rather than we do to save "statistical" lives.  Whether this out-of-line ordering results in the sacrifice of a great number of life years, of course, depends on whether there are large numbers of identified lives that are saved at high prices.  The most frequently cited example is renal dialysis, now estimated to cost around a billion dollars a year. (Though an impressive amount, it is less than 1 percent of our total health expenditures.)  There is now speculation that the availability of an artificial heart could make it possible to save a number of identified, salvageable lives through a disastrously uneconomic process.  If so, we would expect that a public sacrifice of the belief that we will spare no expense to save a life will be required.  At present, it is our impression that the life years lost through undue attention to identified lives do not represent a monumental efficiency problem.  The issues of accountability and prediction have played a larger part in creating the misordering problem.

*Misordering Due to Differential Accountability.*  The problem of differential accountability arises because the penalties and rewards that return to public decisionmakers are far from proportional to the benefits that those decisionmakers generate.  This may be true particularly for life-saving decisions, for these are likely to involve the distinctive reward of a free conscience for the decisionmaker.  In some life-saving circumstances, the chain of causality will be fuzzy and distended, and accountability will be minimal.  In others, particularly where public consciousness has been heightened or identified lives are involved, the consequences of decisions may be highly visible, patterns of causality strongly sketched, and decisionmakers subject to ready penalty if lives are lost in some way that can be tied to their policy choices.

Accountability for expenditures will compete with accountability for lives in either circumstance, but we would expect that the variability in levels of dollar accountability would be somewhat less than that for lives.  This suggests that the stronger the emphasis on accountability in life-saving decisions, the greater

will be the willingness of public decisionmakers to sacrifice
dollars to save quality-adjusted life years. The result will be
uneven trade-off rates across life-saving decisions and both
fewer QALYs and fewer dollars than would be generated by consistent choice procedures.

The differential accountability problem, alas, does not yield
to ready solution. Some progress might be made if we could de-
centralize decisions and let individuals make their own QALY
dollar trade-offs, an approach we shall advocate later. For those
decisions that are inevitably subject to centralized control, proposals for beneficial reform seem more pious than promising. Even
recognizing this, we suggest that accountability for both dollar
expenditures and lives should be strengthened in those areas of
life-saving activity where it is presently weak.

*Misordering Due to Poor Estimates of Benefits.* We may also be
led to save expensive QALYs before cheap ones if our predictions of
how many QALYs will be saved in different areas for the same level
of expenditure are poor. Here, unlike the valuation area, sensitivity to improving estimates could be great. Estimates of benefits
per dollar spent can vary by factors of ten or even one thousand.
This will be particularly the case when there are many intervening
causal steps. (What are the benefits of replacing three Boston
teaching hospitals by the new $118 million, 680-bed Brigham and
Women's Hospital?) Great variations in estimates of benefit also
arise where exceedingly low-level probabilities are involved. (How
likely is a core melt accident in a nuclear reactor?)

By changing estimates of these sorts, we can dramatically shift
the location of a particular life-saving technology along the supply
curve. The problem, we would argue, lies deeper than the persistent difficulty of making informed predictions about how many
lives will be saved with particular technologies. Most decision-
making processes that affect the saving of lives never even attempt
to estimate the end product, the expected number of QALYs that are
added.

## 4. ACCOUNTING

Our discussion of prediction has already provided a limited
introduction to one problem of accounting where lives are involved.
We have defined appropriate units of measurement. The next step,
quite simply, is to add them and subtract them appropriately to
compute an accurate total of the benefits and costs of a program.
We have not had the opportunity to pursue the important task of
developing a complete accounting system. However, we have been able
to identify some significant and frequently recurring errors that
would be eliminated if such a system were developed. We shall
address three of them.

## 4.1 Marginal versus Average Problems

Most policy interventions are proposed as a unit. Thus, we may propose an $X$ million dollar nutrition program, designed to bring all malnourished expectant mothers nutritionally up to an average middle-class American level. An alternative expenditure of the same resources may provide $Y$ home health visits per infant after birth. Sometimes we undertake programs that are defined by a particular level of output, not input. Thus, we may decide to undertake an auto fatalities reduction program to cut traffic fatalities by 10 percent.

A famous early effort of this sort was the U.S. Department of Health, Education and Welfare's Disease Control Memorandum (1966). It outlined a dozen or so interventions, and ranked them in order of the lives saved per dollar expenditure. An advertising program encouraging motorcyclists to wear helmets came out on top.

Unfortunately, the calculations presented in the memorandum do not reveal the information we would like to have. What have we gotten for the last dollars expended on each program? In most cases, though with exceptions, we would expect diminishing returns within each of these alternative programs. That is, the first lives that are saved come cheapest. Maximum efficiency, as is well known, is not achieved by ranking the projects in order of average output per dollar input, but rather by expanding all projects until the output per marginal dollar input is the same. For example, if 98 percent of the benefits of the motorcycle advertising program could be achieved with only half of the proposed expenditure, then it might be undesirable to "waste" the second half of the proposed expenditure. Better to start on a lower average payoff project that may be offering high marginal benefits in the initial range. The principle is a simple one: equate marginal returns across projects.

## 4.2 Worst Case and Best Guess Analyses

Many policy analysts have been indoctrinated in the academic tradition of defending arguments through conservative assumptions. If one is trying to prove a point irrefutably, consideration of the least favorable situation makes sense. This is not, however, an appropriate approach for rational policy analysis. Where policies are being chosen, each possibility should be assigned a probability and an expected performance score should be computed, assuming that we wish to be consistent.

Many variants of worst case analysis are employed frequently in the life valuation area. Two of the most common types are: (a) assume that everything goes wrong and see what the performance will be, and (b) concentrate on the most dangerous aspect of a

process. The second variant, being more subtle, is more likely to
lead to a misinterpretation. For example, when nuclear-based
electricity generation technologies are compared with others,
analysis frequently concentrates on the relatively most dangerous
aspects of the process: the possibility of reactor accidents, diversion of fissionable material, and disposal of radioactive
wastes. Such analyses seldom consider the significant advantage
of nuclear over competitive technologies in air pollution (with
coal), or danger of military confrontations (with oil).

Worst case analyses generate a variety of biases. First, they
lead to distortions in estimates of expected trade-off rates
between lives saved and dollars expended. Second, where alternative approaches are being compared, they favor conventional technologies, whose outputs can be estimated with some precision, as
opposed to those that are less well understood. Third, they tend
to discourage the use of sequential strategies that would enable
society to capitalize on learning.

A more purely logical problem, easily overlooked, is associated
with using best guesses as the basic parameter estimates in complicated systems. The problems that arise are of two sorts.
First, if there are substantial nonlinearities in the system, the
use of means, say, rather than entire distributions may lead to
misestimates. Second, it is possible that in some situations
there will be a correlation of errors, and all parameters will be
under- or over-estimated. In engineering reliability studies,
interrelated failures are termed "common mode failures." These
interrelations increase the probability that an entire system
fails (U.S. Atomic Energy Commission, 1974).

## 4.3  Double Counting

An ideal accounting procedure for life valuation would make
sure that every valued commodity was included and that no valued
commodity was counted twice. In reality, the double counting of
compensated risks is a common error.

Consider a hypothetical cost analysis of a coal-based electri-
city-generator technology. First the dollar costs of producing a
unit of electricity are computed to be $100. One input to this
calculation is the miners' wages, which reflect their valuation of
the risks they run. Let us assume that in deciding what wages to
accept they value their lives implicitly at $300,000 a piece, at
least where small probabilities are involved. If one miner's
life is lost for each hundred thousand units of electricity produced, then $3 of the $100 unit cost of electricity is attributable
to risks of miners' lives.

Next we add up other costs, one of which is expected lives lost. What value should society attach to those miners' lives that are lost in the production process? The first step in arriving at an answer is recognition that those lives have already been counted in the $100. That amount is $97 for inputs other than lives, and $3 as a valuation of the lives lost. To attach an additional value because a particular type of input is used would be double counting, no less inappropriate conceptually than listing wear and tear on mine railroad cars separately, then adding the cost (which is also already included in the $100) to the initial dollar total.

Though we should surely wish to avoid double counting, there are some circumstances in which we might want to attach an explicit cost to lives lost. First and important, the miners may not take all the costs associated with their loss of life into account. Other members of society may feel uncomfortable about allowing them to sacrifice their lives; these others might be willing to pay something so that this would not happen. This is a traditional form of uncompensated externality; the miners cannot charge those "concerned others" for loss of income if they indirectly (by pursuing alternative technologies) or directly (say, through legislation) prevent them from working.

It is worth noting that society presently has rather extensive directed transfer programs whose primary function is the promotion of health, suggesting that the right to bodily well-being is perceived to differ significantly from the right to beer and television enjoyment. If we are more concerned about our fellow citizens' health than about other aspects of their consumption, our societal choices should reflect that fact. But given the rather complex nature of societal decision procedures, and the inherent difficulty of calculating levels of health benefits from various programs, we should not assume too quickly that our "revealed preference" as to the magnitude of this externality is an accurate indicator.

The other members of society may also have a self-interested dollars-and-cents concern for a miner's health. Because there are social welfare programs that provide dependency benefits, health coverage, and the like, miners may not take account of the full resource costs of any health risks. The other members of society will share in some of the costs of their losses. We shall explore this matter in greater detail in our subsequent discussion of appropriate incentives.

## 5.  INCENTIVES AND THE LOCUS OF DECISIONMAKING

Two principles are critical if we are to make appropriate decisions about saving lives.  First, we must place decisionmaking authority in the appropriate hands.  Second, we must also provide decisionmakers with the incentives and information that will enable them to make appropriate decisions.  In the past, most policy discussion on the life-saving issue has taken the locus of decisionmaking as given.  It has focused for the most part on the decision problem of the government or other collective organizations.  The key question that has been examined, as we stressed earlier, is how lives should be valued for such decisions.

It is now widely asserted that future progress in life-saving will depend largely on the actions of individuals.  They will have to drive more safely, eat less, and follow medical regimens more closely than they have to date.  In the words of Victor Fuchs (1967), "the greatest potential for improving the health of the American people ... is to be found in what people do and don't do to and for themselves."  Assuming that this is so, we might inquire why this whole area is suitable for policy investigation.  Our government, after all, is content to stand by while a citizen becomes a daring golfer, overindulgent television viewer, or sloppy homemaker.

Suppose careful investigations reveal that the major health gains available per dollar of expenditure could be achieved by changing the actions that individuals take on their own behalf.  Then if we assume that individuals should be the primary parties concerned with their own health, and that they have the appropriate incentive to care for themselves, present programs for promoting health would seem clearly overextended.

We might come to a quite different conclusion if we started instead from the premise that the competitive market model may not adequately mirror the type of decision that confronts an individual when making choices that affect his or her health.  If so, optimally tailored government policies for saving lives would reflect valuations on those lives different from those implicit in individuals' decisions.  If the society would, on the whole, benefit when individuals sacrificed other goods for survival probability or improved health, then the social valuation would appropriately be higher.  The desirable direction for policy, of course, would be to make decisions that implicitly placed higher values on these individuals' lives than they themselves had done.  This could be accomplished through government regulation.  For example, we can make it hard for an individual to drive a car if not wearing seat belts.  Alternatively, we might try to provide specific incentives to get individuals to pay more attention to their physical well-being, but leaving the final decision to the individual.  For example, a California employer has markedly reduced cigarette smoking through a $7 per week bonus for not smoking on the job (Shepard, 1980). We shall not speculate here on the degree of imperfection in the market for providing health benefits to oneself. We shall argue, however, that if the government does choose to intervene, incentives approaches - as opposed to command-and-control methods - offer numerous advantages.

## 5.1 Why Individuals Might Choose Unhealthy Lifestyles

Before we encourage individuals to pursue more health-promoting lifestyles, we must understand why they might not. We identify four classes of reasons. First, and most important, there are many benefits associated with following unhealthy lifestyles. Most of us eat for recreation as well as for nutrition. We like to drink at parties and then drive home; at times we are in a hurry to get some place, and take liberties with the speed limit. Indeed, we argued earlier that there are relatively few areas where individuals can use money to save their own lives. Most of the trade-offs they make affecting their lifespan relate more to lifestyle than patterns of resource allocation.

Second, there may be externalities among the behaviors of different individuals. Systems of compensation to secure a healthier environment for all may be difficult to establish. It is hard to protect oneself against the drunk driver, the air contaminator, or the person who spreads infection. Moreover, we ourselves may impose some unhealthy externalities on others. The most significant externality is likely to be that between the individual and his or her future self. If one smokes now, one's health is likely to suffer many years later. There is an interesting question, which merits both empirical and philosophical investigation, whether individuals give appropriate weight to their future well-being. To what extent is John Jones at age twenty the same person as John Jones at age forty?

Third, fellow citizens share in the cost of one's illness. This financial relationship probably represents the most significant externality contributing to the divergence between individuals' choices of lifestyles and the pattern of lifestyles that would prove most beneficial for society. Some of this cost sharing arises through private contractual arrangements, whether disability, life, or health insurance plans. Other aspects work indirectly through the market. For example, overall productivity, hence ultimately wages, declines as the absentee rate rises. Finally, government programs that promote health, treat illness, or provide resources to individuals or the families of individuals who are sick, disabled, or dead, create a financial externality through the tax system.

Fourth, individuals who would like to balance health against consumptive pleasure may make suboptimal choices of lifestyle through lack of knowledge. Frequently, the direction of a causal relation is known, but not its magnitude. It is bad to smoke, bad to eat fattening desserts, bad to exceed the speed limit, and bad to eat eggs. For the individual who is willing to give up two of these practices but not all four, it would be nice to know just how bad each is. The trade-offs become much more difficult to estimate when we choose across the spectrum of individual and collective choice. An individual concerned with the risks of nuclear power might be reassured or dismayed to understand that X pounds overweight is equivalent to residence half a mile from a nuclear power plant. This information may help the individual when voting on public issues, as well as when sitting down at the dinner table. Our earlier argument is underscored: substantial gains in life-saving per dollar expended will perhaps be the most readily achieved if we improve our predictions of the benefits from alternative life-promoting interventions.

## 5.2  The Role of Government

There is a role for government in each of the four areas where private behavior may diverge from the optimal behavior for social efficiency. Too many policy analyses, however, leap automatically from diagnosis of an externality or other market imperfection to a prescription for government intervention. Here too, empirical estimation would help. Some of these externalities may be relatively minor. Even if the imperfection is significant, whether the government should intervene or not depends on whether it has effective tools that would enable it to help rather than hurt the matter.

The potential roles for government in encouraging individuals to choose more healthful lifestyles are: establishing incentives to deal with externalities on fellow citizens' health, providing information, and attempting to compensate for the inappropriate incentives created for the most part by private and public compensation programs, such as disability payments for some lower back pains. To the extent that government programs have created these problems, we encounter the problem of creating a pyramid of interventions. Once the government provides compensation for a particular misfortune, it will have an incentive to try to step in to regulate individuals' behaviors so the misfortune is not more likely to occur.

## 5.3  Means to Change Lifestyles

How might government encourage individuals to choose more healthful lifestyles? At the most modest level, the government can merely provide information. If the government can discover, for example, the effects of various diets on health, then it should distribute this information to the general public.  Information of this type, except for the costs of dissemination, is a public good. No private entrepreneur would have an incentive to provide this information efficiently, that is, at a price of zero, the marginal cost of making it available.  (We distinguish availability costs from dissemination costs.)

Information on the likely consequences of different lifestyles is certainly important; yet we should not overrate its impact. Public education campaigns are rarely evaluated; when they are, often no change in behaviors is observed.  A controlled study of television advertisements for wearing seat belts found no increase in the proportion of viewers using them (Robertson et al.,1974). Mandatory warning labels on cigarette packages and ads have apparently had little effect on consumption (U.S. Bureau of the Census, 1975).

Often a second type of information is needed: how to change lifestyles if one wants, or has been convinced, to try.  In a recent national survey on cigarette smoking, 71 percent of male smokers indicated that they had made some effort to stop, yet other data show that only a small proportion had been successful (Horn, 1972).  Data on efforts by individuals to lose excessive poundage (except when hospitalized) show similar failures.  Only a quarter shed as much as twenty pounds; most of these "successes" eventually return to their former weight (Stunkard,1975).  The forces that encourage unhealthy habits in smoking and eating are sturdy; interventions that could empower individuals to control them would be a boon to the national health.

The government can attempt to develop schemes for dealing with externalities among individuals, as might occur with smoking or with driving at excessive speeds.  As we have detailed above, the existence of compensatory programs is likely to provide a strong incentive for individuals to allocate less than sufficient resources to the promotion of their own health.

Understanding the possible justifications for government intervention, we should next estimate the magnitudes of the imperfections in any particular circumstances.  Finally, we should examine alternative forms of government participation, to see which of them, including the possibility of doing nothing, would prove most desirable.

Basically, the government can choose between active and passive roles.  In an active role, the government in effect changes the locus of decisionmaking from the individual to the government. With active role intervention, all individuals will, in effect, have the same decision made for them.  Passive role intervention leaves the decision in the hands of the individual, though the government may change the terms of the decision being made.

Active intervention in the health-promotion area is primarily of two types: direct government provision and government regulation. Thus, the government might make available hospitals, nutrition programs, genetic screening programs, and the like. Regulation includes such measures as making air bags mandatory in new cars, and establishing standards that must be met, such as strictures imposed by EPA and OSHA, or setting minimum requirements for the deliverers of different types of medical care.

Passive modes of government intervention primarily include the provision of information and incentives. In some areas, the provision of information could be an alternative to regulation. The government might publicize the benefits of air bags, or establish certification procedures for medical personnel, allowing the customer to decide whether a physician assistant is capable of delivering certain types of care.

## 5.4  The Provision of Incentives

Where externalities are present, the conceptually appropriate means of government intervention is to provide an incentive for efficient behavior. The magnitude of the incentive should be equal to the net social benefits if an individual takes the desired action. Where health and life-saving are concerned, outcomes will be uncertain; estimates of the externality benefit will have to include some probabilistic modeling. If we know what different outcomes are worth, then we should pay differentially for actions according to the probabilities that they lead to different outcomes.

Let us say Jones can smoke or not. If he smokes he has a .20 chance of contracting the less favorable condition. If he does not smoke, that probability falls to .10. The only externality is that it costs society one thousand dollars if he contracts the less favorable condition. It would be worthwhile for society to charge the individual

$$.2(\$1,000) - .1(\$1,000) = \$100 \text{ for smoking.}$$

A few methodological points may suggest the value of a more spirited investigation of various ways of providing appropriate incentives. First, the actual payments that are made as part of an incentive scheme are lump sum transfers. Unlike the use of X-rays or personnel, they do not involve the use of scarce resources. Second, when individuals are allowed to respond to incentives, rather than forced to follow particular regimens, we tend to influence the individuals who attach the least cost to the change in their action.  For example, if we were trying to encourage individuals to stop smoking and thereby promote their own health, we would want to get only those individuals whose loss of pleasure from smoking was balanced by health gains.  Individuals for whom this pleasure loss is intense - individuals who will incur substantial pains if they stop smoking - should not be required to do so.  In the workplace nonsmoking program mentioned above, five of eleven smokers continued their habit, though they agreed the bonus was "a good idea" (Shepard, 1980).  Third, incentives can be directed where they do the most good.  We can pay fat men to lose weight or change their diets.  We do not need to prohibit everyone, thin or chubby, from enjoying fattening foods. Fourth, incentives can convey information.  In particular, financial incentives can convey information that otherwise fades into the general background noise of government advice.  An individual who is told that five dollars a month will be paid for maintaining his or her weight below a particular level may find that a much more compelling communication of the value of the weight loss than a simple reminder that it is dangerous to be overweight, even if that danger is quantified.  Fifth, in addition to their efficiency effects, incentives programs can redistribute the costs of unhealthy behaviors and unfortunate outcomes.  Taxing cigarettes, for instance, hurts those who are already disadvantaged, in that they are smokers.  If such redistributional consequences are unwelcome, they must be balanced against any efficiency gains an incentives program can expect to achieve.

At present, we have a vast array of public and private programs that reduce individuals' incentives to provide for their own health.  Probably the best way to generate countervailing incentives to induce individuals to behave properly would be to restructure some of our "offending" present programs.  How this can be done is a challenging subject for research study.

## 6.  CONCLUSION

Many of the most important issues confronting society directly or indirectly involve the valuation of lives.  A number of these issues, such as the choice of technologies for generating electricity, the levels of stringency to be applied in environmental protection, or the promotion of transit modes to compete with the automobile, lie somewhat outside the traditional health-care field.  Observing this range of contexts, an economist's first impulse is to attempt to ensure that the values assigned to the preservation of the same life in different areas are constant. This approach may derive from the belief that if risk avoidance could somehow be packaged and sold to individuals, informed consumer sovereigns would reflect such constancy in their own purchases.

Because society does not confront policy issues involving protection against risks on an individual-by-individual basis, results based on hypothetical individual isolated-consumption choices may not apply. This suggests that estimates of the values of lives inferred from market transactions may not be appropriate guides for government decisionmakers. Indeed, this chapter argues that once life valuation is made subject to collective choice, much more than lives and dollars may enter into individuals' valuation functions. Most significantly, the way choices are made may be of vital concern. That is, process may matter. Economists may bemoan this fact, but in view of their professional predilection to take consumers' preferences as given in other contexts, they should not dismiss it.

Procedures for valuing lives must be developed that appropriately reflect not only considerations of process, but also such matters as anxiety, income distribution, and possibilities for compensation. This is a challenging assignment; it should be approached realistically. The search should be for significant insights, useful benchmarks, and helpful guidelines, not unequivocal answers. Present procedures are sufficiently haphazard that even a much qualified analytic approach can provide substantial benefits.

Most analysts would probably agree that we are far from achieving maximum benefit from the resources we are devoting to life-saving activities. With the continued expansion of our expenditures for life-saving, the efficiency loss grows with each passing year.

To curb these losses will require insightful policy reforms. To aid such reforms, we have argued, useful studies might be made: (1) to improve benefit and cost accounting; (2) to refine techniques for predicting the consequences of interventions; and (3) to suggest how incentives should be structured to secure well-considered and efficient life-saving programs.

REFERENCES

Acton, Jan P., 1973. Evaluating Public Programs to Save Lives:
    The Case of Heart Attacks. Research Report R-950-RC.
    Santa Monica, Ca.: The Rand Corporation.

Bailey, Martin J., 1980. Reducing Risks to Life: Measurement
    of the Benefits. Washington, D.C.: American Enterprise Insti-
    tute.

Crouch, Edmund, and Richard Wilson. In Press. "Estimate of
    Risk." University of British Columbia Journal of Business
    Administration, Vol, Nos. 1 and 2 Fall 1979, Spring 1980 p. 299

Friedman, Milton., 1953. Essays in Positive Economics. Chicago:
    University of Chicago Press.

Fuchs, Victor R., 1967. "The Basic Forces Influencing Costs of
    Medical Care." Address given at the National Conference on
    Medical Care Costs. Washington, D.C., June 27.

Horn, D., 1972. "Determinants of Change." In Second World Con-
    ference on Smoking and Health, edited by R. G. Richardson.
    New York: Pitman, pp. 58-77.

Mishan, Ezra J., 1971. "Evaluation of Life and Limb: A Theoreti-
    cal Approach." Journal of Political Economy 79 (December):
    687-705.

Robertson, Lelley, O'Neill, Wixom, Eiswirth, and Haddon, 1974.
    "A Controlled Study of the Effect of Television Messages on
    Safety Belt Use." American Journal Public Health, 64:1071.

Schelling, Thomas, 1968. "The Life You Save May Be Your Own."
    In Prolems in Public Expenditure Analysis, edited by
    S. B. Chase. Washington, D.C.: Brookings Institution.

Shepard, Donald S., 1980. "Incentives for Not Smoking: Experience
    at the Speedcall Corporation - A Preliminary Report." Paper
    presented at the Corporate Commitment to Health, First Execu-
    tive Conference, Washington, D.C., June 9-10.

Starr, Chauncy, 1972. "Benefit-Cost Studies in Sociotechnical
    Systems." In Perspectives on Benefit-Risk Decision Making,
    pp. 17-42. Washington, D.C.: National Academy of Engineering,
    Committee on Public Engineering Policy.

Stunkard, Albert J., 1975. "Presidential Address - 1974: From
    Education to Action in Psychosomatic Medicine: The Case of
    Obesity." Psychosomatic Medicine, 37:195-236.

Thaler, Richard, and Sherwin Rosen, 1974. "The Value of Saving a Live: Evidence from the Labor Market." Discussion Paper 74-2, Department of Economics, University of Rochester.

U.S. Atomic Energy Commission, 1974. "Reactor Safety Study: An Assessment of Accident Risks in U.S. Commercial Nuclear Power Plants." WASH-1400, Washington, D.C.: Government Printing Office, see especially Chapters 6 and 7.

U.S. Bureau of the Census, Department of Commerce, 1975. Statistical Abstract of the United States, 1975, Washington, D.C.: Government Printing Office, p. 751.

U.S. Department of Health, Education, and Welfare, 1966. "Program Analysis - Selected Disease Control Programs." Washington, D.C., Office of the Assistant Secretary for Program Coordination.

Viscusi, W. Kip., 1978. "Labor Market Valuations of Life and Limb: Empirical Evidence and Policy Implications." Public Policy, 26:359-86.

Zeckhauser, Richard J., 1975. "Procedures for Valuing Lives." Public Policy, 23:419-464.

Zeckhauser, Richard J., and Donald S. Shepard., 1976. "Where Now For Saving Lives?" Law and Contemporary Problems, 40, no. 4:5-45

# EIGHT FRAMEWORKS FOR REGULATION*

Lester Lave
Brookings Institution, and
Carnegie-Mellon University

*From, The Strategy of Social Regulation:
Decision Frameworks for Policy,
Brookings Institution, 1981.
Used by Permission.

## 1. INTRODUCTION

The Frameworks used to guide analysis and the decisions made regarding proposed health and safety regulations are inextricably linked. For example, if an agency is required by law to ban any substance shown to be a carcinogen, little or nothing is gained by an elaborate analysis of the economic and other implications of alternative decisions. The decision framework establishes priorities among issues and changes the way both regulators and non-governmental decision makers view health and safety issues. Choosing a decision framework and using it consistently is perhaps the most important device for influencing the billions of decisions governing health and safety that are outside the control of federal regulators. Failure to appreciate the importance of the decision framework is the root of much of the criticism of social regulation. Legislation such as the Toxic Substances Control Act; the Federal Insecticide, Fungicide, and Rodenticide Act; and the Consumer Product Safety Act requires analysis of benefits, costs, and risks in formulating a regulation.[1] Unfortunately, subject areas such as carcinogenicity lack a firm scientific foundation for an analysis for those areas having a scientific foundation. Legislation such as the Delaney Clause is highly specific in requiring a ban,[2] but occasionally this action is so counter to public desires that the agency is condemned for carrying out the legislation, for example, for banning saccharin.

## 2. DECISION FRAMEWORKS

Six frameworks for making regulatory decisions are currently being used and two have been proposed. The frameworks range, roughly, from those requiring the least theory, data, and analysis and offering the least flexibility to those at the opposite pole; they include market regulation, no-risk, technology-based standards, risk-risk (proposed), risk-benefit, cost-effectiveness, regulatory budget (proposed), and benefit-cost.[1]

## 2.1 Market Regulation

Economic theory has formalized the 200-year-old insight of Adam Smith that competitive markets are efficient. In particular (under a set of stringent assumptions including complete information, no transaction costs, rational consumers and producers, no economies of scale in production, and no externalities), a competitive market produces an efficient (or Pareto optimal) equilibrium in the sense that no one can be made better off without making at least one person worse off.[1] This efficiency principle also holds for situations involving risk, such as hazardous products or jobs, although still more stringent assumptions are needed.[5]

Each person in such an economy presumably would decide what is best for him by looking at the array of available products and jobs. Since risk is an undesirable attribute, all risky products and jobs having no compensating attributes would be eliminated, and individuals would scrutinize those risky products and jobs that offered higher pay or some other advantage to determine which should be taken. Under the restrictive assumptions, government regulation would be unnecessary.

Clearly the U.S. economy does not satisfy the host of restrictive assumptions; both buyers and sellers can often influence price, many effects are transmitted outside the marketplace, and often buyers and sellers are woefully ignorant of the health and safety implications of a product. Market equilibrium is inefficient and a case can be made for government intervention. Some economists caution Americans to eschew perfection, arguing that they would be better off in the long run by tolerating these relatively minor evils instead of erecting a huge, self-defeating regulatory structure. Regulation requires resources, but more important, it is virtually impossible to regulate so that incentives are not distorted, and this often leads to even greater inefficiency than in the unregulated market - for example, transportation regulation, particularly of airlines and trains.[6] Many economists argue that regulation is justified only when serious violations of the assumptions occur, and then only if the regulation can be relatively efficient.[7]

An outstanding controversy concerns whether the current U.S. economy is essentially competitive and the consumers well informed. One side claims that the economy has hardly a hint of competition and that most consumers and workers are ignorant.[8] The other side sees intense competition, even within such oligopolistic industries

as automobiles and airlines.[9] Each side can muster persuasive
examples, although general proof is impossible.  Society has tended
to sway with the winds of intellectual discourse, with gusts of
regulation and then deregulation, as in transportation, for example.
There is general agreement, however, that the economy is basically
governed by the free market and that some regulation is necessary.
Successful regulators are pragmatic.  Doctrinaire positions concerning government control or laissez-faire are interpreted within
the context of each case, with the ultimate outcome dependent on
the nature of price control, consumer and worker ignorance, and
magnitude of risk.  Whether to license auto mechanics or to regulate sodium nitrite is decided by consideration of the specific
facts and risks rather than by the doctrinaire view that government should or should not regulate risky situations.

In summary, the decision to use the market to regulate risk
puts faith in consumer information and judgments.  It sees the
costs of bureaucracy constraining private decisions as larger than
costs arising from market imperfections and advises accepting
current imperfections rather than creating a regulatory morass.

## 2.2  No-Risk

The philosphy behind the Delaney Clause of the Food, Drug, and
Cosmetic Act, and food additive amendments generally, is that the
public should be exposed to no additional or unnecessary risk.
Carcinogens cannot be added to foods or remain as residuals in
meat since this might increase the risk of cancer; according to
the Clean Air Act Amendments of 1970, air pollution levels must be
sufficiently low to protect the population from adverse effects,
presumably even the most sensitive members.

This approach has great appeal as rhetoric.  To argue that
carcinogens ought to be permitted in the food supply is to argue
that society must allow higher than necessary risks of cancer.  Why
should any unnecessary exposure be tolerated, even if the risk
appears to be small?

The no-risk framework has the advantage of requiring little
data and analysis and precludes agonizing about the decision to be
made.  According to the Delaney Clause, the only question is
whether a food additive has been shown to be a carcinogen in
humans or animals.  Thus data (on the quality, variety, and price
of food) concerning the consequences of banning may not be considered.  The Delaney Clause has a simple, straightforward answer
to a complicated question:  ban a substance if there is evidence
of carcinogenicity.  Frameworks other than market regulation require answers to a set of complicated questions:  What level of
risks are acceptable?  What benefits would serve to offset the
risks?  Can animal bioassays be relied on to demonstrate human

carcinogenicity?[10]  Can the potency of a substance for humans be demonstrated under current exposure levels?  If one requires simple answers to these questions or distrusts the complicated answers given by experts, no-risk offers an appealing solution.

Unfortunately, the answers are too simple.  Virtually all "natural" foods contain trace elements of carcinogens, including biological contaminants and pesticides.  The Food and Drug Administration treats natural foods differently than food additives; apparently it is less troublesome to die from a cancer induced by a natural food than from one induced by a food additive.  Does anyone seriously propose to ban all foods with trace levels of carcinogens?  Does it make sense to treat those with trace amounts in the same way as those with large amounts of potent carcinogens?

In practice, the Food and Drug Administration considers, at least indirectly, the potency of each substance and the effects on the food supply.[11]  The agency agonizes over decisions concerning carcinogens when banning would reduce the food supply and merely deprive Americans of some cherished food.  Any attempt to impose consistency runs into the impossibility of eliminating risks generally and carcinogens particularly.  It is less obvious, however, that they could not be eliminated from a narrow class of substances such as food additives.

If society were concerned solely or even principally with the safety of food, the no-risk approach would be an appropriate guide for regulation, but society is concerned with many other issues as well.  People eat foods they know to be harmful to their health and they indulge in a range of habits indicating that health is neither their sole objective nor even a very important one.[12]  For example, in addition to overeating and not eating wholesome foods in a balanced diet, people smoke cigarettes, live in cities with polluted air, and work in occupations they know pose risks of accidents and chronic disease.[13]  Attempting to legislate safety by banning food additives that lower cost, enhance flavor and appearance, or increase convenience is like attempting to legislate morality: the rhetorical appeal is evident, but regulation can hope to affect only a tiny proportion of the relevant risk, and at rapidly increasing cost.

The three principal objections to this framework are the current misallocation of resources, the closing of the door to future solutions, and the inconsistency in government policy.  In addition, this framework cannot distinguish between a toxin that is extremely weak and to which few people are exposed and a potent carcinogen to which nearly the entire population is exposed. Insofar as there are many carcinogens and it is costly to ban at least some of them, this framework does not help to develop priorities - which substance should be treated first? - or guidelines - what level of safety ought to be sought where banning is infeasible? Instead, it sends regulators scurrying off to devote much of their attention to relatively benign substances by giving all toxins equal priority.[14] Thus, the framework is a pernicious guide to regulators confronted with complicated problems.

Although Congress has written the no-risk framework into legislation, it is a straw man unworthy of serious consideration. Even the attempt to maintain the facade is increasingly recognized by the regulatory agencies to be impossible. For example, the Food and Drug Administration has attempted to define a "negligible" risk level; any risk below a level of one in one million lifetimes would be considered to be zero for regulatory purposes.[15] The Environmental Protection Agency has taken an even more hostile view of the no-risk framework:

A requirement that the risk from atmospheric carcinogen emissions be reduced to zero would produce massive social dis-locations, given the pervasiveness of at least minimal levels of carcinogenic emissions in key American industries. Since few such industries would soon operate in compliance with zero-emission standards, closure would be the only legal alternative. Among the important activities affected would be the generation of electri-city from either coal-burning or nuclear energy; the manufacturing of steel; the mining, smelting, or refining of virtually any mineral (e.g., copper, iron, lead, zinc, and limestone); the manufacture of synthetic organic chemicals; and the refining, storage, or dispensing of any petroleum product. That Congress had no clear intention of mandating such results seems self-evident.[16]

## 2.3 Technology-Based Standards

Recognizing the difficulty of attempting to estimate the health and safety effects of a proposed standard (much less the problem of quantifying these effects), a number of agencies have placed their reliance on engineering judgments.[17] The best available control technology has been required extensively by the Environmental Protection Agency in regulating air and water pol-lution. This framework has the simplicity of requiring the estimation of neither benefits nor costs. The data and analysis required are for identifying a hazard and then for making the engineering judgment as to the best available control technology. This framework requires a second set of information for determining the best available control technology in addition to the carcino-genicity data required for the no-risk framework.

In practice, however, there is never a best technology but only successively more expensive and stringent technologies. For example, the effectiveness of an electrostatic precipitator in

removing suspended particles from air is proportional to the collector plate area; effectiveness can be increased by increasing
the area.  In practice, engineering judgment defines best available
control technology as a finite collector plate area, even through
further increases in plate area would improve (minutely) the effectiveness of collection.  At some point, additional abatement is
unwarranted because social costs exceed social benefits; but even
then technology is available that would abate emissions further.
In practice, best available control technology embodies implicit
assumptions about the benefits and costs of further abatement.

The crucial issue in implementing this framework at present
is the financial burden each industry can bear.  As long as an
industry is not in danger of bankruptcy, a technology that lowered
emissions would be considered acceptable.  Sufficient uncertainty
exists about what cost level would endanger an industry that
regulators rarely impose standards that come close to doing so.

In summary, the primary advantage of technology-based standards
is that they require no formal evidence on costs or benefits; the
only data required are those necessary for good engineering judgments.  The resulting standard, however, will depend on regulators'
perceptions of industry profitability.  If an area is populated by
an industry teetering on the brink of bankruptcy, best available
control technology will be weak and few emissions will be abated.
If the industry is profitable, it will require large expenditures.
There is more than a theoretical possibility that the first regulation in an industry would press it to the limit of its ability to
afford regulation, leaving no financial resources to handle later
regulations that might be far more important.  Rather than being
a framework for lowering risk or even for using engineering judgments, technology-based standards is a framework for regulating
economic activity through imposing costs arbitrarily among
industries until all are at the same minimal level of profit.

## 2.4  Risk-Risk:  Direct

Even if maximum protection were desired, the Delaney Clause
would be a poor framework because it requires banning carcinogens.
Some toxic substances, such as food additives and fungicides, prevent contamination of food, and thus it is desirable to weigh one
risk against the other, as recognized by the Food and Drug Administration and the Department of Agriculture in the proposed
risk-risk analysis.[18]  Balancing the toxicity of a substance
against the enhanced protection it brings can be done from either
of two perspectives.  The narrow perspective is that of balancing
the risk to the consumer of the additive against the direct health
benefits.  Sodium nitrite may be a carcinogen, but it protects
against botulism; the risk of cancer must be balanced against that
of botulism.  The broad perspective takes account of both producers
and consumers as shown below.

Since the risk-risk framework allows beneficial health effects to be considered along with adverse health effects, it is more flexible than no-risk. It and the remaining frameworks are qualitatively different from no-risk in that they require quantification. of risk and at least partial estimation of benefits. If quantification were impossible, this framework could not be implemented because there would be no method for balancing unmatched risks (for example, chronic respiratory disease versus broken legs). Quantification is particularly difficult for the effects of toxic substances; thus this and the remaining frameworks are subject to the caution of those who contend that potency cannot be estimated from animal bioassays, or at least that potency for humans at low doses cannot be inferred reliably.[19]

All frameworks except the first two allow the possibility of labeling or other action short of banning.[20] This gives wider choice to individuals, permitting them different life-styles. Insofar as choice and diversity are important, alternatives to banning are important.

While the risk-risk framework provides somewhat greater flexibility, it still precludes consideration of nonhealth effects. Conceptually it is a small step since it merely includes both the health risks and health benefits of a proposal. In practice it appears to be a major improvement over the no-risk framework - where it is applicable. Cases, such as sodium nitrite, where the risk-risk framework is invaluable, are the exception. Few substances offer a direct health benefit to the consumer other than drugs, products for which the Food and Drug Administration already uses this framework. The framework is of limited interest because it is of such limited applicability.

## 2.5 Risk-Risk: Indirect

The advantage of the risk-risk framework over the no-risk framework is that it permits wider analysis of risks. One way of stating the objective is that society desires to minimize the adverse health effects associated with a given food such as bacon. Thus society would permit nitrite in bacon if the improvement in the health of consumers from botulism protection exceeded the decrement in health from the risk of cancer. Yet it is evident that the direct risk-risk framework takes only the first step of considering the health of the person consuming the food. People are also associated with the production and distribution of food; society desires to minimize the adverse health effects associated with producing as well as consuming bacon (for a fixed level of production). Workers would not countenance a regulation that offered consumers a small amount of protection at the cost of a large increase in risk to workers.

Since every human activity is risky, a regulation that requires more man-hours to produce a unit of food would increase the exposure and presumably the occupational risk of workers. The indirect risk-risk framework includes occupational risks associated with each additive or contaminant (see the appendix for some methods that might be used to estimate these occupational risks).

One qualification to this approach is that consumers often are unaware of food risks, while workers are likely to have better information and receive a wage premium to take occupational risks. Furthermore, there is some selectivity of workers for a particular job, at least limited flexibility to change jobs or even to quit if the risks are too high, and some ability to remain alert when risks are highest.[21]

The indirect risk-risk framework is an important generalization since it allows consideration of implied health risks to workers. The difficulty is estimating health risks. As a first step in the analysis, assume that the same quantity of a regulated product would be produced as had been produced before. The immediate effect on workers might be estimated by assuming that the average rates of accidents and occupational disease in an industry would apply to the additional effort required by the proposed regulation, for example, additional feed grains to fatten steers because diethylstilbestrol is banned in particular, if banning it required a 10 percent increase in corn production, accidents and occupational disease among corn farmers would be estimated to increase by 10 percent. There are a series of ripple effects, however. The additional farming will require more seed, fertilizer, machinery, and fuel; these in turn will require more steel, coal, and so forth, each of which will involve occupational accidents and disease. Some preliminary notions for quantifying such ripple effects are discussed in the appendix.

## 2.6  Risk-Benefit

Unlike the risk-benefit framework, the three previous ones do not allow consideration of nonhealth effects. The folly of refusing to consider these effects is illustrated by examining one's own choices. For example, most people are willing to risk the minute chance of biological contamination rather than to be bothered with boiling drinking water. They are willing to undertake additional risks in order to get rewards such as additional income and recreational stimulation. For example, there is a risk premium in the pay of workers in hazardous occupations to attract them in the face of the higher risks.[22] These premiums can be extremely high, as for test pilots, steeplejacks, and divers working deep in the ocean. If the effect of a regulation is to lower risk minutely at the cost of a vast increase in price, a lessening of choice or convenience, harm to the environment, or a sacrifice in social goals

generally, society should not be satisfied. The frameworks
previously mentioned suffer from their lack of recognition of
other social goals, such as the ecosystem, endangered species, and
individual freedoms.[23]

Under the risk-benefit framework, regulators would be enjoined
to balance the general benefits of a proposed regulation against
its general risks. This framework is intended to be somewhat
vague, with all effects being enumerated, but with full quantification and valuation being left to the general wisdom of the regulators. The framework may account for cost, convenience, and even
preferences in an attempt to balance benefits against risks.[24]
A vast array of frameworks can come under the risk-benefit heading,
from balancing health risks against health benefits (like the risk-
risk indirect framework) to consideration of all risks, costs, and
benefits. The framework has an immediate appeal to congressmen
and regulators since it is a general instruction to consider all
social factors in arriving at a decision. While no one can oppose
considering all relevant factors, no one has specified precisely
how this is to be done.

The intellectual difficulty with this framework is its lack of
precise definition. Are only health risks to be considered, or are
risks to the present and future environment (air, water, louseworts,
snail darters, and tundra) relevant? If they are not, the framework is no more complete than the previous one, and if they are,
how can the risks to louseworts be added to those to the health of
our great grandchildren and of current workers? Similarly, there
is no guidance about how to quantify benefits: what is the value of
an increase in the supply of food or electricity?

This is the most general and flexible framework, but one
despairs at its implementation. Is it more than an injunction that
decision makers ought to think broadly about the risks and benefits
of their decisions? Insofar as decision makers are suspicious of
quantification or do not believe that it can be done with confidence, this framework serves to broaden their consideration, but it
still relies on their intuitive judgments. While it is desirable
to broaden the scope of matters to be considered, the failure to
define what is irrelevant has lengthened hearings and complicated
the record. The risk-benefit framework makes no pretense at being
an automatic decision-making tool. It forces regulators to consider a broad set of costs and outcomes; they cannot abdicate their
responsibility by examining only a narrow set of effects or appealing to some arbitrary criterion as they can under the no-risk
framework. The risk-benefit framework, however, has produced
decisions and justifications that seem arbitrary and inexplicable;
it has been a step forward, but it is too unsatisfactory to be more
than a transitional step.

## 2.7  Cost-Effectiveness

Many organizations, private and public, find themselves
attempting to increase output even though their current budget is
fixed.  The intellectual contributions in defining this problem
and developing rules to solve it have come from the Department of
Defense.  Although cost-effectiveness is often thought erroneously
to refer to getting some specific project done at lowest cost, the
concept is much broader, referring to accomplishing some general
objective at lowest cost.  President Eisenhower's secretary of
defense, Charles Wilson, described the goal succinctly as an
attempt to "get the most bang for the buck."

How can a goal be achieved within a fixed budget?  For example,
the goal of the National Cancer Institute is to lower the cancer
death rate.  It might achieve this goal by devoting resources to
basic research, clinical trials testing new treatment techniques,
public education, prevention by lowering the amounts of carcinogens in the environment, early detection of cancer, or the provision of more treatment.  How should the fixed budget be allocated
among these competing programs to lower both the incidence of
cancer and the occurrence of death and lesser effects?

Mathematically, this is a problem of maximization under constraints; the solution is to equate the effectiveness of the last
dollar spent on each activity.  For this example, the National
Cancer Institute ought to allocate funds among the programs (taking
care that the most effective projects are done first within each
program) by testing the effectiveness of each dollar.  The first
increment of funds should be given to the program where it would
be estimated to save the most lives.  The second increment of
funds should be allocated by the same criterion, perhaps going to
the same program.  As each successive increment of funds is allocated, the number of lives it saves should fall (since the best
projects were done first).  When all funds have been allocated, it
should be true that the last increment of funds to each program
would be expected to save approximately the same number of lives.
If not, then funds should be reallocated by recalling them from the
program where they are least effective and giving them to the program where they are most effective.  Mathematically, the ratio of
lives saved to dollars expended (for the last increment of funds)
should be equal across programs when all funds have been allocated.
As long as the ratios are not equal, additional lives could be
saved for the same budget by removing funds from the program with
the lowest ratio and adding them to the program with the highest
ratio.

This criterion of the effectiveness of incremental or marginal
dollars is often mistaken for the effectiveness of total dollars.

The confused objective is to equate across programs the ratio of
lives saved per dollar expended with the total expenditures. This
criterion is not efficient since funds could be reallocated so as
to save more lives for the same budget.

If the goal is properly stated and if the budget is appropriate,
cost-effectiveness analysis will lead to the same decisions as the
more elaborate frameworks; this is not true for no-risk, technology-based standards, and risk-risk. Thus, it would be equivalent to the more elaborate frameworks under the proper assumptions,
or it could be a parody of them if the goal or budget is incorrect.

Cost-effectiveness offers a major advantage over benefit-cost
analysis in that it does not require an explicit value for the
social cost of premature death (or other untraded goods). Assumptions about these values are built into the goal and budget (for
example, maximize lives saved for a fixed budget) but need not be
stated explicitly. The flip side of this advantage, however, is
that errors in stating the goal or in determining the budget can
lead to bad decisions, and there is no internal mechanism for
showing the errors in these decisions and the changes in goals or
budget that are necessary.

## 2.8  Regulatory Budget

Cost-effectiveness is a good framework if the relevant costs
are being measured in the analysis. Unfortunately, when the only
costs considered are those of the regulatory agency, the framework
will misallocate resources because only one subset of the total
costs of the regulation to the entire economy is being considered.
The agencies have little or no reason to consider the costs that
their regulations impose on others, unless the costs are so high
that industry bankruptcy is a relevant possibility. The agencies
are instructed to protect the environment, consumers, or workers
without any apparent limits on their ability to impose costs on
others. That the resulting regulations are not universally perceived as desirable, can be judged from the comments of the
affected companies and the fact that the federal government has
often exempted itself from the regulations or has been slow in
implementing them.

An idea originating in the Council of Economic Advisers under
Charles Schultze was to give each regulatory agency an implementation budget in the form of a limit on the total annual costs
that its regulations could impose.[25] For example, the Environmental
Protection Agency might be given an implementation budget of
$10 billion a year, which would mean that the costs of implementing
its air, water, solid waste, radiation, and pesticide regulations
could not exceed $10 billion in that year. Each agency would
develop an implementation budget request, just as it currently

develops its operating budget request. The administration would coordinate and impose priorities on the agencies, and then Congress would react to these requests, modifying them as necessary.

The regulatory budget is one method of implementing cost-effectiveness analysis. The goals needed for the framework are stated in the legislation for each agency, supplemented by whatever informal instructions arise from hearings, appropriations, Office of Management and Budget directives, or presidential intervention. The internal and implementation budgets would be considered and approved by Congress, based on each agency's data on effectiveness. A major advantage of the framework is that it would elicit from the agencies a clearer indication of their priorities and would enable Congress to make more intelligent decisions regarding social values.

The principal difficulties with the framework are in estimating the costs and effects of each regulation. Where a control device must be added to a smokestack, there is debate about the cost of the device and about its expected lifetime, maintenance, and reliability. For a new piece of technology, these difficulties might perhaps introduce a factor-of-two difference in estimated costs. When the regulation will require a change in process or result in banning a substance, the costs become much more uncertain. If there is a factor-of-five-or-ten difference between reasonable high and low estimates of implementation costs, the regulatory budget cannot provide a helpful constraint.

Discipline might be exerted by the use of ex post reviews of previous cost estimates and the resulting experience. Even for regulations that have been implemented, however, it is difficult to estimate the additional costs due to the regulation. In addition, several years would elapse before sufficient experience accumulated to estimate costs retrospectively; disciplining the agency for bad cost estimates during a previous administration would make little sense.

Excluding uncertain or indirect costs (while estimating only direct costs or those that can be confidently quantified) would give a terrible set of incentives to the regulatory agency. For example, banning a substance would minimize direct costs, even though it might impose very substantial indirect costs. Similarly, counting only current costs would lead the agency to design a regulation to impose costs in the future.

Estimating the accomplishments or benefits of a regulation is even more difficult, but this problem is common to all the frameworks from risk-risk to benefit-cost analysis. Good estimates of costs are required for the frameworks encompassed by cost-effectiveness and benefit-cost. There is no easy way to improve the quality of the cost and effectiveness estimates. They will necessarily be uncertain, and agencies will choose estimates from the end of the range that can be justified. This framework cannot be seen as a mechanical way of laying to rest the difficult questions of setting regulations, but it might serve to present more complete information in a useful framework to the correct decision makers.

A good deal of work remains to be done in exploring this framework. Seemingly subtle issues affect the outcome of the analysis. For example, the budget constraint can be stated for all regulatory agencies, for each division or program, or for "discretionary" funds. If trade-offs are made only within narrow programs, the overall result is unlikely to be satisfactory. For example, should the Food and Drug Administration be making trade-offs among food additives or among all activities under its purview that could enhance health? If the Food and Drug Administration were permitted to allocate time and funds among all activities, it might focus on cigarette smoking and ignore food additives. Some groups feel strongly about food additives and would protest a lack of regulatory attention to this area, even if the resources saved more lives by decreasing cigarette consumption.

How would the regulatory budget be determined? Agencies would request large budgets and be opposed by those who must pay implementation costs. The major advantage of this framework is precisely that it puts these issues of social values and of agency efficiency into a forum that facilitates their discussion and resolution and forces Congress to make these judgments. The resulting hearings will raise many relevant issues.

Academicians see the advantages of coherent, well thought out intellectual frameworks. For example, there is a misallocation of resources if some agency assumes, implicitly or explicitly, that the cost to society of premature death is greater than that assumed by another agency. Congressmen express shock at the notion that such a value exists. The political process presents an incremental approach, where a set of social values emerges from dozens of laws and decisions, much like the development of common law. Decision makers feel more confident in answering these questions for particular cases than in giving abstract uniform answers. While the regulatory budget is admirably matched to current American political institutions, the result is unlikely to be an intellectually coherent framework.

## 2.9 Benefit-Cost

This framework is similar to the general balancing of risks against benefits; the principal difference is that it is more quantitative and formal. In addition to enumerating the various benefits of the regulation and then subjectively balancing benefits against costs, this framework would require quantification of the extent to which the benefits and costs vary with the level of regulation, and then would require each of these effects to be translated into dollars.

There are many controversial aspects to its application, including putting an explicit value on prolonging a life, quantifying other benefits, deciding the rate at which effects in the future are discounted to make them equivalent to current effects, and redistributing income.[26] Valuing benefits, or even deciding what is a benefit, runs into the diversity of cultural backgrounds, personal goals, fears, and time horizons. Benefit-cost analysis is the most general and quantitative of the frameworks, and thus elicits the most information and requires the most analysis.[27]

Benefit-cost analysis is a sufficiently broad framework to be adapted to consider virtually any aspect of a regulation or public decision. The implications for those who gain or lose can be folded into the analysis. None of the objections to the framework have the effect of showing an inherent bias or blind spot in the analysis.

In practice, however, the picture is quite different. Benefit-cost analysis is often viewed, correctly, as a tool for defending the status quo. It is rarely used to consider who benefits or pays, and it focuses on the present, giving short shrift to even the near-term future with no importance for events more than a few decades in the future. Adjustment costs are often estimated to be higher than would be observed, reflecting a prejudice that the current situation must be the best one (when adjustment costs are not considered, the analysis is biased toward change). Finally, a number of simplifying assumptions are made that bias the analysis against change.

In a world where data are costly to gather and analyze and are rarely conclusive, one must be content with uncertainty and with making educated guesses. The most important material should be considered first, leaving data and issues of lesser importance for future analysis, if warranted. This framework is especially good at interpreting economic data; economic issues have tended to be of first-order importance and thus the framework has raised the significant issues. The more important non-economic concerns are and the more remote the relationship to economic variables is, the less helpful this framework will be.

For example, when benefit-cost analysis is used within a profit-making corporation to enlighten an investment decision, economic issues are of central concern. In considering government investments in inland waterways, economic considerations are predominant, although a host of other issues must be considered. In

considering whether the United States should purchase oil from a
particular nation in the Middle East, economic considerations are
dominated by political and more general ethical considerations.
In moving from a corporate decision concerning an investment to
purchasing oil from a particular nation, the relative importance
of economic issues declines.  The benefit-cost framework ceases to
be comprehensive and loses any claim to being the sole factor in
making the decision.  Instead, it becomes a framework for raising
issues, organizing information, and deriving quantification where
possible.

Benefit-cost analysis is not the best framework for examining
distributional questions since it offers no way of quantifying the
desirability of transferring income from one individual to another.
While it might be decided that a project has a beneficial redistribution of income, the net benefit cannot be quantified.  Some
ethicists (for example, the utilitarians) were able to deal with
this issue theoretically by defining an optimal distribution of
income; in practice these questions are hotly debated and decisions
are specific to situations.  None of the frameworks can be expected
to handle this set of questions well.

3.  A COMPARISON OF FRAMEWORKS

The eight frameworks stretch from simple solutions (let the
market do it or accept no unnecessary risk) to elaborate ones
(identify all effects and value them in dollars).  The range of
problems is even greater, stretching from purely scientific ones
(is nitrite a carcinogen?) to purely value conflicts (since so few
people buckle their belts, should passive seat belts be required,
even though they are more expensive and less effective than current
belts?).  Only by appreciating the complexity of problems and
frameworks can there be an intelligent analysis of how to improve
standard settings.

The issues of simplicity and the amount of data collection and
analysis required are illustrated by proceeding from the no-risk
to the benefit-cost framework.  Flexibility in finding solutions
and a broader purview of the issues are being purchased at the cost
of collecting and analyzing more data and grappling with myriad
problems, some of which have no solution.  For example, how does
society value a cancer today as against one occurring in twenty
years?  As against one occurring in 300 years?  How is a risk of
death of one in one million to be weighed against the risk of a
broken leg of one in ten?

Four criteria might be used to compare frameworks:

The first is comprehensiveness.  Are all the relevant issues
encompassed within the framework?  No-risk considers only carcino-

genesis (or other health attributes); risk-risk considers all
health consequences either to the consumer (direct) or more
generally (indirect). Cost-effectiveness and the regulatory budget
require examination of costs as well as health, but they can be
considered only within the goals of the agency. Benefit-cost and
risk-benefit are the most encompassing, although even they are not
used in practice to address equity questions.

The second criterion is the intellectual foundation required
of each framework. One can be most certain about the foundation
for the simple frameworks, but drawing in additional considerations
requires more knowledge, assumptions, and value judgments. The
wider coverage comes at a price. In some cases there is insufficient knowledge to be able to quantify or even explore these other
considerations; if so, there is no alternative to a simple framework or an ad hoc decision.

The third criterion is the resources required to implement
the framework. The more complicated frameworks require exploration
of further aspects of the problem, which in turn requires more data
collection and analysis. Generally, the resources available to
analyze alternative regulations constitute a small proportion of
those available for drafting and defending the regulations, and a
minuscule proportion of the cost of carrying out the regulation.
If additional analysis can result in even a tiny improvement in
the quality of the regulation, the reduction in implementation and
other costs should more than pay for the effort.

The fourth criterion is felicitousness. The world is complicated; it changes so rapidly that an agency rarely gets to second-
order priority issues. The most important issues must be treated
first, and they must be raised in easily comprehended fashion. If
the issues are posed in a confused or obscure manner, the decision
is likely to be made on an ad hoc basis. The felicitousness of
the framework is more important than its comprehensiveness.

None of these frameworks is sufficiently complete and sound
to serve as an automatic way of making decisions. The current
Delaney Clause framework would appear to be the most concrete;
even it, however, becomes mired in controversy over proving carcinogenicity - for example, the Newberne study regarding nitrite.[28]

The other frameworks have the more difficult task of quantifying risk and of attempting to quantify other aspects of the
issue (for example, the value of greater choice). In all cases,
judgment is required to examine the suitability of the quantification, the factors that could not be quantified, and the valuation
of the aspects that were quantified. These issues are far too
complicated for a mechanical decision making framework to be appropriate - for example, one of pursuing a project if and only if
estimated benefits exceed costs.

The real question is the extent to which each of these frameworks can prove helpful in informing the decision maker. Must all effects be quantified accurately and all valuations be agreed upon before benefit-cost analysis is helpful?[29] If complete quantification is not possible or if there are difficulties in estimating risk, is it better to slip back to a less demanding framework, possibly back to the no-risk framework?[30] The answer depends on both the amount of uncertainty and the extent to which the general nature of the uncertainty is known. No analysis of health and safety regulations has managed to quantify all aspects of the issue, and it is evident that no future analysis can be expected to be complete. If this lack of completeness is deemed fatal, there is no point in considering benefit-cost analysis further.

Even partial quantification is helpful in making complicated decisions. Knowing that an effect is very important or of no importance is helpful in the analysis. Offsetting this contribution is the tendency of quantified effects to drive out ones that cannot be quantified, but it should be possible to recognize this tendency and attempt to correct it. Even where an effect cannot be quantified, knowing the nature of the effect can be sufficient. For example, the aesthetic effects of air pollution have not been quantified well and are unlikely to be for some time but the sign of the effect is known. Thus, if a benefit-cost analysis finds that the quantified benefits exceed the costs, then adding aesthetic effects can only make the benefits even larger.[31] A more complicated example is spices. They contain toxic substances, including carcinogens, in minute quantities (for example, safarole). The risk of contracting cancer from the use of most spices in cooking food is minuscule and the enjoyment of better-tasting food is great. Should spices be banned in order to eliminate a tiny, controllable source of risk? In general, some of the risks can be computed using standard techniques, but the benefits are more difficult to estimate.

The purpose of regulation is to lower risk or to attain some other social goal. Quantifying the benefits of a proposed or existing regulation is difficult, however. Estimating health effects requires a number of assumptions. Quantifying effects on diversity of choice or consumer satisfaction is even more difficult, but any regulation requires at least an implicit estimate of these effects; they cannot be ignored.

In choosing a framework for each agency, Congress has implicitly met social goals. For example, the market regulation framework keeps the government from regulating health and safety; only by providing facts and analyses can the government influence consumers and businesses. The no-risk framework sets a goal of reducing carcinogenicity without the complications of considering

trade-offs among other social goals. Benefit-cost analysis implicitly elevates the importance of economic efficiency and downplays
political and equity considerations. Selecting a decision framework is a crucial first step in deciding how to regulate health
and safety. The importance of this choice is illustrated in the
next two chapters in an examination of a series of regulatory
decisions.

## REFERENCES

1. Michael S. Baram, "Regulation of Health, Safety and Environmental Quality and the Use of Cost-Benefit Analysis," pp. 46,
   59-60, 75-76; and L. E. Erickson, "Issues and Experiences in
   Applying Benefit Cost Analysis to Health and Safety Standards,"
   app. A and F.

2. The so-called Delaney Clause resulted from hearings held by
   Congressman James Delaney of New York and is found in the Food
   Additives Amendment of 1958 to the Food, Drug, and Cosmetic
   Act of 1938 (72 Stat. 1786). It states in part that "no addi-
   tive shall be deemed to be safe if it is found ... to induce
   cancer in man or animal."

3. Arranging the frameworks is not so simple. For example, risk-
   benefit analysis requires as much information as benefit-cost
   analysis, although it is less formal. While more than a single
   dimension is involved, the ordering is roughly accurate. See
   the bibliography for additional references.

4. Gerard Debreu. Theory of Value: An Axiomatic Analysis of
   Economic Equilibrium; and Kenneth J. Arrow and F. H. Hahn,
   General Competitive Analysis.

5. Kenneth J. Arrow, "Limited Knowledge and Economic Analysis."
   pp. 1-10; and Jacques H. Dreze, ed., Allocation under Uncertainty: Equilibrium and Optimality.

6. Theodore F. Keeler, "Domestic Trunk Airline Deregulation:
   An Economic Evaluation," pp. 75-149; and Paul W. MacAvoy and
   John W. Snow, eds., Railroad Revitalization and Regulatory
   Reform, p. 6.

7. George J. Stigler. The Citizen and the State: Essays on
   Regulation; and Milton Friedman. Capitalism and Freedom.

8.  John Kenneth Galbraith. Economics and the Public Purpose; Galbraith. The New Industrial State; and Robert L. Heilbroner. The Limits of American Capitalism, pp. 1-61.

9.  Friedman. Capitalism and Freedom; and Stigler. The Citizen and the State.

10. Interagency Regulatory Liaison Group, "Scientific Bases for Identifying Potential Carcinogens and Estimating Their Risks," pp. 22-61; National Academy of Sciences. Food Safety Policy: Scientific and Societal Considerations, pp. 5-21 through 5-25; and "Identification, Classification and Regulation of Toxic Substances Posing a Potential Occupational Carcinogenic Risk," Federal Register, vol. 42 (October 4, 1977), pp. 54156-67.

11. Richard A. Merrill, "Regulating Carcinogens in Food: A Legislator's Guide to the Food Safety Provisions of the Federal Food, Drug and Cosmetic Act," pp. 245-46; Peter Barton Hutt, "Food Regulation," pp. 521-24, and 553; and Richard A. Merrill, "Federal Regulation of Cancer-Causing Chemicals," chap. 2: "FDA Regulation of Environmental Contaminants," pp. 21-25.

12. NAS, Food Safety Policy; Victor R. Fuchs, "Some Economic Aspects of Mortality in Developed Countries;" and Fuchs, "Economics, Health, and Post Industrial Society," pp. 153-82.

13. Lester B. Lave and others, "Economic Impact of Preventive Medicine," pp. 675-705; Lester B. Lave and Eugene P. Seskin. Air Pollution and Human Health, chap. 10; and Robert S. Smith, "Compensating Wage Differentials and Public Policy: A Review."

14. Paul F. Deisler, Jr., "Dealing with Industrial Health Risks," p. 8.

15. "Chemical Compounds in Food-Producing Animals," Federal Register, vol. 42 (February 22, 1977), p. 10421.

16. "National Emission Standards for Identifying, Assessing and Regulating Airborne Substances Posing a Risk of Cancer," Federal Register, vol. 42 (October 10, 1978), p. 58660.

17. Larry E. Ruff, "Federal Environmental Regulation," pp. 279-82, 299-303.

18. U. S. Food and Drug Administration and the U. S. Department of Agriculture, "FDA's and USDA's Action Regarding Nitrite."

19. Gio Batta Gori, "The Regulation of Carcinogenic Hazards," pp. 256-61; J. L. Radomski, "Evaluating the Role of Environmental Chemicals in Human Cancer," pp. 27-44; memorandum, Arthur C. Upton, Director of the National Cancer Institute, to the commissioner of the Food and Drug Administration, "Quantitative Risk Assessment;" and National Academy of Sciences. Saccharin: Technical Assessment of Risks and Benefits, p. ES-5.

20. See NAS, Food Safety Policy, pp. 4-17, 8-5; and Oliver E. Williamson, "Public Policy and Saccharin: The Decision Process Approach and Its Alternatives," in Robert Crandall and Lester B. Lave, eds., The Scientific Basis of Health, Safety, and Environmental Regulation.

21. Smith, "Compensating Wage Differentials and Public Policy."

22. Ibid.; and Richard Thaler and Sherwin Rosen, "The Value of Saving a Life: Evidence from the Labor Market," pp. 265-98. The various studies attempt to control for other factors affecting pay, such as years of training.

23. Lester B. Lave and Lester Silverman, "Economic Costs of Energy-Related Environmental Pollution," pp. 619-23.

24. Chauncey Starr, "Benefit-Cost Studies in Sociotechnical Systems;" Starr, "Social Benefit Versus Technological Risk;" NAS, Product Safety; Cyril Comar, "$SO_2$ Regulations;" Richard Wilson, "Direct Testimony in the Matter of Proposed Regulations ... for Toxic Substances Posing a Potential Occupational Carcinogenic Risk;" Bernard L. Cohen, "Society Valuation of Life Saving in Radiation Protection and Other Contents;" Thomas H. Jukes, "Diethylstilbestrol in Beef Production: What Is the Risk?" and Richard C. Schwing, "Expenditures to Reduce Mortality Risk and Increase Longevity."

25. Personal communication with Charles L. Schultze. See U. S. Department of Commerce Regulatory Reform Seminar: Proceedings and Background Paper, pp. 17-31; Christopher C. DeMuth, "Constraining Regulatory Costs," pt. 1: "The White House Review Programs," pp. 13-36; DeMuth, "Constraining Regulatory Costs," pt. 2: "The Regulatory Budget," pp. 29-44; and William Nordhaus and Robert Litan, "A Regulatory Budget for the United States."

26. National Academy of Scienes. Analytical Studies for the U. S. Environmental Protection Agency, vol. 2: Decision Making in the Environmental Protection Agency, app. D; Baruch Fischoff and others, "Approaches to Acceptable Risk," pp. 169-204; and Baram, "Regulation of Health, Safety, and Environmental Quality."

27.  Benefit-cost analysis usually assumes fixed prices, wages, and discount rates.  If the scope of project or projects being analyzed is sufficiently large, however, prices and discount rate must be determined within the analysis.  Thus, applying benefit-cost analysis to all health and safety decisions made by the Food and Drug Administration would require a determination of the appropriate discount rate and would probably specify more projects than the agency could afford to do immediately.  In this sense, the regulatory budget framework is a specialization of benefit-cost analysis.

28.  Paul M. Newberne, "Dietary Nitrite in the Rat."  See also Newberne, "Nitrite Promotes Lymphoma Incidence in Rats, pp. 1079-81;  Council on Agricultural Science and Technology, "Comments on the Newberne Report on the Effect of Dietary Nitrite in the Rat;" Food Safety and Quality. Hearings, pp. 5-28, 176-80, 131-34, 236-38, 356-64; Comptroller General of the United States. Does Nitrite Cause Cancer?  Concerns about Validity of FDA-Sponsored Study Delay Answer.

29.  See, for example, Howard Raiffa. Decision Analysis:  Introductory Lectures on Choice under Uncertainty.

30.  For a discussion of this point see Richard Zeckhauser and Albert Nichols, "The Occupational Safety and Health Administration:  An Overview," app., pp. 161-248; and National Academy of Sciences. Decision Making for Regulating Chemicals in the Environment.

31.  Lave and Seskin. Air Pollution and Human Health, p. 212; and U. S. Environmental Protection Agency. Protecting Visibility: An EPA Report to Congress, pp. 11-12 through 11-17.

A COMPARISON OF AIR POLLUTION CONTROL STANDARDS
AS ADOPTED IN VARIOUS INDUSTRIALIZED CONTRIES


Alfredo Fontanella
Giancarlo Pinchera
ENEA, Direzione Centrale Studi-Roma


## 1. INTRODUCTION

The first laws on the control of atmospheric pollution were passed in the nineteenth century, principally in Great Britain, in order to solve the problems created by the wide and increasing use of coal brought about by the industrial revolution. From then on until the years of the Second World War, the problems faced were essentially those caused by the noxious and malodorous gases of industry. The various laws enacted during this period had a common denominator: the protection of the subjective interests of individuals, of employed workers or of social communities against the 'affronts' of neighboring individuals or companies. They were still a very long way from placing as their objective the protection from pollution, much less the defense of nature as an asset in its own right.

The impetuous urban and industrial development as occurring in a number of countries in the decades that followed, has caused the problem of atmospheric pollution to reemerge in terms that were more serious indeed:

- the sharp increase in energy consumption, with the consequent emission of vast quantities of pollutants into the environment at various stages of the productive cycle and through the use of fossil fuels;

- the development of large industrial zones and the dissemination of potentially noxious productive products;

- the formation of large urban concentrations in which the heating of the dwellings is ensured by myriad small heating plants;

- the enormous expansion of engine-driven vehicular traffic, that brings with it the multiplication of mobile pollution sources.

All these factors have together determined the continuous worsening of the quality of the air as well as the explosion of serious phenomena of smog development in cities such as London, Los Angeles, Tokyo and Milan. The interventions brought about by antismog laws have produced beneficial effects and thereby are permitting a reversal of trend in many areas.

In England, for example, after the dramatic five days in December 1952 during which the London smog caused thousands of deaths, effective steps have been taken. From 1956 to 1976 the average concentration of sulfur dioxide at ground level has been reduced by 75% in spite of the continual increase in fuel consumption. During the same period, smoke emission by industries decreased by over 90%.

At Milan, the concentration of sulfur dioxide at ground level, after having touched an average level of 0.243 ppm during the winter (a critical period due to heating) of 1970, thereby becoming the most polluted city of the year, has recorded a continual decrease, settling in the years from 1976 to 1978 at values fluctuating around 0.137 ppm, still too high nevertheless.

Control laws, standards and organization for the protection of the environment differ greatly from country to country.

The most recent regulations, adopted in industrialized countries for the protection of the atmosphere (like that on the subject of water protection and protection against ionizing radiation and noise), for the most part take the unitary nature of the problem into consideration, thus having as their direct aim the protection of human health and of the environment against pollution more than the mere defense of specific interests. A number of countries have developed integrated laws and standards which accompany the entire cycle of production and of the use of fossil fuels. Nevertheless, only few countries are at present endowed with an organic law on environmental protection which imposes standards and procedures such as to permit an overall approach to prevention and control of the interactions of urban and industrial agglomerations with the various environmental interfaces.

For a number of years now a lively debate has been in course
in the indistrialized countries on the subjects of environmental
protection and a quite tormented process of updating and revision
of the criteria of environmental quality, with regard to contaminants of various sorts, and of the prevention and control measures
intended to ensure compliance with them.  In the light of recent
surveys and studies of the experience gained, regulations and
standards are tending often to become more complete and restrictive.

The standards most widely distributed are those relative
to fuel characteristics and to 'air quality'.  In Germany, Japan
and the U.S.A., there have gradually been added 'emission standards'
(or 'plant performance standards') which regulate the emissions at
the outlet of the plant's smokestack; that is, before they begin to
disperse into the atmosphere.  For the definition of the standards
to which reference is made, please refer to the glossary given in
the Appendix.

The opinion is widely shared that it is urgent to proceed to
the rationalization and to the strengthening of the legislative,
institutional, organizational, regulatory and financial picture
in matters of the prevention and control of the environmental
legislation on the subject of atmospheric pollution are:

-    in 1917, the law on the classification of establishments,
     which imposed the request for authorization on all dan-
     gerous factories;

-    in 1948, the requirement of approval and of periodic in-
     spections for industrial plants;

-    in 1961, the institution of an official body to determine
     emission limits for dusts and toxic substances in the
     various cases;

-    in 1963, the imposition of economic sanctions for plants
     exceeding the emission limits;

-    in 1964, the inclusion of thermoelectric power stations
     among establishments subject to classification;

-    in 1966, the enactment of dust emission limits and technical
     standards for thermoelectric power stations;

-    in 1976, the enactment of regulations for the determination
     of the height of smokestacks for new installations and for
     the reduction of sulfur content in fuels;

-     in 1976, the institution of a special authorization procedure
      for the construction of industrial plants;

-     in 1980, the institution of the national air quality board.

French law on the control of atmospheric pollution establishes
that emissions must not cause disturbance to persons, endanger the
public health and safety, nor harm agriculture, buildings and
monuments.

The concept of environmental quality standards for air has
not been developed in French legislation. It is preferred to control pollution by intervening directly at the level of emissions
from every plant. The definition of emission standards is delegated to local enforcement regulations and to the conditions
appended to the authorization granted.

Law No. 663 of the 19th July 1976 assigns to the Prefect of
the Department involved, with the technical assistance of the
Industry and Mining Interdepartmental Service, the faculty of
authorizing the installation of a number of types of plants, establishing the operating conditions for each case. Among the plants
subject to authorization are thermal power stations with potentials
greater than 800Qth/h. The authorization instrument defines, among
impact. Such urgency derives from the necessity of offering the
operators and entrepreneurs, as also the citizens and adminis-
trators, a sure and unitary framework. It is of course, necessary
to proceed by taking into account the various institutional levels
(national, regional and local) and by beginning with a critical veri-
fication of the existing laws, standards, procedures and structures.
It is necessary, moreover, to bear in mind that the adoption of
regulations which do not permit reasonable flexibility to the
operators and entrepreneurs in the choice of the various technical
options possible to satisfy the legal standards, can introduce
restraints that are not necessary in reality or solutions that
are still technologically immature.

Although bearing the general picture relative to the problem
of atmospheric pollution determined by various types of sources
(domestic heating, automobile traffic, industry and so on) in
mind, it is necessary, in the individual studies, for attention
to be limited to specific systems and subjects.

The purpose of the present memo is to examine the manner in
which various countries are tackling the problem of regulations
for the control of atmospheric pollution generated by combustion
products emitted by industrial plants and, in particular, by
thermoelectric power stations. Taken into consideration are the
technical aspects of the regulations in force in France, Western

Germany, Japan, Great Britain, Italy and the U.S.A., emphasizing
their points of contact and differences of approach. The purpose
of the paper and the competencies of the authors do not include
the treatment of other important aspects of the problem such as,
in particular, its juridical aspects.

The study has been carried out within the scope of the activities of the VESE Project, conducted by ENEA's Central Studies
Office. The VESE (Italian acronym for 'Evaluations of the Effects
of Energy Systems') Project has as its object the systematic study
of the problems of the environmental and socio-economic impact
associated with the various energy sources and technologies with
reference to the entire cycle, from the mines to any waste materials
produced. The sources taken into consideration in the Project are
nuclear energy, oil, coal, methane, solar energy, geothermal and
hydro energy.

## 2. AN OVERVIEW OF NATIONAL REGULATIONS

### 2.1 France

France boasts a long tradition in the field of atmospheric
pollution control. As early as 1810 Napoleon decreed that factories emitting noxious exhalations could not be built without
authorization. The principal landmarks in the evolution of French
other things, the nature and quantity of emissions, the measurement systems, the ways and means of limitation and of control.
The ordinances must be justified on the basis of specific studies
that demonstrate their necessity and effectiveness.

From the zones in which the atmospheric pollution problem
is aggravated by the high density of industry and population,
such as Paris, Lyon and Lille for example, more restrictive norms
are contemplated with regard to the operation of the plants, to
fuel characteristics and to emissions.

The height of smokestacks (see Table 1) is subject to regulation throughout the entire territory, as are fuel characteristics
(see Table 2), and compulsory periodic inspections are contemplated
for steam generators and boilers of high capacity.

## TABLE 1
## REGULATION OF SMOKESTACK HEIGHTS IN FRANCE

Formula for the Calculation of Smokestack Height:

$$h = \sqrt{\frac{A\;q}{C_m}} \quad \sqrt[3]{\frac{1}{R\;T}}$$

in which h is the height in meters, A is a constant (340 for $SO_2$ and 680 for dust ), q is the pollutant flow rate in kg/hr, $\Delta T$ is the annual average difference in temperature between the fumes and the ambient air, in °C, R is the fume emission rate in $m^3$/hr, and $C_m$ is an air quality reference value in mg/$m^3$ of 0.25 for $SO_2$, 0.15 for dust, and 0.20 for $NO_x$.

## TABLE 2
## FUEL STANDARDS IN FORCE IN FRANCE

### A) GENERAL LIMITATIONS

| FUEL | SULFUR CONTENT |
|------|----------------|
| Gasoil and fuel oils for household use | 0.3% |
| Type 1 Heavy Oil | 2 % |
| Type 2 Heavy Oil | 4 % |
| Type 2 BTS Heavy Oil | 2 % |
| Type 2 TBTS Heavy Oil | 1 % |

### B) ZONES OF SPECIAL PROTECTION

  1) Lyon-Villeurbanne

   - Heavy oils are prohibited for plants of capacity Greater than 1,000,000 Kcal/h

   - Maximum sulfur content of fuels is 1%

2) Hauts de Seine, Seine-Saint-Denis, Val de Marne, Lille

- Heavy oils are prohibited for plants of capacity greater than 1,000,000 Kcal/h

- Sulfur content in fuels must not exceed 2gm per 1000 Kcal produced

3) Paris

- Heavy oils are prohibited for plants whose capacity is greater than 1,000,000 Kcal/h

- Fuels having 20% ash content, calorific power less than 5500 Kcal/kg and context of volatile substances of 15% are prohibited.

## 2.2 Federal Germany

The legislation in Federal Germany on atmospheric pollution was revised in 1974. The basic text of the revised legislation is the Federal Law on the Control of Emissions which:

a) Prescribes a license for the construction, alternation and operation of 58 types of dangeous plants;

b) establishes criteria for the design of plants for the storage, transport and processing of dangerous substances;

c) institutes a corps of officials specifically assigned to enforcement.

On the matter of atmospheric pollution, the law refers to Technical Instructions for the Protection of Air Purity. These Instructions establish, in particular:

- input standards for dusts and gases;

- emission standards for dusts and gases;

- height of smokestacks in function of dispersion methods and of surrounding topographical characteristics;

- monitoring instrumentation.

Two types of limitations are indicated for gaseous discharges:

- disturbance limits

- emission limits.

The former are based on objective criteria and take the harm-ful effects produced by the discharges into account. These limits are compulsory. The latter are based on the technological pos-sibilities of achieving the lowest possible level and amount to recommendations. Both types of limits have been established upon indications from the German Society of Engineers (V.D.I).

Fuel standards may be established by local authorities to protect areas of special interest to tourism.

For coal or oil burning plants, the maximum sulfur content is limited to 1%. For plants of capacities greater than 4 TJ/h, equivalent to approximately $10^9$ Kcal, fuels having higher sulfur content may be used provided desulfurization techniques are employed during combustion or afterward, to the fumes.

Table 3 summarizes the main values of the air quality standards and of the emission standards adopted.

## 2.3 Japan

The Japanese legislation on atmospheric pollution control dates back to 1967. But it has since undergone a series of up-datings. A specific text exists that prescribes environmental quality and emission standards by type of plant and by geographic area. More severe environmental quality standards are contemplated, moreover, for the long term. Limits stricter than those in force nationally can be required at the local level.

The examination of the construction of enlargement plans of plants, whose amendment can also be imposed, is required by the competent authorities. Periodic inspections and controls of plants are contemplated with possibilities of enforcement in case of non-compliance.

Table 4 indicates the environmental quality standards relative to a number of pollutants: suspended dusts, nitric oxide, sulfur dioxide and photochemical oxidants. Sulfur dioxide pollution is controlled by limiting its emission. Such limits depend on the height of the smokestack and on the nature of the geographic area in which the plant is located, as illustrated in Table 5.

Tables 6 and 7 respectively show the emission standards for dusts and for oxides of nitrogen, relative to thermoelectric power stations. The values depend on the size of the plant and on the type of fuel utilized.

TABLE 3

FEDERAL GERMANY: QUALITY STANDARDS OF AIR AND OF EMISSIONS

| Pollutant | Environmental Quality Standards for Air (mg/Nm³) | | Emission Standards for Thermo-electric Power Stations (mg/Nm³) |
|---|---|---|---|
| | Long Term | Short Term | |
| Oxides of Sulfur | 0.24 (1yr) | 0.4 (30 min.) | 650 |
| Suspended Dusts | 0.1 (1) (1yr.) | 0.2 (1) (30 min.) | From 50 to 300 depending on the type of fuel and on the size of the plant. |
| Dust Fallout | 0.35 gm/m²day (1 yr.) | 0.65 gm/m² day (24 hours) | -- |
| Nitrogen Oxides | 0.1 (2) | 0.3 (2) | To be reduced as much as poss-ible by available technical means. |
| Carbon Monoxide | 10 (1 year) | 30 (30 min.) | Gaseous fuels: 100 Liquid fuels: 175 Solid fuels: 250 |
| Hydrocarbons | -- | -- | -- |

1) For particles of size less than 10 um. For larger particles the limits are respectively 0.2 and 0.4.

2) As $NO_2$. The limits relative to NO are respectively 0.2 and 0.6

Areas where emission sources are too dense to permit attainment
of the environmental standards as set merely by the aforesaid emission limits are designated as "rerions with local regulation of
the pollutant load". Plans to reduce the pollution are elaborated
for such regions. Such plans are based mainly on more severe
emission limits than those applied nationally (for large installations) and on the disciplining of fuel standards (for smaller installations). The plans are elaborated using calculation models and
make it possible to assess the effect of the limits being required
on the quality of the environment.

The recommended methods of sulfur dioxide emission control
are: importation of fuels having a lower sulfur content, fuel
desulfurization, desulfurization of fumes. The recommendations
to reduce the emission of nitrogen oxides are: two-stage combustion, fuel recirculation, use of special burners, denitrification of fumes. A plan is under study at present for the reduction of the acid rain phenomenon, which includes the control
of sulfate and nitrate ions as those principally responsible.

A survey conducted by a group of experts from the Environmental
Protection Agency in the U.S.A. has shown that, in 1979, environmental regulation was being enforced over 98% of Japanese territory.


## 3.0  GREAT BRITAIN

Great Britain was one of the first countries to enact laws
for the control of atmospheric pollution, as a consequence of its
particular historical situations. The basic laws such as the
'Alkali Act', 'Sanitary Act' and 'Health Act', in fact, date
respectively from 1863, 1866 and 1875. Significant progress has
been made during the last twenty-five years in the reduction of air
pollution. However, important problems still remain to be solved
in this field.

At present, with reference to stationary sources, the principal laws for the control of air pollution are:

- Alkali, etc. Works Regulation Act (1906)
- Health and Safety at Work, etc. Act (1974)
- Public Health Act (1936)
- Public Health Act (1961)
- Public Health (Recurring Nuisances) Act (1969)
- Public Health (Scotland) Act (1897)
- Clean Air Act (1956)
- Clean Air Act (1968)
- Radioactive Substances Act (1960)

TABLE 4

JAPAN: QUALITY STANDARDS OF AIR AND OF EMISSIONS

| Pollutant | Environmental Quality Standards for Air ($mg/Nm^3$) | | Emission Standards for Thermo-electric Power Stations ($mg/Nm^3$) |
|---|---|---|---|
| | Long Term | Short Term | |
| Oxides of Sulfur | 0.1 (1) (24 hours) | 0.26 (1 hour) | See Table 5 |
| Suspended Dusts | 0.1 (24 hours) | 0.2 (1 hour) | From 50 to 800 $mg/Nm^3$ See Table 6 |
| Dust Fallout | -- | -- | -- |
| Oxides of Nitrogen | 0.075 - 0.1 (2) (24 hours) | -- | From 60 to 400 ppm See Table 7 |
| Carbon Monoxide | 11.5 (24 hours) | 23 (8 hours) | -- |
| Hydrocarbons | -- | -- | -- |

1) With a target of 0.04 $mg/Nm^3$ as annual average
2) Ac $NO_2$. With a target of 0.04-0.06 $mg/Nm^3$ as annual average

## TABLE 5

### STANDARDS FOR EMISSIONS OF SULFUR OXIDES

$$q = K \times 10^{-3} \times He^2$$

wherein:

q   =   maximum allowed quantities of sulfur in $Nm^3$ per hour
He  =   effective height of smokestacks in metres
K   =   geographic factor (see example in table)

a)   General standards

| Area No. | Locality | K |
|---|---|---|
| 1 | Tokyo | 3.0 |
| 2 | Chiba | 3.5 |
| 3 | Sapporo | 4.0 |
| 4 | Hitachi | 4.5 |
| 5 | Toyama | 5.0 |
| 6 | Annaka | 6.0 |
| 7 | Tomakomai | 6.42 |
| 8 | Sendai | 7.0 |
| 9 | Asahikawa | 8.0 |
| 10 | Akita | 8.76 |
| 11 | Takasaki | 9.0 |
| 12 | Shizuoka | 10.0 |
| 13 | Hakddate | 11.5 |
| 14 | Mishima | 13.0 |
| 15 | Aomori | 14.5 |
| 16 | Others | 17.5 |

b)   Special standards (2)

| Area No. | Locality (1) | K |
|---|---|---|
| 1 | Tokyo | 1.17 |
| 2 | Chiba | 1.75 |
| 3 | Kashima | 2.34 |

Note:   (1)   only the first of the localities indicated are given
        (2)   for new installations

TABLE 6
JAPAN: DUST EMISSION STANDARDS

| Fuel | Standards (mg/Nm$^3$) Ordinary | Special |
|---|---|---|
| Liquid or Gas | 100 | 50 |
| Lignites | 800 | 400 |
| Others | 400 | 200 |

TABLE 7
JAPAN: EMISSION STANDARDS FOR OXIDES OF NITROGEN

| FUEL | Fumes Emitted (x 1000 Nm$^3$/h) (1) | Standards (ppm) |
|---|---|---|
| Gaseous | 500 and over | 60 |
| | 40 - 500 | 100 |
| | Less than 10 | 150 |
| Solid | All | 400 |
| Liquid | 500 and over | 130 |
| | 10 - 500 | 150 |
| | Less than 10 | 180 |

1) Plants built after 19.VIII.1979. Plants already in existance must gradually reduce their emissions.

-    Control of Pollution Act (1974)
-    Local Government Planning and Land Act (1980)

The most recent legislation in England and Wales differs in some
cases from that of Scotland and Northern Ireland.

The law requires the adoption of the most suitable measures
to impede the discharge into the atmosphere of noxious or troublesome substances unless in a form such that the discharge does not
result in any damage.

At the national level, no concentration limits exist with
regard to air quality and emissions except for dusts, for which
the emission limit varies from 115 to 460 mg/m$^3$ depending on the
case.

The emission limits, together with the manner in which they
are enforced, are established case by case by an inspectorate for
the purpose (in England the 'Alkali and Clean Air Inspectorate')
at the time of the plant's construction and may vary subsequently
as a result of technical and scientific progress and in relation
to general pollution conditions.  The Inspectorate's ordinances
are compulsory in nature.  Periodic inspections and checks are
contemplated and local authorities are empowered to casue discharges to the air to be interrupted should they prove to be
deleterious to health or to the environment.

The Inspectorate bases its ordinances on criteria drawn
from the most up-to-date knowledge on the effects of the pollutants in question, especially on the effects on human health,
on animals, on agriculture and on the natural environment.
Account is taken, moreover, of the pollution level previously
existing in the area.  For example, it is not permitted for a
new discharge to cause the concentration in the environment to
increase more than 0.40 - 0.57 mg/m$^3$ (as an average value over
3 minutes).

In the case of coal- or oil-burning thermoelectric power
stations, it is believed that the best pollution control policy
consists of the dispersion of fumes by means of properly designed high stacks.  In the calculation of smokestack heights
the procedure is followed of not exceeding, on the ground, a
fraction (generally 1/40) of the maximum permissible concentration in working environments for the same substances.  The desulfurization of fumes is not believed to be necessary and its
cost unjustified.

The use of fuels has recently been regulated on a national
scale (see Table 8).

TABLE 8
GREAT BRITAIN:  FUEL STANDARDS

| Fuel | Sulfur Content |
|------|----------------|
| Gasoil | 0.8% |
| Fuel Oil | 0.5% |
| **Starting 1.X.1980** | |
| Fuel Oil | |
| Light (12.5 cS/180$^0$F) | 3.5% |
| Medium (30 cS/180$^0$F) | 4 % |
| Heavy (70 cS/180$^0$F) | 4.5% |
| Extra Heavy (115 cS/180$^0$F) | 5 % |
| All Fuels | |
| London Area | |
| New Installations | 1 % |
| Effective 1987 | |
| Existing Installations | 1% |

- authorization by the Minister of Industry, after having consulted the Central Commission on Air Pollution, the Ministers of Health and of the Environment, the Superintendency of Monuments and the Chairman of the region.

Any request to build or alter a plant must be accompanied by information concerning precautionary measures for the protection of the public health and of the environment, as well as the pollution monitoring system.

The legislation in force at the present time contemplates the limitation of emissions; that is, regarding the total contribution to the concentration of pollutants at ground level due to industrial sources.  The limits set by the enforcing regulations of Law 615/1966 for industrial plants are given in Table 9. They apply to industrial plants situated in zones A and B as defined in the same law and do not represent environmental quality standards (see Glossary) because they take into consideration neither the background level nor the controbution of nonindustrial sources.

For ENEL's new thermoelectric power stations, Law 880/1973 Article 6 prescribes the adoption over the entire national territory, of "a double chemical and meteorological survey network with double terminals" and sets a limit value for the concentration at ground level of sulfur dioxide (0.25 ppm as the average over 30 minutes and 0.10 ppm as the average over 24 hours, equivalent respectively to 0.657 and to 0.263 mg/m$^3$). Such values represnet a ceiling below which to maintain pollution. When exceeded, ENEL is required to "take every immediate measure for the pollution to be brought back within the said limits".

## 4.0  ITALY

The Italian legislation on air pollution control is comprised by:

- Law No. 615 of 13.VII.1966, which is the basic text;
- D.P.R. No. 139 of 22.XII.70, which regulates household heating plants;
- D.P.R. No. 322 of 15.IV.71, relative to industrial plants;
- D.P.R. No. 323 of 22.II.71 and Law No. 437 of 3.VI.71 with regard to motor vehicles;
- Law No. 880 of 1973, updating for thermoelectric power stations;
- Law No. 833 of 23.XII.1978, instituting the National Health Service.

The objective of the legislation is to control the gaseous emissions which jeopardize the public health or damage property. The problem of protecting aesthetic qualities of air is not addressed.

In addition to the laws just listed, further instruments for the control of atmospheric pollution are:

- urban and rural regulatory plans;
- specific provisions on the siting of industrial plants;
- laws for the protection of natural and artistic amenities.

The control of air pollution by industrial plants is effected by means of:

- pinpointing potentially polluting plants (over the entire territory);
- limitation of concentrations at ground level of pollutants (zones A and B defined by the law);
- disciplining fuel use (zones A and B separately).

TABLE 9

ITALY: EMISSION LIMITS AT GROUND LEVEL OF INDUSTRIAL ORIGIN AS PER THE ENFORCEMENT REGULATIONS OF LAW 615/1966

| POLLUTANTS | Peak Concentrations (1) 1013 millibar 25°C | | | Average Concentrations (1) 1013 millibar 25°C | |
| --- | --- | --- | --- | --- | --- |
| | p.p.m. (mg/m³) | sampling interval | freq. in 8 hours | p.p.m. (mg/m³) | sampling interval in hours |
| Oxides of sulfur expressed as $SO_2$ (2) | 0.30 (0.79) | 30 | 1 | 0.15 (0.39) | 24 |
| Chlorine ($Cl_2$) | 0.20 (0.58) | 30 | 1 | | |
| Hydrochloric acid | 0.20 (0.30) | 30 | 1 | 0.03 (0.05) | 24 |
| Fluorine compounds expressed as fluorine | 0.06 | 30 | 1 | (0.02) | 24 |
| Hydrogen sulfide (hydrogen sulfydrate) | 0.07 (0.10) | 30 | 1 | 0.03 (0.04) | 24 |
| Total organic substances expressed as hexane. Refinery derivatives | 80.00 | 30 | 1 | 40.00 | 24 |
| Oxides of nitrogen ($NO_2$) | 0.30 (0.56) | 30 | 1 | 0.10 (0.19) | 24 |
| Carbon monoxide | 50.00 (57.24) | 30 | 1 | 20.00 (22.89) | 8 |
| Lead compounds (Pb) | (0.05) | 30 | 1 | (0.01) | 8 |
| Inert suspended dusts | (0.75) | 120 | 1 | (0.30) | 24 |
| Free crystalline silica contained in the dusts expressed as $SiO_2$ | (0.10) | 120 | 1 | (0.02) | 24 |

1) At ground level, off the plant premises. Applied to industrial plants located in both A and B zones.

2) For ENEL's new thermoelectric plants the peak value has been reduced to 0.25 ppm (equal to 0.657 mg/m³), the 24 hour average to 0.10 ppm (equal to 0.263 mg/m³). Applied over the entire territory. Law 830/1973.

Thermoelectric power stations using fossil fuels are included among the industrial activities which, due to their potential to pollute, must be placed outside urban areas. A special procedure is contemplated for siting them which includes:

-   approval on the part of CIPE (an interministerial committee), in agreement with an interregional consultive committee, of ENEL's construction programs;

-   designation of sites by the regions involved, in agreement with ENEL and the local governments. If the region cannot identify sites by the deadlines set, the designation is made by CIPE;

Article 4 of the law that instituted the National Health Service establishes that "by decree of the Prime Minister, upon a proposal of the Minister of Health, after having heard the opinion of the National Health Board, maximum limits of acceptability are set and periodically subjected to revision for the concentrations and maximum exposure limits relative to pollutants of chemical, physical and biological natures and of sound emissions in the enclosed spaces of work and dwelling and in the outdoor environment".

The National Health Board, in its sitting of the 26th January 1982, approved the "limit values for pollutants of the unconfined air" elaborated by a study group created by the CCIA (italian acronym for 'Central Commission against Atmospheric Pollution') and operative at the Higher Health Institute. The limit values, given in Table 10, before going into force as the quality standards for unconfined air (enforcement methods to be established) must be subjected to the foregoing procedure.

The informatory spirit of the limits, according to those who proposed them, sprang from the consideration that the objective of the limits themselves, understood as air quality standards, must be the protection of human health in the widest sense this term can have pursuant to the World Health Organization's definition: not only the absence of illness, but a complete state of physical, mental and social well-being for all individuals, including particularly sensitive population groups. Account was also taken, in the elaboration of the values proposed, of EEC directives and proposals.

TABLE 10
UNCONFINED AIR QUALITY STANDARDS FOR THE HEALTH AND SANITATION
PROTECTION·OF THE POPULATION, PROPOSED BY THE
NATIONAL HEALTH BOARD (JANUARY 1982).

| POLLUTANT | MAXIMUM CONCENTRATION ALLOWED FOR A SET PERIOD OF EXPOSURE | |
|---|---|---|
| Sulfur dioxide (expressed as $SO_2$) | Median of the 24-hour average concentrations detected over a 1 year period | $80\mu g/m^3$ |
| | A percentile of 98 of the 24-hour average concentrations detected over a 1-year period | $250\mu g/m^3$ |
| Nitrogen dioxide (expressed as $NO_2$) | 1-hour average concentrations must not exceed more than once per day | $200\mu g/m^3$ |
| Ozone (expressed as $O_3$) | 1-hour average concentration not to be reached oftener than once per month | $200\mu g/m^3$ |
| Carbons monoxide (expressed as CO) | 8-hour average concentration 1-hour average concentration | $10\mu g/m^3$ $40\mu g/m^3$ |
| Lead (expressed as Pb) | Arithmetic average of the 24 hour average concentration detected over 1 yr. | $2\mu g/m^3$ |
| Fluorine (epxressed as F) | 24-hour average concentration Average of the 24 hour avg. concentrations detected over 1 month | $20\mu g/m^3$ $10\mu g/m^3$ |
| Suspended particles (determined by the gravimetric method given in the Directive) | Arithmetic average of the 24-hour average concentrations detected over a 1-year period | $300\mu g/m^3$ |

| Feedstock | Concentration limit values | Conditions under which limit applies |
|---|---|---|
| Total hydrocarbons except for methane | Average concentration over 3 consecutive hours of the day to be specified according areas of the competent regional authorities: 200 $\mu g/m^3$ | To be adopted only in the areas and uring the periods of the year in which the air standards for ozone indicated in Table 1 were significantly exceeded. |

TABLE 11
ITALY:  DISCIPLINE OF FUELS

a)  Exempt of limitations

- Gaseous fuels (methane and similar)
- Distillates of oil (kerosene, gasoil, etc.) with sulfur content no greater than 1.10%
- Metallurgical and gas cokes containing volatile substances up to 2% and sulfur up to 1%.
- Anthracite and derivates containing volatile substances up to 13% and sulfur up to 2%.
- Wood and charcoal.

b)  Exempt of limitations in Zone A; limited to thermal and industrial plants of capacities greater than 500,000 Kcal/h in Zone B:

- Fuel oils having viscosities up to $5^0$ Engler at $50^0$C and sulfur content, up to 3%.

c)  Permitted in both zones only for thermal and industrial plants with capacities greater than 1,000,000 Kcal/h and with municipal authorization:

- Fuel oils with viscosities greater than $5^0$ Engler at $50^0$ and with sulfur contents up to 4%.

d)  Subject to special limitations:

- Steam coals having volatile substances up to 23% and sulfur up to 1% - only for extremely large plants.
- Pressed fuels (such as bricks or pellets) containing volatile substances up to 13% and sulfur up to 2% - only for the heating of single rooms.
- Lignite and peat - forbidden in Zone B.

e)  Permitted by way of exception:

- Steam coals containing sulfur up to 2%.
- Sulcis coal, in the same basin.

CCIA's elaboration is based on the "conviction regarding the necessity of establishing exposure limits to pollutants of the air outside, regardless of the origin of the pollution, whose validity may be extended over the entire national territory". As a consequence the study group has recognized the need to revise the present limitation system. The latter, in fact, contemplates only the industrial contribution to air pollution, which must be summed to that derived from other sources, thus leaving the total air pollution level undetermined.

Of course the revision of the present regulation system cannot but take into account the legislation in force and the experience gained in other countries, such as those taken into consideration in this paper, on the subject of values of air quality standards and of the relative measurement time intervals, as well as of other types of standards.

The fuel-use discipline is summarized in Table 11. With reference to thermoelectric power stations, the following are permitted:

- natural gas, without any limitation;
- fuel oil, with sulfur content up to 3% or 4% depending on cases;
- coal, with sulfur content up to 1%;
- lignites and peat, outside Zone b.

Also permitted exceptionally, that is for limited periods and in the absence of a contrary view of the Minister of Health, are coals having sulfur contents up to 2%.


5.   UNITED STATES OF AMERICA

Environmental legislation in the U.S.A. has, in recent years, undergone a considerable evolution, including the approval, the amendment and the more updated interpretation of laws, regulations and standards.   One of the objectives of this evolution is the elaboration of an organic system of principles by which to regulate the entire cycle of every energy system.   Table 12 provides the list of federal legislation relative to the production and utilization of coal, by way of an example.

The control of air pollution is regulated by a specific text of law, the Clean Air Act of 1963, which has undergone a series of amendments in subsequent years, among which the most important are those of 1970 and 1977.

The law institutes a specific body, the EPA (Environmental Protection Agency) whose duty, among other things, is to enforce the control of air pollution at the federal level.

It is EPA's task to identify pollutants, define criteria and to promulgate the ambient levels of the NAAQS (National Ambient Air Quality Standards), permitted for the purpose of protecting health and the environment, on a national scale.

The standards are established on the basis of the pertinent scientific literature, whose assessment and specific studies promoted for the purpose are looked after by the EPA. EPA periodically reexamines and updates the list of pollutants and the relative standards.

TABLE 12

U.S.A.: LIST OF FEDERAL LEGISLATION ON THE PRODUCTION AND UTILIZATION OF COAL

1) Regulations on open mines (Department of the Interior);
2) Laws on mining concessions (1920 and 1975);
3) Law on the policy and management of federal territory, of 1976;
4) Laws on species in danger of extinction (1973 and subsequent amendments);
5) Law of 1977 on safety and health (which amends the federal law on safety and health in coal mines of 1969);
6) Law on safety and health in mines, of 1977;
7) Laws for protection against lung diseases, of 1977 and 1978;
8) Law on air purity (Clean Air Act)
9) Law on water purity (Clean Water Act
10) Law on the protection of potable waters, of 1974)
11) Law on the conservation of resources and on recovery of 1976;
12) Law on the control of noise, of 1972;
13) Law for the coordination of energy production and for protection of the environment, of 1977;
14) Law on the utilization of fuels in thermoelectric power stations and in industry.

A distinction is made between primary and secondary standards. The former regard health protection and must be enforced on the short term (within three years of their promulgation). The secondary standards concern the protection of material assets and of other productive activities and have longer enforcement terms.

TABLE 13

| Pollutant | Environmental Quality Standards for Air (1) (mg/Nm3) | | Emission Standards for Thermo-Thermoelectric Power Stations NSPS (ng/J) |
|---|---|---|---|
| | Long Term | Short Term | |
| Oxides of Sulfur | 0.8 (1 year) | 0.37 (24 hours) | Liquid and gaseous fuels: 340, Solid Fuels: 520 |
| Suspended Dusts | 0.075 | 0.26 | 13 (3) |
| Dust Fallout | -- | -- | -- |
| Oxides of Nitrogen | 0.1 (4) (1 year) | -- | Gaseous fuels: 86, Liquid fuels: 130, Solid fuels: 215 |
| Carbon Monoxide | 10 (8 hours) | 40 (1 hour) | -- |
| Hydrocarbons | -- | 0.16 (6) (3 hours) | |

1) Only the primary standards are shown.
2) Also required is an abatement of 70 to 90% (see text).
3) For both solid and liquid fuels.
4) As $NO_2$
5) With reference to the most common types of fuels. Also required is an abatement of 25 to 65%.
6) Is not binding. For photochemical oxidants the limit is 0.16 $mg/Nm^3$ as a 1-hour average.

USA: Air Quality and Emission Standards

With regard to thermoelectric power stations using fossil
fuels, the following pollutants are subject to regulation:  sulfur dioxide, total suspended dusts, oxides of nitrogen, carbon
monoxide, hydrocarbons and photochemical oxidents.  The relative
standards were promulgated in 1971 (see Table 13) and are under
revision at present.

The enforcement of the standards is normally delegated to
the administrations of the individual States which draw up local
enforcement plans (SIP, State Implementation Plan) in which measures are established to maintain the quality of the environment
within the range prescribed by the federal standards.

Local implementation plans (SIP), among other things, con-
template emission limits for various plants and require approval
by EPA, which must ascertain that the plans of the various states
are compatible with federal standards.  Table 14 shows, by way
of example, the emission limits set by Illinois for coal-fired
power plants.

Emission standards federal in nature have been established
for plants built after 1975.  They are generally more restrictive
than those contemplated by SIP.  Such standards, called 'New
Source Performance Standards' (NSPS), are laid down for the various
types of plants in function of the power productd and are based
on the 'best available control technology' (BACT).  They specify
the percentage of removal of the various pollutants present in
smoke.  Table 15 shows the 1979 NSPS for thermoelectric power
stations using fossil fuels.  The limitations are generally sufficiently restrictive to require the adoption of abatement systems
(smoke desulfurization and denitrification).  They are applied to
plants whose construction begain after the 18th September 1978.

These standards limit the emission of sulfur dioxide to
1.20 lb/million Btu (520 ng/J) and an abatement of 90% of the
pollution potential present in smokes (which may be reduced to
70% when this leads to emissions of less than 0.6 lb/million
Btu).  The degree of abatement required is stated in function of
fuel's calorific power and of its sulfur content; for example,
for a fuel producing 10,000 Btu/lb, the required abatement is:

-       70% for sulfur contents below 1%;
-       70 to 90% for sulfur contents from 1 to 3%;
-       90% for sulfur contents from 3 to 6%;
-       over 90% for sulfur contents over 6%.

The regulations also prescribe the methods and procedures
for checking compliance with the limitations.

## TABLE 14

### U.S.A.: S.I.P. EXAMPLE: STANDARDS RELATIVE TO COAL UTILIZATION IN ILLINOIS

| POLLUTANT | ENVIRONMENTAL STANDARDS ($mg/m^3$) | | | | Emissions ($lb/million/Btu$) | | |
|---|---|---|---|---|---|---|---|
| | Primary | | Secondary | | Existing Plants | | New Plants |
| | Ann. Avg. | Once Per Yr. Maximum | Ann. Avg. | Once Per Yr. Maximum | In Cities | Rural Areas | |
| $SO_2$ | 0.08 | 0.365 | 0.60 | 0.260 | 1.8 | 6.0 | 1.2 |
| Suspended Dusts | 0.074 | 0.260 | 0.060 | 0.150 | 2.0 | 2.0 | 1.0 |
| $NO_2$ | 0.100 | -- | 0.100 | -- | 0.9 | -- | 0.9 |

216

*A. Fontanella and G. Pinchera*

TABLE 15

U.S.A.: NSPS FOR COAL-FIRED THERMOELECTRIC POWER STATIONS BUILT AFTER 1978

| Sulfur Dioxide | Maximum Emission | 1.2 lb/million Btu |
|---|---|---|
| | Abatement Required | 70% if this leads to a maximum emission of 0.6 lb/million Btu |
| | | 90% in other cases |
| | | NONE if anthracite is used to 100% |
| | Checking Criterion | Average value of 30 successive days of operation |
| | Monitoring | Continuous |
| Dusts | Maximum Emission | 0.03 lb/million Btu |
| | Abatement Required | 99% |
| | Max. Opacity of Smoke | 20% |
| | Tolerance on Max. Emission Emission Allowed | Opacity up to 27% for not more than 6 minutes 6 minutes in any hour. |
| | Checking Criterion | Manual sampling at smokestack for maximum emission |
| | | Average over 6 minutes for max. opacity |
| | Monitoring | Continuous for Opacity |
| Oxides of Nitrogen | Maximum Emission lb/million Btu) | Sub-bituminous coal 0.5 |
| | | Fuel containing more than 25% by weight of a number of types of lignate 0.8 |
| | | Other types of coal 0.6 |
| | Abatement Required | 65% |
| | Checking Criterion | Average value of 30 successive days of operation |
| | Monitoring | Continuous |

For purposes of siting the plants, the law identifies two types of areas in which further restrictions apply. The areas in which the quality of the air is better than that prescribed in the NAAQS are defined as 'Prevention of Significant Deterioration Areas' (P.S.D.). They include national parks and monumonts, deserts, etc. To install plants in such areas it is necessary to demonstrate by means of a study for the purpose based on monitoring techniques and on mathematical simulation, that the limits set by NAAQS and SIP will in no case be exceeded It is necessary, moreover, to demonstrate that the best available control technologies, based on a specific study that takes economic, environmental, energy, etc. factors into account, are being adopted.

The areas in which the basic level of pollution is higher than the NAAQS are defined as 'Nonattainment Areas'. The States are required to elaborate a pollution reduction plan for purposes of returning to values within the limits set by NAAQS by the end of 1982. In areas defined in this way, no new plants may be built unless the strictest emission control measures possible (L.A.E.R.) are adopted or the emissions of the new plant will be more than compensated by a reduction in the existing sources.

No precise standards exist with regard to the height of smokestacks. The law limits itself to making reference to good engineering practice.

## 6.　CONCLUSIONS

Two different criteria of regulation and control of atmospheric pollution emerge from the review of national regulatory instruments. A first group of countries, including the U.S.A., Federal Germany (and, to a degree, Italy) regulate by imposing minimum measures and restrictions which apply nationally, with additional geographic or other requirements (type of plant, fuel utilized, etc.). A second group, comprising Great Britain and France, mainly applies general principles to develop regulations and modes of implementation on case by case basis.

More stringent control generally occurs with the first type of approach although the example of Great Britain demonstrates the validity of the second.

In the regulations of the first group, four principal types of standard limitations are discerned:

-   air quality standards[1]
-   smokestack emission standards[1]
-   standards[1] on the potential of fuels
-   the prescribing of specific technologies to abate or prevent the formation of pollutants.

Table 16 indicates in which countries ordinances exist at present that are valid at the national level for the four types of limitations. Not included are the cases in which such limitations have the nature of recommendations, usage or provisions of local or temporary validity. Table 17 indicates the directives of the European Economic Community emanated in 1980 and the recommendations of the World Health Organization. Table 18 summarizes the limits of emission at the smokestack in force in a number of countries, for various pollutants.

All in all, the standards adopted for the control of atmospheric pollution in the U.S.A., Japan and Federal Germany appear, although the presence of significant differences, to be the most complete and effective. Also these countries, however, are involved in the lively debate which, on the scientific, technical, economic, juridic and sociopolitical levels, concerns the criteria of definition of the standards and other aspects of the regulations on the control of atmospheric pollution.

Of course, in order to have a complete enough picture for a profound comparison between the various countries under consideration, it is necessary to identify and examine also other aspects of the regulations and of the control organization, such as the manner in which data are obtained (measurement points, analytic methods, specifications for the instrumentation, etc.), how control is effected, as well as the degree of effective application of the regulations themselves.

In conclusion, it may be affirmed that it is forseeable and to be hoped for that the existing standards undergo updatings and amendments, also in depth, as a consequence of the improvements in knowledge about the effects of pollution, of the progress of technology and of the refinement of analyses of risk and of benefits deriving from energy production.

Also to be hoped for is an effort towards the achievement of a uniformity which, although taking into due account the various geographic, meteorological, demographic, and economic and cultural situations of the various countries, will make more effective defense of the environment possible.

[1]See the glossary also for the meaning of "maximum admissible value."

TABLE 16

THE ADOPTION OF NATIONAL STANDARDS FOR THE CONTROL OF ATMOSPHERIC POLLUTION IN A NUMBER OF COUNTRIES

| | USA | Japan | Federal Germany | Italy | Great Britain | France |
|---|---|---|---|---|---|---|
| Air Quality Standards | Yes | Yes | Yes | Yes[2] | No | No |
| Emission Standards | Yes | Yes | Yes | No | No | No[1] |
| Requirement that Smoke be Desulfurized | Yes | Yes | Yes | No | No | No |
| Fuel Standards | Yes | Yes | Yes | Yes | Yes | Yes |

[1]Except for Dust
[2]Limits for emissions into the air due to total contributions of industrial plants

## TABLE 17

### AIR QUALITY STANDARDS
DIRECTIVES OF THE EUROPEAN ECONOMIC COMMUNITY AND RECOMMENDATIONS OF THE WORLD HEALTH ORGANIZATION

| Pollutant | E.E.C. Directives | | | W.H.O. Recommendations | |
|---|---|---|---|---|---|
| | Air Quality Standards (mg/Nm³) | | | | |
| | (1) | (2) | (3) | Long Term | Short Term |
| Oxides of Sulfur | 0.12 (4) | 0.18 (5) | 0.35 (6) | 0.06 (1 Year) | 0.20 (7) |
| Dusts | 0.08 (°) | 0.13 (°) | 0.25 (°) | 0.04 (1 Year) | 0.12 (7) |
| Carbon Monoxide | -- | -- | -- | 0.01 (8 hours) | 0.04 (1 hour) |
| Photocmehical Oxidants | -- | -- | -- | -- | 0.12 (1 hour) |

1. Median of average daily values detected during year.
2. Median of average daily values detected during winter (1st October to 31st March).
3. 98% of all daily average values detected during year.
4. Reduced to 0.08 for a value associated with dust larger than 0.04(°).
5. Reduced to 0.3 for a value associated with dust larger than 0.06(°).
6. Reduced to 0.25 for a value associated with dust larger than 0.15(°).

(°) Values referred to the black smoke method.

## TABLE 18

### EMISSION STANDARDS FOR THERMOELECTRIC POWER STATIONS

| Pollutant | U.S.A. (ng/J) | Japan | Federal Germany (Mg/Nm$^3$) |
|---|---|---|---|
| Oxides of Sulfur | Liquid and gaseous fuels: 340 <br> Solid fuels: 520 (1) | Regulation by geographic parameter | 650 (3) |
| Suspended Dusts | Liquid and solid fuels: 13 | From 50 to 800 mg/Nm$^3$ depending on the fuel and size of the plant | From 50 to 300, depending on the fuel and on the size of the plant |
| Nitrogen Oxides | Gaseous fuels: 86 <br> Liquid fuels: 130 <br> Solid fuels: 215 | From 60 to 400 ppm, depending on the fuel and on the size of the plant | Reduced to the narrowest limits permitted by technology |
| Carbon Monoxide | -- | -- | Gaseous fuels: 100 <br> Liquid fuels: 175 <br> Solid fuels: 250 |

1. Also abatement from 79 to 90% required.
2. Also abatement from 25 to 65% required.
3. Limit recommended by the Federal Government.

7. Glossary

The definitions of a number of words used in the text are given here as a memory aid. The sometimes controversial nature of some definitions and the different meanings terms can take in the various legislations are emphasized.

Pollutant:  any substance which, in the light of chemical, bio-logical, toxicological and epidemiological knowledge, may be deemed to be capable of altering the characteristics of the environment in a manner harmful to the public health and well-being, to the environment itself and to the socio-economic activities present in it.  This definition has been taken from the legislation of the United States of American and is the one most widely applied.  In other legislations dominance is frequently given to the aspect of 'disturbance to neighbors' or the pro-tection of the environment is subordinated to the damage to other productive activities.

The classification of pollutants is often controversial, as in the case of carbon dioxide.

Photochemical pollutants (also photochemical oxidants):  organic substances, among which principally the hydrocarbons, which under-g o particular transformations in the atmosphere by effect of the sun's rays.  These transformations lead to the production of ozone and the formation of aerosols that are particularly harmful to the eyes and respiratory passages, to vegetation and to visibility.

Not all hydrocarbons undergo transformation to the same de-gree.  For example, if the speed of oxidation of NO to $NO_2$ is taken as the photochemical reactivity index, under equal conditions a negligible effect is observed for methane and acetylene; while reactivity is maximum for a number of olefines such as 2-Butene and 2-Pentene.

Maximum admissible concentrations (MAC):  the concentration values of specific substances that must not be exceeded in the working environment in order to safeguard the health of the workers present. They are established on the basis of toxicological studies, con-sidering an exposure of 8 hours per day and 40 hours per week.

These values are normally utilized to calculate the minimum ventilagion required in rooms to prevent toxic effects.  In some cases they are utilized to assess dilution factors for the dis-charge of pollutants into the environment or as reference values in the calculation of the height of chimneys.

Environmental quality standards[1]:  the concentrations of specific substances that must not be exceeded in the environment and
in the air and water in particular, a distinction is made between
primary and secondary standards:  the former are directed at the
protection of human health and their implementation is more urgent.  The latter, on the other hand, concern the protection of the
natural environment, of economic and cultural, etc., assets.

It is necessary, in the definition of standards, to specify
the measurement techniques (instrumentation, number of measurements, duration of each sampling, etc.) and the points at which
to carry out the sampling (e.g. at the level of a man's head or at
a certain distance from the plant).

Environmental standards are not linked to the discharges of a
specific plant alone, but represent the total effect of all the
sources distributed over the area and of the background level (due
to natural sources and to long-distance transport.

Discharge limits:  maximum concentrations at ground level of pollutant substances attributable to a specific source or to a uniform
type of sources such as, for example, industrial plants.  Even if
from the prevention and control points of view it is important to
discriminate between the contributions from various polluting
sources, in practice the adoption of discharge limits leads to
difficulties and controversies precisely in the control action.
So that they tend to be substituted by environmental quality
standards and by emission standards.

Emission standards (or plant performance standards):  regulations
on the emissions from the smokestack of a specific source of
air pollutants.  They are generally expressed as the maximum admissible concentrations of specific polluting substances present in
the fumes at the chimney outlet.  The values of emission standards
depend on the type of plant being considered.  For thermoelectric
power stations they may be expressed in function of the height of
the smokestack, of the type of fuel employed, of the total quantity
of fumes diacharged and of the geographic area in which the plant
is operating.

Another way of expressing emission standards is that of limiting
the quantity of pollutants that may be dischaged per unit of energy
produced or utilized in the plant.  This magnitude, as opposed to
the concentration of pollutants present in fumes, does not depend
on the surplus air employed for combustion.

---

[1]The term 'standard' when meaning the maximum admissible value has
been taken from U.S.A. legislation.  In other legislations terms
such as 'maximum value' and 'reference value' are frequently
utilized.

<u>Regulation of the use of fuels</u>:  reduction of polluting potential upstream of the combustion process.  A typical case is the limitation of the sulfur content in the fuels burned in thermoelectric power stations.  The extent of these limitations can depend on the type of fuel, on the type of plant, on its dimensions, on the geographic area in which it operates and on meteorological conditions.

<u>Technical apparatus for the control of pollution</u>:  systems for the continuous reduction of polluting emissions such as, for example, the treatment of fuels prior to combustion.  The control authorities can require the application of a particular technology, however they generally prescribe the adoption of the 'best applicable technology', or of the 'best available technology'.

<u>Best available control technology</u>[2]:  the technology that makes possible the greatest limitation of polluting emissions, among those tested on an industrial scale, from both the technical and economic points of view.

<u>Best practicable control technology</u>[2]:  the technology that makes possible the greatest limitation of polluting emissions, among those tested on an industrial scale, from both the technical and economic points of view.

<u>Monitoring systems</u>:  networks for the detection of polluting emissions and of the pollution level of the ambient air.  It may have various functions, among which:

- checking for compliance with the terms of law;
- regulation of the plant and of emission-reduction apparatus;
- checking the validity of the measures prescribed for the reduction of pollution.

---

[2]The concept derives from the British legislation and has found a stricter definition and application in the legislation of the United States of America.

ACTUAL AND PERCEIVED RISK: A REVIEW OF THE LITERATURE[*]

Vincent T. Covello

National Science Foundation
Washington, D.C.

A truly unexpected result came out of
the Kemeny Commission's study of Three
Mile Island. A group that set out to
investigate a technology ended up
talking about people ... In the commission's own words, "It became clear
that the fundamental problems were
people-related problems." (Editorial,
The Washington Post, October 31, 1979).

## 1. INTRODUCTION

Behavioral and social scientists in several countries are
currently grappling with several people-related questions: What
factors influence individual perceptions of risk? What accounts
for anomalies in the way individuals and groups behave when faced
with ostensibly comparable risks--such as the risks of nuclear
power, dam failures, or earthquakes? What weight should decisionmakers attach to public perceptions of risk in determining how
safe is safe enough? Are there ways to increase our capacity for
dealing with technological risks in a rational manner?

[*]This review draws heavily on the work of Paul Slovic,
Baruch Fischhoff, and Sarah Lichtenstein, and I would
like to acknowledge this contribution. I would also like
to thank Jeryl Mumpower, Mark Abernathy, Joshua Menkes,
and Jiri Nehnevajsa for their help. The views expressed in
this paper are exclusively my own and do not necessarily represent the views of the National Science Foundation.

What follows is a review of the behavioral and social science
literature pertaining to these questions [See Refs. (1)-(148)].
Before beginning, however, several points need to be made concerning
the significance and quality of the data.  First, most of the reported findings are based on surveys of small, highly specialized,
and unrepresentative groups.  An important set of risk perception
studies undertaken by Paul Slovic and his colleagues, for example,
relied almost entirely on data gathered from residents of Eugene,
Oregon, a small university town located in a state with a high
level of environmental concern and a progressive environmental
protection program [See for example Refs. (24), (113), (115)].
The respondents included 40 members of the Eugene League of Women
Voters, 40 college students at the University of Oregon, 25 Eugene
businessmen, and 15 persons selected nationwide for their expertise
in risk analysis.

   Second, little attempt has been made by researchers to analyze the effects of ethnicity, religion, sex, region of the
country, age, occupation, education, income, marital status, and
other background, group, or contextural variables on risk perceptions.  Few studies in the field challenge the prevailing
assumption that different segments of the population share the
same risk perceptions.

   Third, the risk perception literature suffers, in an exaggerated form, from shortcomings common to nearly all survey research (23):  (a) people typically respond to survey questions
with the first thing that comes to mind, and then become committed
to their answer; (b) people typically provide an answer to any
question that is posed, even when they have no opinion, when they
do not understand the question, or when they hold inconsistent
beliefs; (c) survey responses can be influenced by the order in
which the questions are posed, by whether the emphasis is on
speed or accuracy, by whether the question is closed or open, by
whether respondents are asked for a verbal or numerical answer,
by interviewer prompting, and by how the question is posed.
Risk perception surveys are especially vulnerable to these type.
of biases, since people are often unfamiliar with the activity or
substance being assessed and because they may not understand the
technical and methodological issues under debate.

   Fourth, few studies have examined the relationship between
perceptions of technological hazards and the behavior of people
in actual situations.  Empirical studies done in other social and
behavioral fields suggest that the linkages between perception
and behavior are highly complex and appear to be mediated by
several factors [See for example Refs. (9), (4), (53)].  Researchers have shown, for example, that activist behavior is related to a willingness to participate in group activities, a
positive identification with potential group leaders, a belief in

the efficacy of social action, and physical proximity to arenas
of social conflict [See Refs. (9),(25)]. With only rare exceptions
[See for example Refs. (68), (76)], risk perception researchers
have not examined these and other variables.

The findings reported in this paper are confounded even
further by several additional problems:  risk perceptions may
change rapidly; people may not understand how their perceptions
and preferences translate into policy; and people may prefer
alternatives that are not realistically obtainable. With these
reservations and qualifications in mind, some of the major
findings of the risk perception literature are discussed below.

## 2.  HUMAN INTELLECTUAL LIMITATIONS

Research suggests that people do not cope well when confronted
with risk problems and decisions.  Intellectual limitations and
the need to reduce anxiety often lead to the denial that risk and
uncertainty exist and to unrealistic oversimplifications of
essentially complex problems [See Refs. (106), (113), (136)].  To
simplify risk problems, people use a number of inferential or
judgmental rules, known technically as heuristics [See Refs. (46),
(47), (133), (134), (135), (136)]. Two of the most important are
(1) information availability, or the tendency for people to judge
an event more frequent if instances of it are easy to imagine or
recall; and (2) representativeness, or the tendency of people to
assume that roughly similar activities and events (such as nuclear
power technologies and nuclear war) have the same characteristics
and risks.  These judgmental operations enable people to reduce
difficult probabilistic and assessment tasks to simpler tasks;
however, these  judgmental operations also lead to severe and
systematic biases and errors.

One bias associated with information availability is that
people have difficulty imagining low probability/high consequence
events happening to themselves [See for example Refs. (7), (113),
(106)].  Unless people have been made graphically aware of the
risks, typically through past experience, they are unlikely to
take protective action [See Refs. (54), (55), (147)].  A classic
example is the observed reluctance of flood plain residents to
purchase low cost flood insurance (55).  Compounding the problem
is the difficulty people have understanding and interpreting
probabilistic information.  People residing in 100 year flood
plains, for example, typically believe that a recent severe flood
precludes the possibility of another severe flood in the near
future [See Refs. (7), (147)].  According to folk wisdom but not
according to probability theory, lightning never strikes the same
place twice.

Biases associated with information availability have also been used to explain the results of studies in which people were asked to judge the frequency of various causes of death, such as accidents, tornados, and diseases (113). These studies show that the risks of low frequency events tend to be overestimated, and that the risks of high frequency events are underestimated. People underestimate fatalities caused by asthma, stroke, and diabetes, and overestimate fatalities from homicides, botulism, fires, snake bites, tornados, and abortion. Overestimated causes of death tend to be dramatic, sensational, and of interest to the media, whereas underestimated causes of death tend to be un-spectacular, undramatic, and of little interest to the media [See for example Refs. (10), (113)].

Researchers have also pointed out that information availability biases may cause risk information campaigns and educational efforts to work at cross-purposes (119). Information may heighten the imaginability and consequently the perceived probability of a rare event, even when the information is designed to assure individuals that the event is unlikely. A package insert listing all the risks of taking a drug, or a published report describing safety precautions taken at a DNA research laboratory, may only serve to increase concern about the substance or activity. By identifying previously unknown ways in which things can go wrong, the information provider takes the chance that people will in-correctly assess the information--i.e., that they will consider the event more likely as a result of increased knowledge. As one observer notes (67):

> ...We generally assume that informed
> scientific advice is valuable to poli-
> tical policy-makers. However, in the
> context of a controversial political
> issue, and when the relevant technical
> analysis is ambiguous, then the value
> of scientific advice becomes questionable.
> A technical controversy sometimes creates
> confusion rather than clarity, and it
> is possible that the dispute itself may
> become so divisive and widespread that
> scientific advice becomes more of a
> cost than a benefit to the policy-maker
> and society. (p. 61)

Unfortunately, few researchers have critically examined the controversial hypothesis implicit in this work--that the very dis-cussion of a low-probability hazard increases the judged probabili-ty of the hazard, regardless of what the evidence indicates.

Disputes and controversies about risk are made all the more difficult by another psychological mechanism: once beliefs are formed, individuals will frequently structure and distort the interpretation of new evidence, and will often resist disconfirming information [See for example Refs. (113), (102)]. People tend to dismiss evidence contradicting their beliefs as unreliable, erroneous, and unrepresentative. The accident at Three Mile Island, for example, provided confirming evidence for those already convinced that nuclear power technology is safe (76). The accident also reinforced the beliefs of those that believed that nuclear power technology is dangerous. Convincing people that a hazard they fear is safe is extremely difficult even under the best conditions. Any accident or mishap, no matter how small, is seen as proof of high risk (119).

## 3.   OVERCONFIDENCE

A second set of research findings address the problem of overconfidence. Researchers have shown that experts and laypersons are typically overconfident about their risk estimates. In one study participants were asked to state the odds that they were correct in judging which of two lethal events was more frequent (113). Most people claimed that the odds of their being wrong were 100:1 or greater. In actuality, people were wrong about one out of every eight times. Such overconfidence can produce serious judgmental errors, including judgments about how much is known about the hazard and about how much needs to be known. Of equal or greater importance, overconfidence leads people to believe that they are comparatively immune to common hazards. Studies show (1) that most people rate themselves among the most skillful and safe drivers in the population; (2) that people rate their own personal risk from several common household hazards as lower than the risk for others in society; (3) that people judge themselves average or above average in their ability to avoid bicycle and power lawnmower accidents; and (4) that people underestimate and are extremely unrealistic about their chances of having a heart attack [See for example Refs. (126), (101), (144)]. In general, people underestimate the risks of activities that they perceive to be familar and under their personal control, such as automobile driving.

Overconfidence has also been used to explain in part the observed reluctance of about 80-90 percent of the U.S. driving population to wear seat belts (110). Unfortunately, few empirical studies have examined this issue or the more general relationship between perceived risk and protective behavior. One new study by Slovic, Lichtenstein, and Fischhoff of seat belt usage is, however, exploring and empirically testing the hypothesis that seat belt usage will increase if the public is presented with information about the lifetime risks of driving instead of information about the risks of taking a single trip.

## 4. EXPERT AND NON-EXPERT ESTIMATES OF RISK

A third set of findings bear on expert and non-expert estimates of risk. A consistent research result is that technical experts and non-experts differ substantially in their risk estimates (49). Risk estimates of technical experts are closely correlated with annual fatality rates, whereas the risk estimates of non-experts are only moderately to poorly correlated with annual fatality rates (115). In explaining these differences, researchers have identified several factors other than annual fatality rates that influence public perceptions of risk [See for example Refs. (62), (113), (139), (140)]. Risks are perceived to be higher if the activity is perceived to be involuntary, catastrophic, not personally controllable, inequitable in the distribution of its risks and benefits, unfamiliar, and highly complex. Other factors influencing risk perceptions are whether the adverse effects are immediate or delayed, whether exposure to the hazard is continuous or occasional, whether the technology is perceived to be necessary or a luxury, whether the adverse effects are well-known or uncertain, and whether the activity is certain to be fatal.

Several studies have shown that these dimensions of risk are closely related to one another [See for example Refs. (112), (139), (140)]. Such correlations have prompted several research groups to reduce the various dimensions of risk to a smaller number of factors. One study identified at least two factors (24): the level of technological complexity and the hazard's severity or catastrophic potential. In a follow-up study that examined a larger set of hazards and risk characteristics, Slovic, Fischhoff, and Lichtenstein found three factors (113): familiarity, dread, and the number of people exposed to the hazard. In an on-going European study of risk perception, Vlek and Stallen identified several additional factors influencing risk perception and risk acceptability, including the beneficiality of the technology and the degree to which protection is provided by institutional means [See for example Refs. (139), (140)]. In spite of these different findings, it is clear that a hazard's catastrophic potential is uppermost in the minds of people. Since catastrophic events may threaten the survival of individuals, families, societies, and the species as a whole, such concern may be quite justifiable.

Analysis of inter-correlations between the various dimensions of risk have also led researchers to challenge Starr's (121) well-known proposition that the risks of voluntary activities are more acceptable to the general public than the risks of involuntary activities [See Refs. (113), (3), (85)]. One problem with this proposition is that voluntary risks are also perceived by the public to be controllable, equitable, familiar, and non-

catastrophic.  These correlations suggest in turn that the observed
greater willingness of the public to accept voluntary risks may
be due to these other factors and not to the voluntary nature of
the activity.

5.  RISK PERCEPTION AND NUCLEAR POWER:  A CASE STUDY

To date, most of the research on risk perception has focused
on nuclear power [See for example Refs. (6), (13), (27), (29),
(44), (50), (66), (70), (73), (74), (76), (78), (79), (80), (87),
(89), (91), (92), (94), (119), (120), (128), (129), (142), (143)].
These studies have produced several important findings.  First,
researchers have shown that nuclear power has nearly all the
characteristics associated with high perceived risk.  The risks
of nuclear power are perceived to be involuntary, delayed in their
consequences, unknown, uncontrollable, unfamiliar, potentially
catastrophic, inequitable, and certain to be fatal (119).  Public
perceptions of nuclear power contrast sharply with non-nuclear
sources of electric power, which are perceived to be non-cata-
strophic, familiar, controllable, and comparatively safe.

Second, researchers have shown that disputes about nuclear
power are often about values and goals that far transcend issues
of health and safety [See for example Refs. (70), (76), (78),
(86), (87), (120)].  Many people are concerned about nuclear
power not because of its specific risks but because of its
associations with nuclear weapons, with highly centralized political and economic systems, and with technological elitism.  The
debate about nuclear power is also colored by social class--people
with lower socioeconomic status are less supportive of nuclear
power than those with higher socioeconomic status; by sex--women
are less supportive of nuclear power than men; and by concerns
about the credibility of institutions charged with estimating,
evaluating, and managing the risks (70).

Despite these concerns, research studies consistently show
that the public, by a margin of 2 and sometimes 3 to 1, supports
nuclear power, even in the aftermath of Three Mile Island [See
Refs. (70), (43)].  Somewhat counterintuitively, researchers
have also found that people living within the vicinity of a
nuclear power plant (and therefore presumably subject to the
greatest objective risk) are more supportive of nuclear power
than those living farther away (70).  In explaining this finding
it has been proposed that people living near power plants receive
greater economic benefits, that they experience greater cognitive
dissonance, and that they have had their worst fears assuaged by
a history of accident-free operations.  Interestingly, those who
are least supportive of nuclear power live in areas where power
plants are under construction or being planned.  One policy implication arises from these findings.  In several countries, including

France, proposals are currently being considered to compensate those who live in the vicinity of nuclear power plants. If the intention is to win wider public acceptance, then the policy is misdirected. Those living nearest to the power plant are already supportive and little would be gained by compensating them. By comparison, compensating those who are least supportive--i.e., those living in areas where power plants are under construction or being planned--might have a major impact. Such a policy could, of course, also backfire by providing support for the belief that the risks of nuclear power are indeed substantial.

## 6. CONCLUSION

One conclusion that can be drawn from this literature review is that conflicts about technological risks appear to be rooted in the different risk analysis methods and approaches used by technical experts and non-experts (1). Technical experts often implicitly and sometimes explicitly assign equal weight to hazards that take many lives at one time and to hazards that take many lives one at a time; non-experts typically assign greater weight to hazards that take many lives at one time, i.e., to catastrophes. Technical experts often implicitly and sometimes explicitly assign equal weight to statistical and known deaths; non-experts typically assign greater weight to known deaths. It is interesting in this regard to note the high levels of public concern and massive allocations of resources devoted to rescuing an identifiable person lost at sea. Technical experts often implicitly and sometimes explicitly assign equal weight to voluntary and involuntary risks; non-experts typically assign greater weight to involuntary risks. Technical experts typically express risks in quantitative terms, and use computational and experimental methods to identify, estimate, and evaluate the risks; non-experts typically express risks in qualitative terms, and use intuitive and impressionistic methods to identify, estimate, and evaluate the risks. Technical experts typically believe that quantitative estimates of risk should be the prime consideration in risk acceptability decisions; non-experts typically believe that quantitative estimates of risk should be only one among several quantitative and qualitative considerations in risk acceptability decisions. Technical experts often implicitly and sometimes explicitly assign the same weight to different ways of dying; non-experts typically feel that some ways of dying are worse than others. How one dies, and with how much suffering, is as important as where and when. Public risk acceptance and the success of risk management policies are likely to hinge on whether these differences are taken into account in the decision-making process.

REFERENCES

1.  Allison, A., Carnesale, A., Zigman, P., and DeRosa, F., Governance of Nuclear Power Report submitted to the President's Nuclear Safety Oversight Committee (September 1981).

2.  Atkinson, J.W., "Motivational Determinants of Risk-Taking Behavior." Psychology Review 64 (1957):359-372.

3.  Becker, G. M., and McClintock, C. G., "Value: Behavioral Decision Theory." Annual Review of Psychology 18 (1967): 239-286.

4.  Bem, D. J., Wallach, M., and Kogan, N., "Group Decision Making Under Risk of Aversive Consequences." Journal of Personality and Social Psychology 1 (1965):453-460.

5.  Bowen, J., "The Choice of Criteria for Individual Risk, for Statistical Risks, and for Public Risk." Risk-Benefit Methodology and Application (UCLA-ENG-7598). Edited by D. Okrent. Los Angeles: University of California, December 1975.

6.  Bowman, C. H., et al., The Prediction of Voting Behavior on a Nuclear Energy Referendum (IIASA RM-78-8). Laxenburg, Austria: International Institute for Applied Systems Analysis Research, February 1978.

7.  Burton, I. and Kates, R. W., "The Perception of Natural Hazard in Resource Management." Natural Resource Journal, 3 (1964):412-441.

8.  Buttel, F., and Flinn, W., "The Politics of Environmental Concern: The Impacts of Party Identification and Political Ideology on Environmental Attitudes." Environment and Behavior 10 (March 1978):17-35.

9.  Cole, G., and Withey, S., "Perspectives on Risk Perceptions." Risk Analysis: An International Journal 1:2 (1982).

10. Combs, B., and Slovic, P., "Causes of Death: Biased Newspaper Coverage and Biased Judgments." Journalism Quarterly 56 (1979):837-843.

11. Craik, K. H., "Environmental Psychology." New Directions in Psychology. Edited by T. M. Newcomb. New York: Holt, Rinehart, and Winston, 1970.

12. Crowe, M. J., "Toward a 'Definitional Model' of Public Per-
    ceptions of Air Pollution." Journal of the Air Pollution
    Control Association 18 (March 1968):154-157.

13. de Boer, C., "The Polls: Nuclear Energy." Public Opinion
    Quarterly (Fall 1977):402-411.

14. Delcoigne, G., "Education and Public Acceptance of Nuclear
    Plants." Nuclear Safety 20 (November-December 1979):
    655-664.

15. Downs, A., "Up and Down with Ecology--The Issue Attention
    Cycle." The Public Interest 28 (1972):38-50.

16. Edwards, W., "Behavioral Decision Theory." Annual Review
    of Psychology 12 (1961):473-498.

17. Edwards, W., and Tversky, A., Decision Making, Selected
    Readings. Middlesex: Penguin Books, 1967.

18. Englemann, P. A., and Renn, O. "On the Methodology of Cost-
    Benefit Analysis and Risk Perception." Directions in
    Energy Policy. Edited by B. Kursunoglo and A. Perlmutter.
    Cambridge, Massachusetts: Ballinger Publishing Company,
    1979, pp. 357-364.

19. Falk, H., "The Effect of Personal Characteristics on Atti-
    tudes Toward Risk." Journal of Risk and Insurance 43
    (June 1976):215-241.

20. Fischhoff, B., "Behavioral Aspects of Cost Benefit Analysis."
    Impacts and Risks of Energy Strategies: Their Analysis
    and Role in Management. Edited by G. Goodman. London:
    Academic Press, 1979.

21. Fischhoff, B., "Hindsight/Foresight: The Effect of Outcome
    Knowledge on Judgment Under Uncertainty." Journal of
    Experimental Psychology: Human Perception and Per-
    formance 1 (1975):288-299.

22. Fischhoff, B., "Informed Consent in Societal Risk--Benefit
    Decisions." Technological Forecasting and Social Change
    13 (May 1979):347-357.

23. Fischhoff, B., Slovic, P., and Lichtenstein, S., "Labile
    Values: A Challenge for Risk Assessment." Society,
    Technology and Risk Assessment. Edited by J. Conrad.
    London: Academic Press, 1980, pp. 57-66.

24. Fischhoff, B., Slovic, P., Lichtenstein, S., Read, S., and Combs, B., "How Safe is Safe Enough? A Psychometric Study of Attitudes Toward Technological Risks and Benefits." Policy Sciences 9 (1978):127-152.

25. Fishbein, M., and Ajezen, I., Belief, Attitude, Intention and Behavior: An Introduction to Theory and Research. Reading, Massachusetts: Addison-Wesley, 1975.

26. Flanders, J. P., and Thistlewaite, D. L., "Effects of Familiarization and Group Discussion Upon Risk Taking." Journal of Personality and Social Psychology 5 (1967):91-97.

27. Foreman, H., ed., Nuclear Power and the Public. Minneapolis: University of Minnesota Press, 1970.

28. Friedman, M., and Savage, L. J., "The Utility Analysis of Choices Involving Risks." Journal of Political Economy 56 (1948):279-304.

29. Gould, L., and Walker, C. A., eds., Too Hot to Handle: Public Policy Issues in Nuclear Waste Management. New Haven, Connecticut: Yale University Press, 1981.

30. Green, C. H., "Revealed Preference Theory: Assumptions and Presumptions." Society, Technology and Risk Assessment. Edited by J. Conrad. London: Academic Press, 1980, pp. 49-56.

31. Green, C. H., "Risk: Attitudes and Beliefs." Behaviour in Fires. Edited by D. V. Canter. Chichester: Milay, 1980.

32. Green, C. H., and Brown, R. A., "Counting Lives." Journal of Occupational Accidents (1978).

33. Green, C. H., and Brown, R. A., Life Safety: What is it and How Much is it Worth? (CP52/78). Borehamwood, Hertfordshire, U.K.: Department of the Environment, Building Research Establishment, 1978.

34. Green, C. H., and Brown, R. A., Metrics for Societal Safety (Note N 144/78). Borehamwood, Hertfordshire, U.K.: Department of the Environment, Building Research Establishment, 1978.

35. Green, C. H., and Brown, R. A., Perceived Safety as an Indifference Function (Note N 156/78). Borehamwood, Hertfordshire, U.K.: Department of the Environment, Building Research Establishment, 1978.

36.  Green, C. H., and Brown, R. A., The Perception of, and
     Attitudes Towards, Risk: Final Report: Vol. 2.
     Measure of Safety (FRO/028/68). Dundee, Scotland:
     School of Architecture, Duncan of Jordanstone College
     of Art, University of Dundee, April 1977.

37.  Green, C. H., and Brown, R. A., The Perception of, and
     Attitudes Towards, Risk: Final Report: Vol. 3.
     Stability of Perception under Time and Data (FRO/028/68).
     Dundee, Scotland: School of Architecture, Duncan of
     Jordanstone College of Art, University of Dundee,
     April 1977.

38.  Green, C. H., and Brown, R. A., The Perception of, and
     Attitudes Towards, Risk: Final Report: Vol. 4.
     Initial Experiments on Determining Satisfaction with
     Safety Levels (FRO/028/68). Dundee, Scotland: School
     of Architecture, Duncan of Jordanstone College of Art,
     University of Dundee, April 1977.

39.  Green, C. H., and Brown, R. A., Problems of Valuing Safety
     (Note N 70/78). Borehamwood, Hertfordshire, U.K.:
     Department of the Environment, Building Research Establish-
     ment, 1978.

40.  Greenberg, P. F., "The Thrill Seekers." Human Behavior 6
     (April 1977):17-21.

41.  Hammond, K. R., and Adleman, L., "Science, Values and Human
     Judgment." Science 194 (October 22, 1976):389-396.

42.  Harris, Louis and Associates, Inc., Harris Perspective 1979:
     A Survey of the Public and Environmental Activists on the
     Environment (59). New York: Louis Harris and Associates,
     1979.

43.  Harris, Louis and Associates, Inc., Risk in a Complex Society.
     Chicago: March & McClennon Public Opinion Survey, 1980.

44.  Harris, Louis and Associates, Inc., A Second Survey of
     Public and Leadership Attitudes Toward Nuclear Power De-
     velopment in the United States. New York: EBASCO, 1976.

45.  Kahan, J. P., How Psychologists Talk About Risk (P-6403).
     Santa Monica, California: Rand Corporation, October 1979.

46.  Kahneman, D., and Tversky, A., "On the Psychology of Pre-
     diction." Psychological Review 80 (July 1973):237-251.

47.  Kahneman, D., and Tversky, A., "Prospect Theory: An Analysis
     of Decision Under Risk." Econometrica 47 (March 1979):
     263-291.

48.  Kasper, R. G., "Perceived Risk: Implications for Policy."
     Impacts and Risks of Energy Strategies: Their Analysis
     and Role in Management. London: Academic Press, 1979.

49.  Kasper, R., "Perceptions of Risk and Their Effects on
     Decision-Making." Societal Risk Assessment: How Safe
     is Safe Enough? Edited by R. Schwing and W. Albers.
     New York: Plenum Press, 1980, pp. 71-80.

50.  Kasperson, R. E., Berk, G., Pijaka, D., Sharaf, A., and
     Wood, J., "Public Opposition to Nuclear Energy: Retro-
     spect and Prospect." Science, Technology and Human
     Values 5 (Spring 1980):11-23.

51.  Keeney, R. L., and Kirkwood, C. W., "Group Decision Making
     Using Cardinal Social Welfare Functions." Management
     Science 22 (1975):430-437.

52.  Keeney, R. L., and Raiffa, H., Decisions with Multiple Ob-
     jectives: Preferences and Value Tradeoffs. New York:
     John Wiley & Sons, 1976.

53.  Klausner, S., ed., Why Man Takes Chances: Studies in Stress-
     Seeking. Garden City, New York: Doubleday, 1968.

54.  Kunreuther, H., "Limited Knowledge and Insurance Protection."
     Public Policy 24 (1976):227-61.

55.  Kunreuther, H., Ginsberg, R., Miller, L., Sagi, P., Slovic, P.,
     Borkan, B., and Katz, N. Disaster Insurance Protection:
     Public Policy Lessons. New York: John Wiley & Sons, 1978.

56.  La Porte, T. R., "Public Attitudes Toward Present and Future
     Technology." Social Studies of Science 5 (1975):373-391.

57.  La Porte, T. R., and Metlay, D., "Technology Observed:
     Attitudes of a Wary Public." Science 188 (April 11, 1975):
     121-127.

58.  La Porte, T. R., and Metlay, D., They Watch and Wonder:
     Public Attitudes Toward Advanced Technology. University
     of California Institute of Governmental Studies,
     Berkeley, 1975.

59.  Lerch, I., "Risk and Fear." New Scientist 185 (January 3,
     1980):8-11.

60.  Lichtenstein, S., Fischhoff, B., and Phillips, L. D.,
     "Calibration of Probabilities: The State of the Art."
     Decision Making and Change in Human Affairs. Edited by
     H. Jungermann and G. de Zeeuw. Dordrecht: Reidel, 1977.

61.  Lichtenstein, S., Slovic, P., Fischhoff, B., Layman, M.,
     and Combs, B., "Judged Frequency of Lethal Events."
     Journal of Experimental Psychology: Human Learning and
     Memory 4 (1978):551-578.

62.  Lowrance, W., Of Acceptable Risk: Science and the Determi-
     nation of Safety. Los Altos, California: Kaufman, 1976.

63.  Maderthaner, R., Guttman, G., Swaton, E., and Otway, H. J.,
     "Effect of Distance on Risk Perception." Journal of
     Applied Psychology 63:3 (1978):380-382.

64.  Maderthaner, R., Pahner, P., Guttman, G., and Otway, H. J.,
     "Perceptions of Technological Risks: The Effect of Con-
     frontation" (NASA RM-76-53). Laxenburg, Austria: Inter-
     national Institute for Applied Systems Analysis, 1976.

65.  Marquis, D. G., and Reitz, H. J., "Effects of Uncertainty on
     Risk Taking in Individual and Group Decisions." Behavioral
     Science 14 (July 1969):281-288.

66.  Maynard, W. S., Nealey, S. M., Hebart, J. A., and Lindell, M. K.
     Public Values Associated with Nuclear Waste Disposal
     (BNWL-1997). Seattle: Battelle Human Affairs Research
     Centers, 1976.

67.  Mazur, A., "Disputes Between Experts." Minerva 11 (1973):
     55-81.

68.  Mazur, A., "Opposition to Technological Innovation."
     Minerva 13 (1975):58-81.

69.  McEnvoy, J., "The American Concern with the Environment."
     Natural Resources and the Environment. Edited by
     W. R. Burch, Jr., et al. New York: Harper and Row, 1972.

70.  Melber, B. D., Nealey, S. M., Hammersla, J., and Rankin, W. L.,
     Nuclear Power and the Public: Analysis of Collected Sur-
     vey Research (PNL-2430). Seattle: Battelle Human Affairs
     Research Centers, 1977.

71.  Mitchell, R. C., Public Opinion on Environmental Issues:
     Results of a National Public Opinion Survey. Washington,
     D. C.: Council on Environmental Quality, Department of
     Agriculture, Department of Energy, and Environmental Pro-
     tection Agency, 1980.

72. Mitchell, R. C., "Silent Spring/Solid Majorities." Public
    Opinion, 2 (August-September 1979).

73. National Council on Radiation Protection and Measurements,
    Perceptions of Risk: Proceedings of the Fifteenth Annual
    Meeting, March 14-15, 1979. Washington, D. C.: National
    Council on Radiation Protection and Measurements, March 1980.

74. Nelkin, D., Nuclear Power and its Critics. Ithaca, New York:
    Cornell University Press, 1971.

75. Nelkin, D., "The Political Impact of Technical Expertise."
    Social Studies of Science 5 (1975):35-54.

76. Nelkin, D., "Some Social and Political Dimensions of Nuclear
    Power: Examples from Three Mile Island." American Poli-
    tical Science Review 75 (March 1981):132-145.

77. Nelkin, D., Technological Decisions and Democracy. Beverly
    Hills, California: Sage Publications, 1977.

78. Nelkin, D., and Pollack, M., "Political Parties and the
    Nuclear Energy Debate in France and Germany." Comparative
    Politics (January 1980).

79. O'Hare, M., "Not On My Block You Don't: Facility Siting and
    the Strategic Importance of Compensation." Public Policy
    25 (Fall 1977).

80. Okrent, D., and Whipple, C., An Approach to Societal Risk
    Acceptance Criteria and Risk Management (UCLA-ENG-7746).
    Los Angeles: University of California, School of Engineer-
    ing and Applied Science, June 1977.

81. Opinion Research Corporation, "Public Attitudes Toward En-
    vironmental Trade-Offs." ORC Public Opinion Index 33
    (August 1975):1-8.

82. Otway, H. J., "The Perception of Technological Risks: A
    Psychological Perspective." Technological Risk: Its Per-
    ception and Handling in the European Community. Edited by
    M. Dierkes, S. Edwards, and R. Coppock. Cambridge,
    Massachusetts: Oelgeschlager, Gunn and Hain, Publishers,
    Inc., 1980, pp. 35-45.

83. Otway, H. J., Risk Assessment and Societal Choices (IIASA
    RM-75-2). Laxenburg, Austria: International Institute
    for Applied Systems Analysis, February 1975.

84. Otway, H. J., et al., "On the Social Aspects of Risk Assessment." Journal of the Society for Industrial and Applied Mathematics (1977).

85. Otway, H. J., and Cohen, J. J., Revealed Preferences: Comments on the Starr Benefit-Risk Relationships (IIASA RM-75-5). Laxenburg, Austria: International Institute for Applied Systems Analysis, 1975.

86. Otway, H. J., and Fishbein, M., Public Attitudes and Decision Making (IIASA RM-77-54). Laxenburg, Austria: International Institute for Applied Systems Analysis, 1977.

87. Otway, H. J., and Fishbein, M., The Determinants of Attitude Formation: An Application to Nuclear Power (IIASA RM-76-80). Laxenburg, Austria: International Institute for Applied Systems Analysis, 1976.

88. Otway, H. J., Maderthaner, R., and Guttman, G., Avoidance Response to the Risk Environment: A Cross-Cultural Comparison (IIASA RR-75-14). Laxenburg, Austria: International Institute for Applied Systems Analysis, 1975.

89. Otway, H. J., Maurer, D., and Thomas, K., "Nuclear Power, The Question of Public Acceptance." Futures 10 (April 1978):109-118.

90. Otway, H. J., Pahner, P. D., and Linnerooth, Social Values in Risk Acceptance (IIASA RM-75-54). Laxenburg, Austria: International Institute for Applied Systems Analysis, November 1975.

91. Pahner, P. D., "The Psychological Displacement of Anxiety: An Application to Nuclear Energy." Risk-Benefit Methodology and Application (UCLA-ENG-7598). Edited by D. Okrent. Los Angeles: University of California, December 1975.

92. Pahner, P. D., A Psychological Perspective of the Nuclear Energy Controversy (IIASA RM-76-67). Laxenburg, Austria: International Institute for Applied Systems Analysis, 1976.

93. Payne, J. W., "Relation of Perceived Risk to Preferences Among Gamblers." Journal of Experimental Psychology: Human Perception and Performance 104 (1975):86-94.

94. Pearce, D. W., "The Nuclear Debate is About Values." Nature 274 (1978):200.

95. Pearce, D. W., "The Preconditions for Achieving Consensus in the Context of Technological Risk." Technological Risk: Its Perception and Handling in the European Community. Edited by M. Dierkes, S. Edwards, and R. Coppack. Cambridge, Massachusetts: Oelgeschlager, Gunn, and Hain, Publishers, Inc., 1980. p. 58.

96. Powers, W. T., Behavior: The Control of Perception. Chicago: Aldine, 1973.

97. Pratt, J. W., Raiffa, H., and Schlaifer, R., "The Foundations of Decision Under Uncertainty." The American Statistical Association Journal 59 (1964):353-376.

98. Raiffa, H., Decision Analysis: Introductory Lectures on Choices Under Uncertainty. Reading, Massachusetts: Addison-Wesley, 1968.

99. Rapoport, A., and Wallsten, T. S., "Individual Decision Behavior." Annual Review of Psychology 23 (1972):131-175.

100. Ravetz, J. R., "Public Perceptions of Acceptable Risks as Evidence for Their Cognitive, Technical, and Social Structure." Technological Risk: Its Perception and Handling in the European Community. Edited by M. Dierkes, S. Edwards, and R. Coppack. Cambridge, Massachusetts: Oelgeschlager, Gunn, and Hain, Publishers, Inc., 1980, pp. 46-57.

101. Rethans, A., An Investigation of Consumer Perceptions of Product Hazards. Unpublished Ph.D. dissertation. University of Oregon, 1979.

102. Ross, L., "The Intuitive Psychologist and His Shortcomings." Advances in Social Psychology. Edited by L. Berkowitz. New York: Academic Press, 1977.

103. Rowe, W. E., An Anatomy of Risk. New York: John Wiley & Sons, 1977.

104. Sapolsky, H. M., "Science, Voters, and the Fluoridation Controversy." Science 162 (1968):427-433.

105. Sjoberg, L., "Risk Generation and Risk Assessment in a Social Perspective." Foresight, the Journal of Risk Management 3 (1978):4-12.

106. Sjoberg, L., "Strength of Belief and Risk." Policy Sciences 2 (August 1979):39-52.

107.  Slovic, P., "Assessment of Risk-Taking Behavior."
      Psychological Bulletin 61 (1964):220-233.

108.  Slovic, P., "Choice Between Equally Valued Alternatives."
      Journal of Experimental Psychology: Human Perception
      and Performance 1 (1975):280-287.

109.  Slovic, P., and Fischhoff, B., "Cognitive Process and
      Societal Risk Taking." Cognition and Societal Behavior.
      Edited by J. S. Carroll and J. W. Payne. Potomac,
      Maryland: Lawrence Erlbaum Associates, 1976.

110.  Slovic, P., Fischhoff, B. and Lichtenstein, S., "Accident
      Probabilities and Seat Belt Usage: A Psychological Per-
      spective." Accident Analysis and Prevention, (1978) 10,
      281-285.

111.  Slovic, P., Fischhoff, B., and Lichtenstein, S., "Behavioral
      Decision Theory." Annual Review of Psychology 28 (1977):
      1-39.

112.  Slovic, P., Fischhoff, B., and Lichtenstein, S., "Charac-
      terizing Perceived Risk." Technological Hazard Management.
      Edited by R. W. Kates and C. Hohenemser. Cambridge,
      Massachusetts: Oelgeschlager, Gunn, and Hain, 1981.

113.  Slovic, P., Fischhoff, B., and Lichtenstein, S., "Facts and
      Fears: Understanding Perceived Risk." Societal Risk
      Assessment: How Safe is Safe Enough? Edited by R. Schwing
      and W. Albers, Jr. New York: Plenum, 1980, pp. 181-216.

114.  Slovic, P., Fischhoff, B., and Lichtenstein, S., "Informing
      People about Risk." Product Labeling and Health Risks
      (Banbury Report 6). Edited by L. Morris, M. Marsis, and
      I. Barofksy. Cold Spring Harbor, New York: Cold Spring
      Harbor Laboratory, 1980.

115.  Slovic, P., Fischhoff, B., and Lichtenstein, S., "Rating
      the Risks." Environment 21:3 (April 1979):14-39.

116.  Slovic, P., Fischhoff, B., and Lichtenstein, S., "Risky
      Assumptions." Psychology Today 14 (June 1980):44-45,
      47-48.

117.  Slovic, P., Fischhoff, B., Lichtenstein, S., Corrigan, B.,
      and Combs, B., "Preference for Insuring Against Probable
      Small Losses: Insurance Implications." Journal of Risk
      and Insurance 45 (June 1977):237-258.

118. Slovic, P., Kunreuther, H., and White, G., "Decision Processes, Rationality and Adjustments to Natural Hazards." Natural Hazards: Local, National, and Global. Edited by G. F. White. New York: Oxford Press, 1974.

119. Slovic, P., Lichtenstein, S., and Fischhoff, B., "Images of Disaster: Perception and Acceptance of Risks from Nuclear Power." Energy Risk Management. Edited by G. Goodman and W. Rowe. London: Academic Press, 1979.

120. Spangler, M. B., "Risks and Psychic Costs of Alternative Energy Sources for Generating Electricity." The Energy Journal (January 1981).

121. Starr, C., "Social Benefit versus Technological Risk." Science 165 (September 19, 1969):1232-38.

122. Starr, C., "Some Comments on the Public Perception of Personal Risk and Benefit." Risk vs. Benefit: Solution or Dream? Edited by H. J. Otway. Los Alamos, New Mexico: Los Alamos National Laboratory, 1971.

123. Starr, C., and Whipple, C., "Risk of Risk Decisions." Science 208 (June 1980):1114-1119.

124. Stumpf, S. E., "Culture, Values, and Food Safety." BioScience 28 (March 1978):186-190.

125. Swaton, E., Maderthaner, R., Pahner, P. D., Guttman, G., and Otway, H. J. The Determinants of Risk Perception: A Survey (IIASA RM-76-XX). Laxenburg, Austria: International Institute for Applied Systems Analysis, 1976.

126. Svenson, O., "Are We All Among the Better Drivers?" Unpublished report, Department of Psychology, University of Stockholm, 1979.

127. Tamerin, T., and Resnick, L. P., "Risk Taking by Individual Option--Case Study: Cigarette Smoking." Perspectives on Benefit-Risk Decision Making. Washington, D.C.: National Academy of Engineering, 1972, pp. 73-84.

128. Thomas, K., Maurer, D., Fishbein, M., Otway, H., Hinkle, R., and Simpson, D., A Comparative Study of Public Beliefs About Five Energy Systems (IIASA RR 80-1). Laxenburg, Austria: International Institute for Applied Systems Analysis, 1979.

129. Thomas K., Swaton, E., Fishbein, M., and Otway, H., Nuclear
     Energy: The Accuracy of Policy Makers' Perceptions of
     Public Beliefs (IIASA RR 80-2). Laxenburg, Austria:
     International Institute for Applied Systems Analysis, 1979.

130. Thompson, M., "Aesthetics of Risk: Context or Culture?"
     Societal Risk Assessment: How Safe is Safe Enough?
     Edited by R. Schwing and W. Albers. New York: Plenum
     Press, 1980, pp. 273-286.

131. Thorngate, W., "Efficient Decision Heuristics." Behavioral
     Science 25 (1980):219-225.

132. Tubiana, M., "One Approach to the Study of Public Acceptance."
     Directions in Energy Policy. Edited by B. Kursunoglu and
     A. Perlmutter. Cambridge, Massachusetts: Ballinger Pub-
     lishing Company, 1979.

133. Tversky, A., "Elimination by Aspects: A Theory of Choice."
     Psychological Review 79 (1972):281-299.

134. Tversky, A., and Kahneman, D., "The Framing of Decisions
     and the Psychology of Choice." Science 211 (1981):
     1453-1458.

135. Tversky, A. and Kahneman, D., "Availability: A Heuristic
     for Judging Frequency and Probability." Cognitive
     Psychology 4 (1973):207-232.

136. Tversky, A., and Kahneman, D., "Judgment Under Uncertainty:
     Heuristics and Biases." Science 185 (September 27, 1974):
     1124-1131.

137. Tversky, A., and Sattath, S., "Preferences Trees."
     Psychological Review 86 (1979):542-573.

138. Velimirovic, H., An Anthropological View of Risk Phenomena
     (IIASA RM-75-XX). Laxenburg, Austria: International
     Institute for Applied Systems Analysis, 1975.

139. Vlek, C., and Stallen, P. J., "Judging Risks and Benefits
     in the Small and in the Large." Organizational Behavior
     and Human Performance 28 (October, 1981).

140. Vlek, C., and Stallen, P. J., "Rational and Personal Aspects
     of Risk." Acta Psychologica 45 (1980).

141. Von Neuman, J., and Morgenstern, O., Theory of Games and
     Economic Behavior. Princeton: Princeton University
     Press, 1944.

142. Von Winterfeldt, D., Edwards, W., Anson, J., Stillwell, W., and Slovic, P., Development of a Methodology to Evaluate Risks From Nuclear Electric Power Plants: Phase I-- Identifying Social Groups and Structuring Their Values and Concerns. Final Report to Sandia National Laboratories, Albuquerque, New Mexico, May, 1980.

143. Von Winterfeldt, D., and Rios, M., "Conflicts about Nuclear Power Safety: A Decision Theoretic Approach." Proceedings of the ANS/ENS Topical Meeting on Thermal Reactor Safety. Edited by M. H. Fontana and D. R. Patterson. Springfield, Virginia: National Technical Information Service, 1980, pp. 696-709.

144. Weinstein, N. D., "It Won't Happen to Me: Cognitive and Motivational Sources of Unrealistic Optimism." Unpublished paper, Department of Psychology, Rutgers University (1979).

145. Wendt, D., and Vlek, C., eds., Subjective Probability, Utility and Human Decision Making. Dordrecht: Reidel, 1974.

146. White, G. F., "Formation and Role of Public Attitudes." Environmental Quality in a Growing Economy. Edited by H. Jarret. Baltimore: Johns Hopkins University Press, 1966.

147. White, G. F., "Human Response to Natural Hazard." Perspectives on Benefit-Risk Decision Making. Washington, D.C.: National Academy of Engineering, Committee on Public Engineering Policy, 1972.

148. Zebroski, E. L., "Attainment of Balance in Risk-Benefit Perceptions." Risk-Benefit Methodology and Application. Some Papers Presented at the Foundation Workshop, Asilomar, California (UCLA-ENG-7598). Edited by D. Okrent. Los Angeles: University of California, 1975.

PSYCHOLOGICAL ASPECTS OF RISK:
THE ASSESSMENT OF THREAT AND CONTROL*

P. J. M. Stallen
TNO-Apeldoorn, The Netherlands

A. Tomas
University of Nijmegen, The Netherlands

## 1. INTRODUCTION

"Technological hazards are big business," Harriss, Hohenemser and Kates (1978) conclude. Compared to such major sectors of national efforts as social welfare programs, transportation and national defense, technological hazards and their management require increasingly large amounts of money. Tuller (1978) estimated the total control and damage costs due to technological hazards to be 98,0 - 180,0 billions of dollars in U.S. fiscal year 1974. Harriss et al. (1978) convincingly argue that the hazards of technology have in industrial nations replaced natural hazards of floods, pestilence and disease (see Table 1).

If for no other reasons, the expenditures and losses due to technology are deserving the attention they are receiving. Especially from the side of government and industry there is an urgent need for a better scientific understanding of safety problems of complex technological systems. Let us assume that such a scientific interest offers the most appropriate approach to model - and so control - technological uncertainty. Our interest in

---

*This paper reports major parts of the theoretical foundation of an ongoing extensive in-depth study of people's reaction to technological threat. This study is sponsored by the Dutch Ministry of Health and Environmental Hygiene, Ministry of Social Affairs, TNO and the Openbaar Lichaam Rijnmond. We benefited greatly by discussing our ideas with Roel Meertens, Peter Stringer, Pieter Defares and Charles Vlek, who all made extensive comments on an earlier draft.

risk assessment then concerns two different levels of decision
making about technology: <u>social choice</u> and <u>personal choice</u>.


Table 1.   Comparative hazard sources in U.S. and developing
           countries.   (Source: Harriss et al. 1978).

| | <u>Principal Causal Agent</u>[a] | | | |
| --- | --- | --- | --- | --- |
| | <u>Natural</u>[b] | | <u>Technological</u>[c] | |
| | Social cost[d] (% of GNP) | Mortality (% of total) | Social cost[d] (% of GNP) | Mortality (% of total) |
| United States | 2-4 | 3-5 | 5-15 | 15-25 |
| Developing countries | 15-40[e] | 10-25 | n.a.[f] | n.a.[f] |

[a]Nature and technology are both implicated in most hazards.  The
division that is made here is made by the principal causal agent
which, particularly for natural hazards, can usually be identified unambiguously.
[b]Consists of geophysical events (floods, drought, tropical cyclones,
earthquakes and soil erosion); organisms that attack crops,
forests, livestock; and bacteria and viruses which infect humans.
In the U.S. the social cost of each of these sources is roughly
equal.
[c]Based on a broad definition of technological causation, as discussed
in the text.
[d]Social costs include property damage, losses of productivity from
illness or death, and the costs of control adjustments for preventing damage, mitigating consequences, or sharing losses.
[e]Excludes estimates of productivity loss by illness, disablement,
or death.
[f]No systematic study of technological hazards in developing countries
is known to us, but we expect them to approach or exceed U.S. levels
in heavily urbanized areas.


        As psychologists, we feel more competent in dealing with the
latter and leave the area of collective decision making about technology to other disciplines like policy science, sociology, and
welfare economy.[1)]   However, in as far as politicians and experts

too, are individual decision makers, our paper may well contain.
relevant data to their way of expressing preferences and handling
technological risks. As Slovic, Fischhoff and Lichtenstein (1976)
noted, policy makers when asked to "weigh the benefits against the
risks"

>"often have highly sophisticated methods at their dis-
>posal for gathering information about problems or
>constructing technological solutions. When it comes
>to making decisions, however, they typically fall
>back upon the technique which has been relied upon
>since antiquity:intuition. The quality of their in-
>tuitions sets an upper limit on the quality of the en-
>tire decision-making process and, perhaps, the quality
>of our lives. There is an urgent need to link the
>study of man's judgmental and decision-making capa-
>bilities to the making of decisions that affect the
>health and safety of the public."

In summary, in this paper we will concentrate on how individual persons evaluate technological risks. To us, this means
studying how they judge an attribute (i.e., risk) of an activity
(i.e., exposure to hazard). The important questions are: How does
their representation of the risky activity look like? Or, how have
they constructed an image of technological risk? Are there systematic biases involved? In the next section we will discuss two
models that have recently been proposed to facilitate an answer
to such questions.

## 1.1  Models of "Risk Perception"

Vlek and Stallen (1980) have attempted to develop an ordering
of aspects of risk which is meaningful from a decision-theoretic
point of view. Figure 1 shows how they have decomposed the concept of acceptable risk and how its analytical components relate.

These authors argue that patterns of personal risk experience
(see bottom, figure 1) may be systematically related to fundamental
characteristics of risky decision situations, such as one's relative freedom of choice of exposure to hazard, the controllability
of decision consequences and the type of need that is either ful-
filled or frustrated by the possible consequences of the exposure.
Most aspects of risk that have been discussed in the recent literature on personal risk assessment can be subsumed under one of their
conceptual categories. For example, Rowe's "four most significant
factors" (1977, p. 119): a. voluntariness, b. controllability,
c. discounting in time, and d. discounting in space, are regarded
as composed of elements 1-4, 5-8, 9-10 and 11-12, respectively
(see Figure 1).

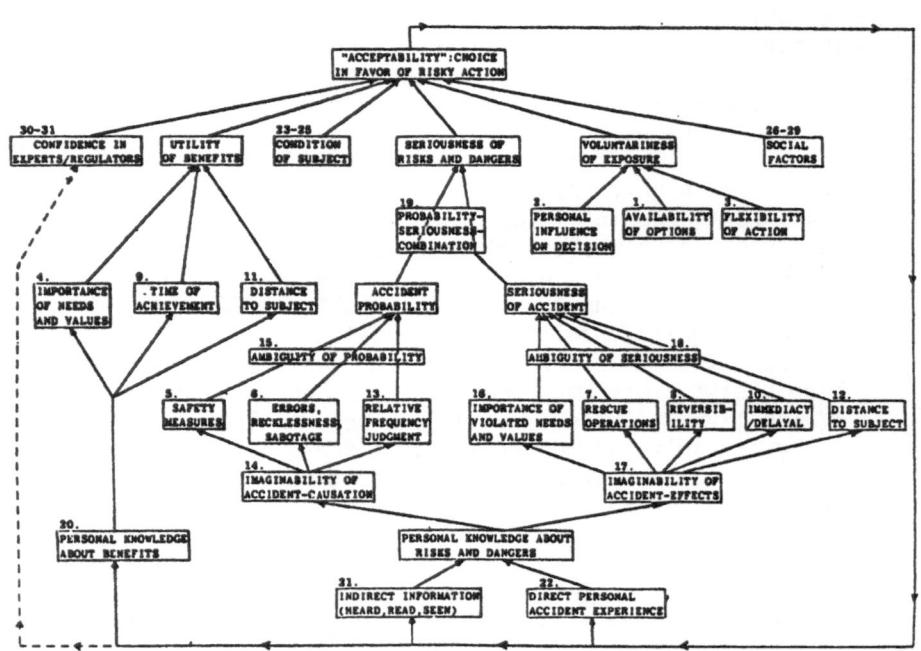

Figure 1. Rational ordering of the various aspects of risk.
(Source: Vlek & Stallen, 1980)

Figure 1 is clearly based on the assumption of man as a rational decision maker who weighs all the risks and benefits, integrates them, compares alternatives and chooses the one with the maximum net benefit. The fact that most decisions bear upon ill-defined problems, however, has led Vlek and Stallen to propose a rational structure with less global, less abstract and less formal variables. Their <u>conceptual</u> model has the advantage of providing a framework for ordering many studies on personal risk assessment. Moreover, it may function as a conceptual bridge between, on the one hand, the quantitative approach of most risk analysts who combine probabilities with losses in calculating levels of risks, and, on the other hand, psychologists who want to show that people necessarily use various heuristics in estimating how much risk is involved in some activity.

For studying the dynamics of such latter decision making, Tomas and Stallen (1981) have proposed to use a model that gives more credit to the <u>functional</u> relationship between various psychological processes that mediate between the uncertain (technological) environment and behavior. Figure 2 shows the basic structure of their process-model. The internal processes of appraisal and coping determine how a person will react to threatening technological uncertainty. This reaction is not necessarily a passive response, it may often as well be seen as an active process of structuring one's environment so as to make it more compatible with one's needs.

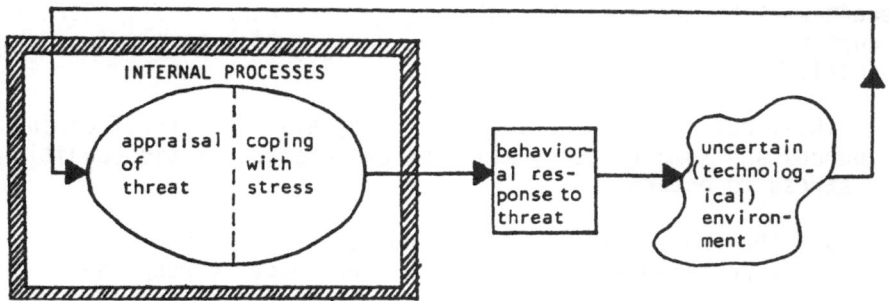

Figure 2.   General model of person/environment relationship.


This more explicit emphasis on the often need-based motivational character of people's evaluation of threat (and their behavioral response) may be seen as a second characteristic of their model.   In his analysis of environments as perceptual targets, Ittelson (1973) has asserted that "the first response to the environment is affective.  The direct emotional impact of the situation, perhaps largely a global response to the ambiance, very generally governs the directions taken by subsequent relations with the environment.  It sets the motivational tone and delimits the kinds of experiences one expects and seeks" (p. 16). Most contemporary psychological studies of personal risk assessment underestimate this role of affective or emotional processes in assessing the risk of exposure to the hazardous environment.[2] This is somewhat surprising, because most people's response to technological risks seems to be more in terms (!) of "feelings of (in)-security" than in terms of "(in)sufficiently elaborated cognitive representations."  From the first point of view, one major area of psychological theory which seems to be very relevant to such feelings is the literature of stress and coping.  The following elaboration of our general model may both serve as a short introduction to one dominant approach within this area and show its relevance for studying how people react to (technological) threat.  In this latter respect it primarily has a heuristic function, like the aforementioned conceptual model of Vlek and Stallen.

## 1.2  Threat, Stress and Behavior

One of the most influential programs of psychological stress research is the work done by Richard Lazarus and his associates

(Lazarus, 1966; 1976; 1980). Typical of Lazarus' approach is his view on stress as an intermediary variable between the stressor and behavior. In 1966 he defines stress as "a psychological condition involving the anticipation on the part of the organism of his inability to cope with some future stimulus." In his more recent work he stressed the <u>interactive</u> character of it: "Stress occurs when there are demands on the person which tax on (...) his adjustive resources" (Lazarus, 1976, p. 47).

Thus, both the emphasis on the ready availability of coping resources and some (future) stimulus are characteristic conditions for stress to occur:

   a.  the individual is called upon to respond under
       circumstances in which he has no adequate response
       available, and
   b.  the consequences of not responding are important to
       the individual.

A final characteristic of Lazarus' view on stress as well as on coping is his cognitive orientation.

With respect to this latter emphasis on cognition, we take a slightly different position. Lazarus' views stress as transactional between the person and environment (see Kanner, Coyne, Schaefer and Lazarus, in press), and not - at least not clearly - as transactional between internal cognitive and emotional states. This may foreclose seeing stress as possibly aroused by changes in the person's emotions or his motivational system. His "demands" indeed can easily be interpreted as exclusively cognitive demands. Aside from this difference in point of view, however, Lazarus' model sufficiently suits the problems that we must address ourselves to. To a large extent, the threatening stimuli in technological assessment are likely to be future <u>expected</u> effects of present technological developments. Often it is a rather distant future (like in the case of delayed environmental effects). It is hard to see how such stimuli could generate stress without a subject who believes them to take place. This focus on stress as related to in some sense ordinary every day exposure and often chronic threat, is also close to Lazarus' current research interests (see Kanner et al., in press; Folkman and Lazarus, in press). Far too little attention has been given to the ways most people cope with stressful events in their day-to-day lives, i.e., when not subject to extreme circumstances like tornadoes, parachute jumping, doctoral examinations or spinal surgeries. In our opinion, technological threat has increasingly become a "normal" condition of our life.

In theory, the processes that mediate between the threatening environment and the response can be differentiated into two classes: appraisal and coping (see Figure 2). These two kinds of internal processes can be distinguished in most models within the stress literature. However, in a temporal sense (with appraisal preceding the coping) a sharp distinction is often hard to make, particularly when the individual is facing acute threat (cf. e.g., Krohne, 1978). In general, the processes that will be described below will operate at degenerate levels in cases where threat is acute as compared to chronic and the individual is hypervigilant instead of vigilant (cf. Janis and Mann, 1977).

1.2.1 Appraisal. Folkman and Lazarus (in press) define appraisal as "the cognitive process through which an event is evaluated with respect to what is at stake (primary appraisal) and coping re- sources and options (secondary appraisal)." The primary appraisal process is seen as based on the assessment of two elements:

1. the threatening event;
2. the possible state of not having reached pursued goals, and of having lost one's stakes at the same time.

Because of its embeddedness in the subject's entire motiva- tional and goal structure, the evaluation of this latter state already entails an implicit anticipation of future coping oppor- tunities. The more important the goals or the larger the stakes, the more difficult it will be to substitute them for other goals and, consequently, the more the subject will feel forced to look for alternative ways of fulfilling his needs. Thus, the questions that always lurk behind the assessment of threat are: is there anything I should do to reduce the threat? The answer to these questions is the result of the secondary appraisal.

In Figure 3 we have shown how the various aspects of the appraisal and coping processes can be represented.[3] Some conse- quence of the exposure of the person to a given technological en- vironment called the external stimulus, is selectively perceived as a threat.[4] This selection mechanism is dependent upon the general psychological state of the person. For example, pre- judices (source I) or fear (source II) may bias the perception. The cognitive resource primarily contains all knowledge that the person has ready at hand or can retrieve from memory. It is both factual knowledge like base rates and subjective knowing like scenarios or scripts. This all can be geared to the assessment of the threatening event. Examples are specific inferences that can be made about what to expect from the threat, or attributions about what caused or who is responsible for the event. Under affective processes we subsume the emotional involvement in the threatened situation, commitment to past behavior and the like. The relevant evaluations or affects are drawn from the affective resource. This

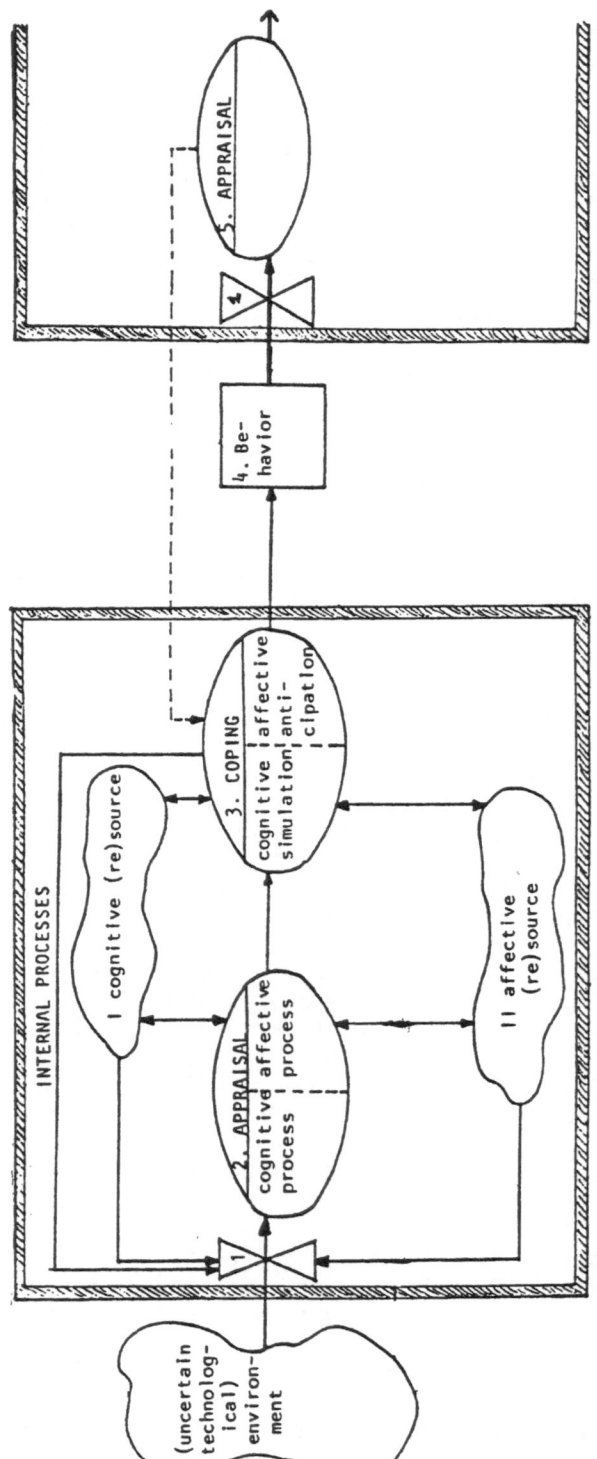

Figure 3. Schematic representation of the various psychological processes involved in dealing with a threatening uncertain environment.

we consider to be the value base for directing behavior.  Motivational constructs like level of self-esteem and conformity are
to be represented in this resource.  Although affective reactions
can occur without extensive cognitive encoding, and it even can be
argued that affect and cognition are under the control of separate
and partially independent systems (Zajonc, 1980).  Particularly when
the threatening event is cognitively complex and affectively am-
biguous (as is likely with most technological threat), the interaction of cognitive and affective processes will gain fundamental
importance.

1.2.2  Stress.  In line with Lazarus' definition we define stress
as follows: it is the psychological state of an individual who
experiences threat in his environment in such a way that the
demands on him that follow from the threat are in conflict.
Feelings of insecurity with respect to technological or industrial
developments are so regarded as one specific form of stress.  The
notion of demand in this definition of stress, or the notions of
something "at stake" and of "options" in the definition of appraisal
by Folkman and Lazarus, all refer to the fact that stress arises
in the context of goal directed and motivation-based behavior.  It
is the very relevance of the threatening event to the possibility
of not attaining one's goals and not satisfying one's needs that
makes the event threatening.  Thus, at the same time, any threatening event does generate two hypothetical future states: one with
and one without the activity.  In Figure 4 we have graphically
represented this argument.

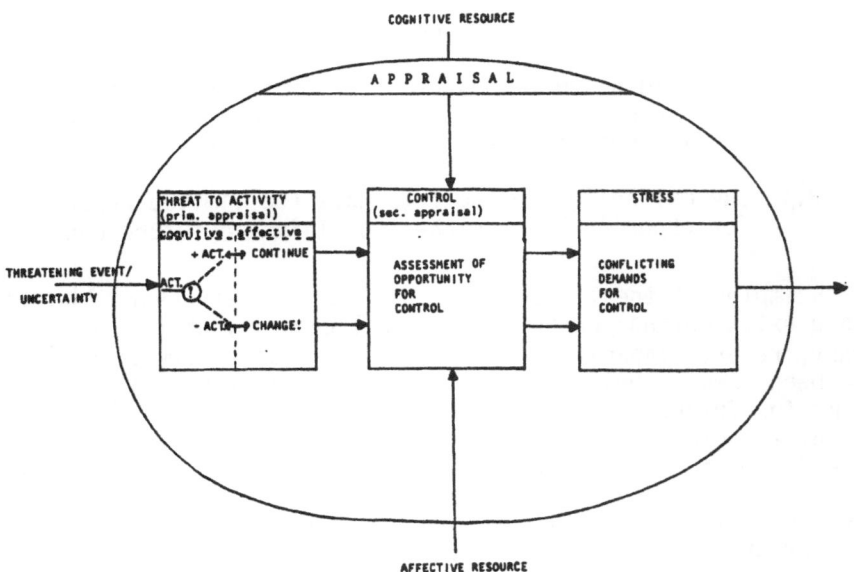

Figure 4. Structure of the appraisal process leading to stress.

In cognitive terms, the threat is represented by way of possible future consequences of the threatened activity: the activity will (+ ACT.) or will not (- ACT.) lead to the intended outcomes. As the activity has been directed to specific goals, and (quite a lot of) personal energy has been invested in it, one will generally be motivated to continue the activity. On the other hand, the threatening event calls for a change or modification of the activity. So the individual faces two mutually exclusive demands for future behavior.

As such there are demands for control of behavioral outcomes. Whether the option "change" can be realized or not will, of course, depend on available opportunities to do so. On the one hand, this is a matter of (re)sources that can be used to undertake a new or safeguarded activity. On the other hand, it requires a sufficiently well defined threat to know exactly how the activity should be changed. From a rational decision theoretic point of view one would like the person to have information about the relevant levels of all the variables of Vlek and Stallen's model (see Figure 1). Both types of information - concerning the nature of the threat and the nature of the available resources - together define whether the two mutlally exclusive demands make for a psychological conflict. This conflict we call stress.

There are at least three conditions that may oppose an easy change, and which will foster stress or aggravate the conflict.[5]

1.   The threat could apply to other (substitute) activities. In that case it is hard to respond to the threat by changing the activity, assuming that one does not alter one's goals. Again a conflict exists.

2.   It could be that the threatened activity is directed at or closely linked to central values. These it will be difficult to forgo, so a conflict arises.

3.   One may believe himself to have too limited capabilities for choosing between the two claims to future behavior.

Examples of the latter condition are when stressors like noise impose extra demands on the attentional capacity of the subject. Hereby, reserve capacity will be occupied that otherwise could have been used to help resolving the conflict between the two claims for future action. The stressfulness of the second condition, for example, is evident from research indicating the need for enduring social support in overcoming family crises. The knowledge that most food has artificial additives or the fear that at many other alternative places for settlement the soil will be polluted by chemical waste, will generate feelings of insecurity with respect to one's living condition. As such it exemplifies the first of the afore mentioned three conditions.

1.2.3 Coping. With continuously high levels of stress one would get psychologically or physically harmed. One therefore must cope with the stress. Folkman and Lazarus (in press) define coping as "the cognitive and behavior efforts to master, tolerate or re- duce external and internal demands and conflicts between them."[6] They continue: "Such coping efforts serve two main functions: the management or alteration of the person-environment relation- ship that is the source of stress (...) and the regulation of stressful emotions (...)." The first type of coping they call problem focused coping, the second emotion focused coping. In most circumstances of no-acute threat this distinction can only be gradual. There are practically no social phenomena that do not implicate affect in some important way, and probably very few per- ceptions and cognitions in everyday life are not "hot." Conse- quently, problem focused coping will most often have a significant emotional or affective component as well. However, as Zajonc (1980) notes, "Affect is always a companion to thought whereas the converse is not true" (p. 154). Particularly when threat is acute, it is likely that the problem-focused coping is entirely dominated by emotion-focused coping.

Since in our terms coping essentially involves the reappraisal of the threatening person-environment relationship, it has the same components as the appraisal process. Because of its orientation to future options, the components are labelled: cognitive simulation and affective anticipation (see Figure 3).

In coping with stress, one could seek to reduce the stress by

1.  exploring one's opportunities for altering the threatening environment,
2.  looking for ways to move away from the situation,
3.  reappraising the threat.

As Figure 3 shows, the person will generally search for new or additional information in his environment before he decides what to do. Essential is the ordering of the issues and accompa- nying choices, i.e., the strategy of the coping process. Janis and Mann (1977) have argued that certain questions do not come up at all before certain others are addressed.

The person generally will first explore the possibility to continue the present activity in a slightly modified (protected) way. In this way the person tries to achieve the same goals with essentially the same means. When there is no acceptable way to continue the present activity with precaution, he is forced to search for alternatives. He will do so only when he feels confi- dent in finding an acceptable solution. If the person also cannot imagine a realistic and acceptable option (considering one's own

and other available resources), and it is also not possible to
forget about the activity itself, there will be a tendency to
deny the necessity of making a decision. This may happen through
a variety of defense mechanisms. There are three main defensive
strategies:

Type A: procrastination,
Type B: shifting of responsibility,
Type C: bolstering of the least aversive alternative.

A postponement of the decision (type A) is possible when one
thinks that the risks of doing so are not too serious. This will
typically be accompanied by a (temporary) lack of interest in the
issue. Type B means a denial of one's own responsibility and
laying the decision into someone else's hands. As an example of
this one might think of the generally observed tendency within
the public to believe that the authorities should act to mitigate
the consequences of a natural disaster. In Section 2.2.2 we will
return to this type of defense as based upon a shift in locus
of causality. Finally, one might bolster the intention to continue
the threatened activity by exaggerating its positive consequences
and/or minimizing the negative (type C).[7] This may be facili-
tated by a selective attention to the threatening environment.
Figure 3 shows this mechanism by the arrow from the coping back to
the attention rate.

All three forms of defensive avoidance are attempts to escape
from the decisional conflict which is typical of a condition of
(high) stress: one cannot afford to change and one cannot afford
not to change. However, these kinds of conflict resolutions are
characterized by a high level of vulnerability to unanticipated
challenges. Only when there is enough time to work out all
options and enough confidence to find an acceptable solution,
one engages in what we call rational searching behavior. It is
only to this kind of searching processes that one could apply
Janis and Mann's procedural criteria for a high quality decision
(Janis and Mann, 1977), or the 9 stages for the resolution of a
poorly structured decision problem (Vlek and Wagenaar, 1979).

Finally, any action means a change in the environment. One
normally would like to know whether the action has resulted in a
subjectively safe(r) situation. In Figure 3, this opportunity to
learn from one's own behavior is represented by connecting the
appraisal of the threat in the new situation (post choice) (box 5)
to the reappraisal of the threat that one - prior to choice -
had expected to be present after one had performed the intended
behavior (box 3).

## 2. PSYCHOLOGICAL STUDIES OF TECHNOLOGICAL RISK ASSESSMENT

To what extent do the results of empirical studies that deal with natural and/or technological hazards support the foregoing analyses and where do they fit into the structure of the postulated model? Unfortunately, there are not many studies directly relevant to this topic, even with lenient criteria. For example, only about 10 studies can be selected from Rowe's (1977) reference list that pertain to the area of interest. Of these, only 5 (2%!) deal specifically with technological threat. Presumably, this rate truly reflected the efforts of studying the influence of the risk-generating system on people's judgments, attitudes and behavior in proportion to the efforts spent on studying the risk-generating system itself (risk estimation). At best the situation has only slightly improved since.[8] Technological hazards are still no big business for social psychologists.[9] Thus, it does not surprise to see that the relatively few empirical investigations of how people deal with an uncertain technological environment are not embedded in a common larger theoretical framework. Fortunately, outside the specific scope of technology there are interesting results of fundamental research which we believe are highly relevant to our topic. In the next section we will discuss both types of studies in as far as they can be grouped under the common denominator of the assessment of threat (Section 2.1) and the assessment of control (Section 2.2). Studies that explicitly address themselves to ways of coping and acting will be discussed elsewhere (Tomas and Stallen, 1981).

### 2.1 The Assessment of Threat

Within the social sciences, technological risk assessment has mainly been studied from the point of view of the psychologist[10] (e.g., Slovic and Lichtenstein, 1980; Green and Brown, 1980; Vlek and Stallen, 1981).[11] The last authors studied how inhabitants of a heavily industrialized area (Rotterdam) judge the acceptability of and the riskiness of various risky activities, as diverse as "smoking in bed before sleeping," "landing liquiffed natural gas" and "receiving full anaesthesia before a medical operation." They concluded that subjects base their judgment of acceptable risk primarily on the believed personal necessity of the activity. According to the authors, "necessity" of the activity derives from the need for its associated benefits. As a secondary dimension, judgments about "acceptability" seem to be based upon the assessment of the "scale of production and/or distribution of benefits and of potential accidents." Figure 6 shows these two major conceptual dimensions as the horizontal (= primary dim.) and vertical axis (= secondary dim.), respectively, of a two-dimensional space. The vectors indicate group averages in the weighing of the contributions of both dimensions to the overall judgment of acceptability.[12] Average group rank

orders of the 26 stimuli according to acceptability can be found by
projecting the stimulus points of Figure 6 onto the respective
(extended) vectors.  Clearly, these results show that persons living
closer to the industry tend to consider 7 "large scale" activities (all lying in the upper half of Figure 6) as relatively more
<u>acceptable</u> than persons living farther away from the industry.

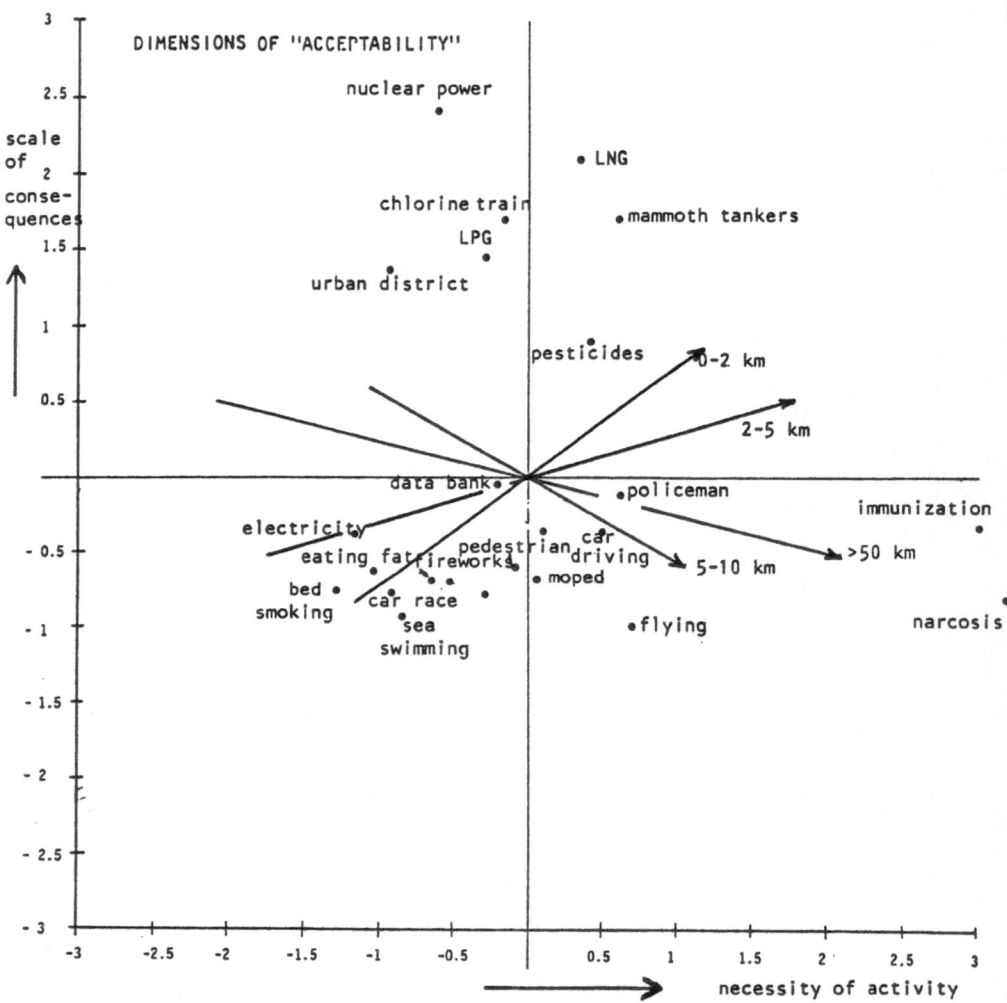

Figure 6. Two-dimensional stimulus configuration based on
          analysis of rank orders of acceptable risks.
          Arrows represent average judgments for each of
          4 groups living at varying distances from the
          industrialized area.  (Source: Vlek and Stallen,/1981)

A possible explanation for this may be that the former group
(0-2 km) judges the benefits that accrue to them and/or to society
in general to be greater. We can test for this hypothesis, as
Vlek and Stallen also analyzed judgments about the benefits of the
activity. The dimensional structure recovered was virtually
identical to Figure 6. However, in this two-dimensional cognitive
space of "beneficiality of risky activities" no such differences
between distance-groups in weighing the two-dimensions as shown in
Figure 6 were found. Because Vlek and Stallen also did not find
any such group-specific differences in the judgments of riskiness
(see below), the following explanation can be put forth. When
confronted with the difficult question whether they believe the
risks are acceptable or not, those subjects who in fact are most
exposed to the large-scale industrial risks bolster their judgment
by exaggerating the benefits associated with those activities.

The cognitive map of the riskiness of the various activities,
as constructed by Vlek and Stallen, is represented in Figure 7.
It shows that subjects probably base their judgments of riskiness
of an activity primarily on its level of catastrophic potential
(= horizontal axis) and, secondary, upon the degree to which protection is provided by institutional means (= vertical axis).
Vlek and Stallen labelled the latter dimension "degree of
organized safety," with positive values on the Y-axis indicating
a high degree of organized safety. However, as they have suggested,
an alternative interpretation may be in terms of personal avoid-
ability as this, too, is in close agreement with their data. Their
data also allow for an interpretation in terms of "imaginability
of negative consequences." As can be seen from Figure 7, persons
high in feelings of insecurity with respect to industrial risks,
base their judgments of riskiness almost exclusively upon the
degree of catastrophic potential.

Regarding the cognitive determinants of judgments about risks,
Slovic et al. (1980) have reported similar results. With a
different set of risk-stimuli and a different technique of
analysis,[13] They interpreted their two dimensions as:

1.  dread, i.e., catastrophic, hard to prevent,
    fatal, etc., and
2.  familiarity, i.e., observability, knowledge, immediacy
    of consequences.

Slovic et al. (1980) did try to further define the precise
nature of their "dread-factor." Results of a small exploratory
study led them to put forth the following explanation: "An accident
that takes many lives may have little or no impact on perceived
risk if it occurs as part of a familiar, well understood and self-
limiting process." In contrast, a small accident may greatly enhance perceived risk and trigger strong corrective action because

it may signal either a possible breakdown in safety control
systems or the possibility that the mishap might proliferate.
Thus, the number of people killed may be relatively unimportant
in determining the degree of dread or catastrophic potential of
risky activities.

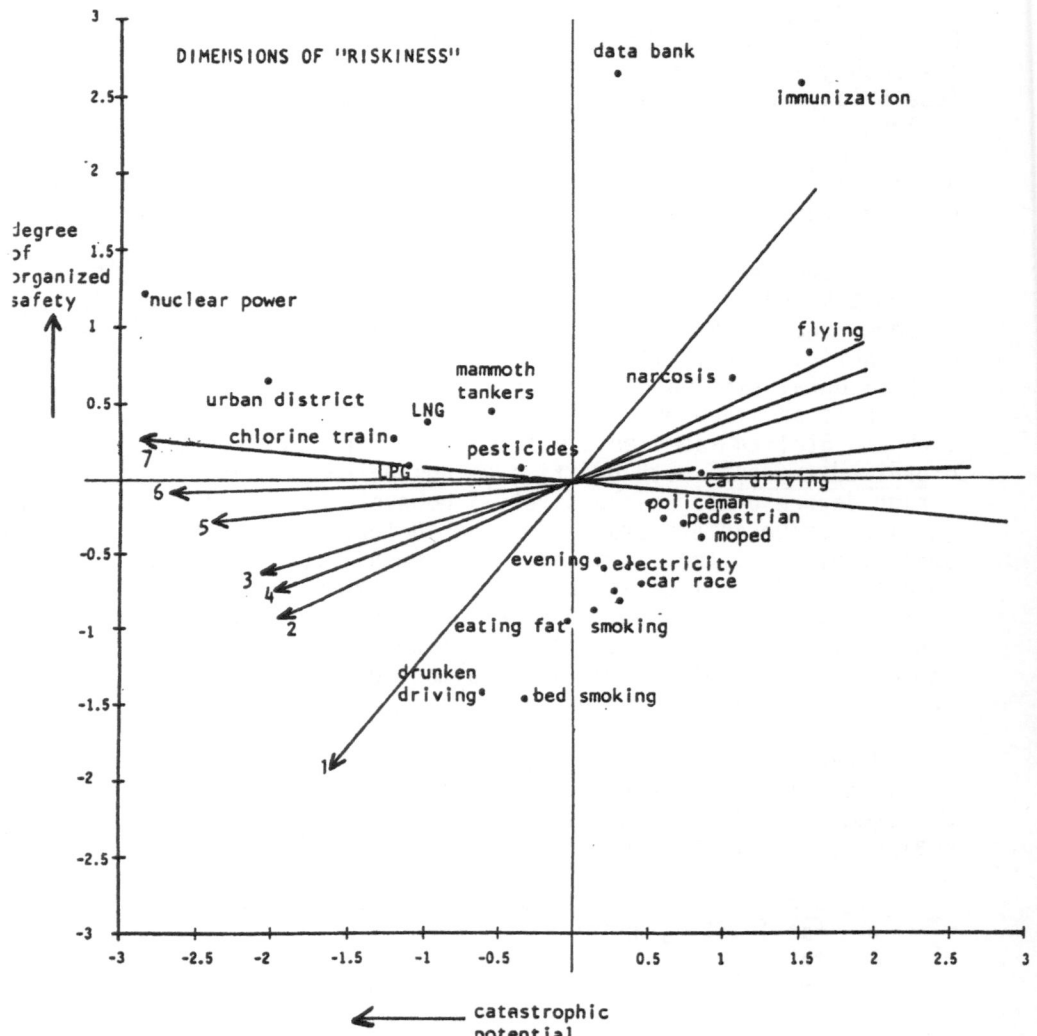

Figure 7. Two-dimensional stimulus configuration based on
          analysis of rank orders of riskiness. Arrows
          represent average judgments for each of 7 groups
          of persons in the extent to which they feel in-
          secure with respect to industrial risks.
          (Source: Vlek and Stallen, 1981)

However, the vast majority of research efforts in technical risk analysis is devoted to the calculation of numerical estimates of probable fatalities which in most cases are derived from figures of relative frequencies. Within the technical community, such estimates are generally regarded as to provide the rational measure of risk. The question that immediately arises, is how well represented are such relative frequencies in layman's mind? Using different samples of causes of death and approaching different groups of people, Lichtenstein, Slovic, Fischhoff, Layman and Combs (1978), Green and Brown (1980) and Stallen and Vlek (Apeldoorn: TNO - mimeo) all report highly similar results: overestimation of low frequencies and underestimation of high frequencies. Moreover, the results of Stallen and Vlek show (see Figure 8) that the elicitation of like estimates from experts in the field of risk analysis or risk regulation shows essentially the same bias. As a possible explanation of the general finding, Lichtenstein et al. suggest that people make use of the availability heuristic (Tversky and Kahneman, 1974): assessment of frequency or probability is based on the number and ease with which instances come to mind. As the number recalled per category (= cause of death) is relatively independent of the total number within that category, a flattening of responses is likely to occur.[14]

If, as Slovic et al. (1980) seem to suggest, uncertainty is experienced as threat depending upon its implications for future behavior, and if the number of fatalities due to an accident is not seen as a major determinant of judged seriousness of accidents, then it is worthwhile to explore in more detail what is known about people's abilities in predicting future and explaining present (or past) threat. We will discuss these abilities under the heads "foresight" and "hindsights," respectively.

2.1.1 Foresight. Most work on prediction has typically been done in areas called human information processing and judgment under conditions of uncertainty or risk (here shortly referred to as judgment). It generally reveals people to be quite inept at all but the simplest inferential tasks - and sometimes even at them - Fischhoff (1976) writes. The following examples are taken from the specific field of hazard research and may illustrate this finding.

In an almost classical study, Kates (1962; see also Burton and Kates, 1964) found that floodplain dwellers misjudged the probabilistic nature of renewed flooding of their residential areas. Many viewed floods as repetitive and even cyclical phenomena, thus replacing randomness by a determinate order in nature. Another common view among residents was that the occurrence of severe flood in one year was seen as making it unlikely to have another severe flood the following year. Such biases resemble the better known gambler's fallacy. They may be caused by the

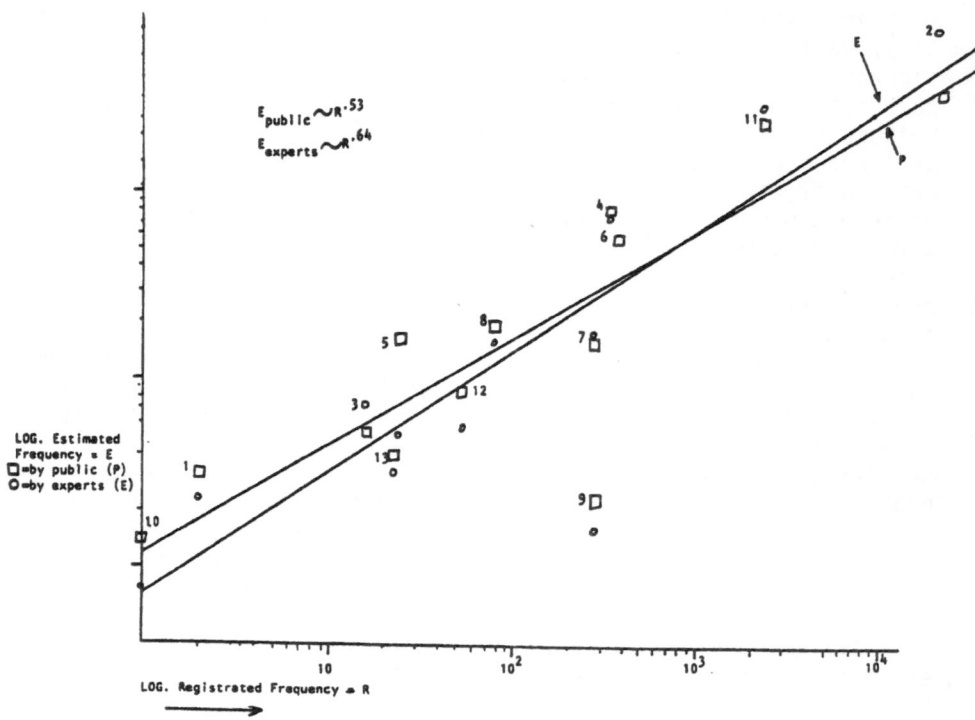

Figure 8. Relationship between registered frequency and
         subjectively estimated frequency for 13 causes
         of death.

1. Poisoning as caused by spraying
   of agricultural crops.
2. Heart disease.
3. Do-it-yourself repairment of
   electrical wiring and installations.
4. Motorbicycle accident.
5. Murder with gun or bomb.
6. Pedestrian accident in traffic
7. Drowning.
8. Fire in house or building.
9. Faulty anaesthetical treatment.
10. Lightening
11. Car accident.
12. Collision with train on crossing.
13. A leaking gas pipe at home.

use of simplifying judgmental strategies (heuristics). For
example, when given evidence regarding the specific case
(= flooding), people consistently ignore base rate information,
thus giving too much weight to the vivid evidence at the expense
of adequately adjusting earlier stored statistics. However, as
we will argue below, an alternative explanation for Kates'
finding may be given in more explicit motivational terms.

With respect to uncertainty about (consequences of) future
events, Monat, Averill and Lazarus (1972) made the distinction
between event-uncertainty and temporal-uncertainty. The former
applies to cases where one knows when an accident may happen but
not whether it actually will happen or not. Temporal uncertainty
exists if it is only unknown when a threatening event may find
place (and not whether it will occur or not). Their experiment -
exposing subjects to the threatening event, i.e., a painful
electrical shock - suggests that temporal uncertainty is experienced as the more threatening of the two types of uncertainty.
As time wore on, subjects under the condition of temporal uncertainty engaged in avoidance-like activities like attention diversion. Although the authors wrote "To date, we know of no literature attempting to assess intrapsychic coping strategies employed
under conditions of temporal uncertainty" (Monat et al., 1972,
p. 238), as a matter of fact the studies in natural hazard management at the Department of Geography, University of Chicago, may be
seen as dealing with such strategies. The judgmental strategies
that the flood plain residents in Kates' study used, for example,
just had the effect of making order out of disorder, thereby de-
picting the future as more controllable than in fact it is. We
will discuss this motive in more detail further on in section 2.2.

Another example in studying human judgment is its insensitivity
to varying sample size when estimating population parameters. Particularly when information about sample bias is pallid and information about the nature of the sample is vivid, people make unwarranted generalizations from samples to populations (see e.g.
Hamill, DeCamp, Wilson and Nisbett, 1980). Not only lay people
are subject to such biasing information processes. Several studies
indicate that such misperceptions also operate among experts like
psychometricians, well trained statisticians or people when making
decisions with grave social consequences (see e.g. Goldberg, 1968;
Tversky and Kahneman, 1974; Slovic et al., 1976).

Although people's probability estimates apparently violate
basic laws of probability theory, people tend to view such laws as
rational when they are questioned about their estimation competence
or when asked to retrospect onto their own judgmental processes.
Under these conditions, they exaggerate their own information processing sophistication. For example, they underestimate their
actual reliance on only a few major variables or their use of

heuristics in performing complicated multivariate judgments
(Shepard, 1964; Slovic and Lichtenstein, 1971). Nisbett and
Borgida (1975) have suggested that perhaps only when we have
rather well rehearsed schemata for dealing with certain types of
abstract, data summary information, we process information in a
fashion that the scientist would describe as rational. This may
be particularly relevant given the generally low frequency of
most natural and technological hazards within the human lifetime.
Here, information at a higher level than the individual cases
hardly exists and, if only for this reason, it will be difficult
for the individual to fall back on such schemata. Thus, it would
appear rational to see people rely more heavily on other resources
in trying to come to grips with both past and future events like
technological hazards.

2.1.2 Hindsight. Results of laboratory studies suggest that
people process information about past events in such a way as to
systematically reduce its perceived surprisingness. Telling
people that an event has occurred has been shown to increase their
subjective probability that it was going to happen (Fischhoff,
1975). Subjects, told the correct solution to earlier problems
(like the answer to almanac-type questions, inadmissible evidence
in court or debriefing information after participation in a psychological experiment), overestimated how much they actually knew before being told (Fischhoff, 1977). Walster (1967) reported that
when confronted with news of an accident, people tended to exaggerate in retrospect its predictability, and more so with more
serious accidents. This knew-it-all-along effect appears to be
rather robust: debiasing instructions do not produce increased
adjustment nor do they reduce bias. These findings also shed some
light on important social aspects of the dynamics of appraisal.
If, by exaggerating the predictability of the past, people underestimate how much they should learn from it, they will find it
even more difficult to reconstruct the uncertainties that actually
have faced others. As an example, they may be very insensitive
for the uncertainties that risk-regulators have to face. Thus,
when confronted in such a social context with "unexpected" accidents, people may act on the basis of a "they-could-have-known-
it-all" belief. At their turn, the decision makers will come to
fear such overreacting and, for example, are likely to develop an
even more reserved attitude with respect to early informing of
the public about impending disasters.

A similar effect has been reported in studying people's
response to the cancellation of threatening alarms. Breznitz
(1976) observed lowering heart rates, skin conductances as well
as diminishing beliefs in the threat, in short, a generally lower
level of emergency preparation when subjects were exposed to new
alarms after their first exposure to a false alarm. Breznitz
hints at an I-knew-it-all-along explanation of the FAE (= False

Alarm Effect) when he suggests that "on the subjective level, people (...) might be particularly disturbed to discover that it was all unnecessary - that, in fact, they made fools of themselves" (p. 131, emphasis ours). Here, Breznitz, too, seems to assume that his respondents reacted in a rather deterministic way to the cancellation and, by implication, to the alarm. In hindsight, they may have come to view the alarm as signaling "almost certainly nothing to happen." Clearly then, a second similar alarm will not be as much fearful. To Breznitz, the credibility loss of the source of the threatening information is FAE's chief feature. In our opinion, however, it may be but one of the *a posteriori* arguments that support people's expectation of whether or not a new alarm will materialize.

## 2.2 The Assessment of Control

"Perceived Choice" and "Perceived Control" are major themes of modern psychology (see, e.g., Harvey, 1976; Perlmutter and Monty, 1980). In section 1.2 we have argued that they are also essential aspects of the notion "stress." In Figure 4 we have tried to illustrate that it is essentially the availability of psychological (re)sources that determines whether the perception of a threatening event will make for a psychological conflict. When the two demands for future behavior are conflicting, the individual experiences stress. In his review of experimental research on personal control over impending harm, Averill (1973) shows that when subjects are given opportunities for control, decreased stress reactions have generally been observed in comparison to conditions where no such control was possible. Thus, one would expect that any attempt to resolve that conflict by personally resuming control over outcomes of own behavior will (more or less effectively) reduce the stress.

2.2.1 Modes of Control. In section 2.1 we have already given examples of the need to control future outcomes, viz. the very act of interpreting (through inferences etc.) the threatening event. This motivation of responding to threat can be called cognitive control. To have explained events apparently builds up to the self-image of being knowledgeable and, as such, it satisfies the need to be in control of one's environment, to know what. Paradoxically, the hindsight experiments indicate that this very feeling that we have made sense out of an event may be a good guarantee that we will not improve our foresight or predictive efficacy. The role of cognitive control in facing threats is emphasized in Jordan's (1968) speculations about behavior under risk. Using Lewin's terminology, he argued that whether locomotion from one mental region to another will take place or not depends on the certainty characteristics of those regions. These are:

1.  knowing whether negative events might occur to which
    I do not wish to be exposed when being in that region,
    and
2.  knowing whether I can avoid undesirable consequences
    of such an event might it still occur.

If a region has both characteristics, the subject can move
with psychological certainty even though not all the consequences
of his act will be known. "All he has to know is that certain un-
wanted outcomes are excluded. When certainty exists, the taking
of risks and of chances becomes exciting, otherwise threatening,"
(Jordan, 1968, p. 135). Burger and Arkin (1980), for example,
report that those subjects who could not predict the onset and who
knew that they could not control the termination of loud noise,
felt significantly more depressed in solving problems than those
who could either control or predict, or could do both.

Of interest to the study of cognitive control in determining
the level of stress is the assessment of warning signals. Experiments with both humans and animals have generally found that a
warning signal by itself has little effect on the experience of
stress. One of the reasons seems to be that warning signals often
do not provide sufficient information regarding the onset of the
direct threat. As a result, subjects must remain vigilant and
responsive to all cues in their environment. Because the meaning
of a warning signal is so imparted to cues, in fact not associated
with the threat, the stressfulness of the entire environment may
have been increased. Thus, one would expect to see people strive
for temporal certainty. Indeed, this preference has been observed
when subjects were presented with relatively simple stimuli like
electric shocks (cf. Monat et al., 1972) or bursts of loud noise.
The mere knowledge of the conditions for a threat to materialize
even makes subjects select higher levels of shock than when such
information is not given (Bowers, 1968). On the basis of a review
of experimental research, Averill (1973) concluded that stress is
primarily a function of the meaning of the control response for
the individual and not just of its effectiveness in preventing or
mitigating the impact of a potentially harmful stimulus. If this
holds, believing one has meaningful control may still result in
ineffective behavior. Weinstein (1980), for example, discussed
various studies which indicate that people are generally unreal-
istically optimistic about the future. For a range of positive
and negative events he examined the conditions under which such
optimistic biases occur. Among others, the following hypotheses
relating to the role of perceived control were supported: The
greater the perceived controllability of a negative event, the
greater the tendency for people to believe that their own chances
for not being harmed are better than average; also, the greater
the perceived controllability of a positive event, the greater the
tendency for people to believe that their own chances are greater

than average. Weinstein suggests the following explanation for this effect. If an event is perceived to be controllable, it signifies that people believe there are steps one can take to increase the likelihood of a desired outcome. Assuming in addition that people can more easily bring to mind their own actions than the actions of others, people are likely to conclude that desired outcomes are more likely to happen to them than to others.

In summary, available evidence suggests that, compared to behavioral control, the need to reduce the uncertainty, which often accompanies such control, is the more potent determinant of stress. The general preference for clear expectancies, however inaccurate, seems to facilitate adaption and to reduce long-term stress even though they may initially lead to increased reactivity.

If the question were put to most people, there are good reasons to expect that they would prefer to have a choice among alternatives rather than to have decisions made for them (cf. Brehm, 1972). Yet, as Averill (1973) notes, social commentators from Hobbes to Fromm have emphasized man's willingness to relinquish such control, to "escape from freedom." In this and the following paragraphs we will devote our attention to this paradoxical relationship between decisional control and stress. Our starting point will be Kelly's (1955) fundamental observation that a man controls his destiny "to the extent that he can develop a construction system with which he identifies himself and which is sufficiently comprehensive to subsume the world around him. If he is unable to identify himself with this system [...] but he can experience no personal control."

Kelly's above statement leads us to expect that the individual must be able to agree with the degree of structuring that exists. On the one hand, as Kaufmann (1971) has emphasized, this requires us to look at the environmental determinants of personal control: the individual can only develop a sufficiently coherent construction system if his (social) environment is sufficiently stable. On the other hand, it suggests the importance of perceived congruity between individual belief systems and behavior constraints in determining the experience of control and, consequently, of choice. Some evidence for this is to be found in an experiment by Lewis and Blanchard (1971), who reported that stronger identification with a role leads to the experience of greater decisional control. If subjects are placed in completely novel or unstructured settings, i.e., when behavior is not constrained by situation-bound cues, they will generally experience considerable stress.

The relevance of congruity between, on the one hand, expectations or beliefs and, on the other hand, external conditions for behavioral control is also evident from studies of the

"illusion of control." According to Langer (1975), an illusion
of control is an expectancy of a personal success probability in-
appropriately higher than the objective probability would warrant.
Langer found that people who had been allowed to draw their
lottery-ticket themselves were only willing to sell their ticket
at a price which was considerably higher than of those people who
had simply been given a ticket. Wortman (1975) conducted an experiment in which she asked crapshooters either before or after
they had thrown the dice to bet on the number of eyes they were to
throw and had thrown, respectively. The results strongly supported
her hypothesis: those players who "caused" their own outcome and
knew beforehand what they hoped to attain, perceived themselves to
have more control over the outcome, more choice about which outcome they received, and more responsibility for their outcome than
players in the remaining conditions. Wortman argues that only if
people know what they hope to attain, can they exert control. It
is likely that Langer's results are similarly mediated by her subjects' differential attempts to exert control.

Related to the concept of role-identification is the notion
of commitment, which has been defined as "pledging or binding of
an individual to an act" (Kiesler, 1971). It means that the individual identifies with his action or, in other words, believes
that he acts on the basis of a proper appraisal of both internal
and external demands or constraints. As such, he believes his
appraisal is causal of his action. Experiments have indeed supported the idea that attribution of self-responsibility for an act
increases commitment to that act (see, e.g., Mayer, Duval and
Hensly-Duval, 1980). One major variable that influences this
attributional process and, consequently, determines commitment,
is perceived choice. It presumably does so, Mayer et al. argue,
primarily by affecting the salience of the self as a plausible
cause. Harvey (1976) has summarized how some variables influence
perceived choice (see Table 2).[15]

In summary, we may perceive personal control if we intend a
certain outcome, if personal choice is salient to us, and, because
of hindsight effects, perhaps mostly so if the intended outcome
in fact occurs. In retrospect, the perception of such (illusory)
control may lead via the attribution of causality to self, to
commit ourselves to the act. Thus, commitment implies the perception of responsibility for the act. The major effect of commitment is to make that act resistant to change.

In a final paragraph, we will elaborate the conceptual relationship between causality of and responsibility for an accident.
These variables - often confounded - may be important in explaining
differential reactions to (technological) threat.

Table 2.    Some determinants of perceived choice with their
            respective direction of influence.
            (Source: Harvey, 1976, p. 90).

| Variable | Finding for Variable |
|---|---|
| Similarity in attractiveness of options | Greater perceived choice for small than for large or zero difference in attractiveness |
| Uncertainty about outcomes of options | Greater perceived choice the more uncertain the decision maker about the outcomes of the options |
| Valence of options | Greater perceived choice when the outcomes of the options are positive than when they are negative in attractiveness |
| Number of options | Greater perceived choice the greater the number of options when decision maker thinks he has expeditiously evaluated options; greater perceived choice for moderate than for small or large number of options when decision maker thinks he has spent more than normal amount of time in evaluating options |
| Locus of control | Greater differences for internals than externals in response to options varying along dimensions of similarity in attractiveness and valence of outcomes |
| Variability in reinforcement | Greater freedom attributed to reinforcing agent who rewards and punishes on an intermittent schedule than to one who reinforces on a continuous schedule |
| Costs in taking an action | Less freedom is attributed to reinforcing agent or to decision maker the more costly the action taken or decision made |
| Acting in line with predisposition | Greater freedom attributed to person, the more the behavior was consistent with a relevant predisposition |
| Consequences of a decision | Relatively high self-attribution of choice for decisions having positive consequences and relatively low self-attribution of choice for decisions having negative consequences |
| Magnitude of negative consequences of an action | Less freedom attributed to self by actors and more freedom attributed to actor by observer the more serious the consequence of an action |

2.2.2 Accident Causation and Responsibility for Accidents.
In the foregoing sections we have tried to argue that the
answer to the question "Do I feel that I have opportunities to
control the threat?" is an important determinant of stress or
feelings of insecurity. These opportunities were supposed to be
both cognitive and affective in kind, as illustrated by Figure 4.
They concern ways to understand one's technological environment,
to protect against anxiety, to structure vague feelings and beliefs
for a decision, to get compensation from a person or industry at
fault, or to provide a sense of identity. All such functions are
specific types of control and serve a common goal which could be
called systems - maintenance: they are "ways of keeping oneself
going in a difficult and unpredictable world" (Fischhoff, 1976).
As such, the assessment of a threatening event is very much an
active process. Of special interest to the process of how people
appraise threatening events like industrial alarms and emission
of pollutants is the area of research on attribution of responsibility for an accident. This is particularly important as the
assignment of responsibility might be closely related to take
action or demanding it from others.

The major impetus to the research on attribution of responsibility for an accident is a provocative study by Walster (1966).
She presented subjects with a description of an accident in which
some act of a stimulus person (perpetrator) led to harm to some
other person (victim). Her experiment demonstrated that outcome
severity determined the amount of responsibility assigned to the
perpetrator. Walster explained this effect by reasoning that her
subjects had attributed more responsibility in the serious accident to protect themselves from the idea that their lives, too,
could be affected by such unfortunate events. To believe in
chance happenings would be too threatening. This "defensive
attribution hypothesis" has since stimulated a large number of
studies. Despite its continued popularity, however, there has
been observed a worrisome unevenness in growth of data collection
and conceptual development (see, e.g., Fischhoff, 1976; Fincham
and Jaspars, 1980).

In an earlier critical review of theoretical notions and empirical results, Vidmar and Crinklaw (1974) had already pointed at
the absence of "any rationale that predicts whose fate subjects
will identify with when both a perpetrator and victim are involved
in an accident" (p. 114). One possible answer has been suggested
by Pryor and Kriss (1977): with the fate of the subject who is
most similar to the perceiver. If it is the victim with whom one
identifies, the perpetrator will have the most dissimilar characteristics. Because of their saliency they will give rise to the
attribution of causality to this act. If the perceiver identifies
more with the perpetrator-role, he may attribute causality to environmental factors (chance). Indeed, experimental results

(Shaver, 1970; Crinklaw and Vidmar, 1971) indicated that the more similar the subject and the stimulus person are, the less is the responsibility assigned for the mishap. However, this only appeared true when the first of the following three specific orientations was present.[16] This condition is that the perceiver has to be primarily interested in what the observed situation means affectively to him (value-maintenance set). When his orientation is causal-genetic, i.e., he is primarily interested in what the social, physical etc. conditions of the behavior/act are, or situation-matching, i.e., his main concern is the evaluation of the behavior of the stimulus person as it relates to norms appropriate to that situation, no differential assignment of responsibility with varying similarity was found. In conducting one of the few experiments that has not at the outset confounded the above three inferential sets, Chaikin and Darley (1973) found that with a value-maintenance set:

1. In general, less responsibility is attributed to change (= environmental factor) when the consequences of the accident become more serious.
2. Perpetrator-like perceivers attribute responsibility to environmental factors (like chance; but they also derogated the victim!) more so than do those subjects who have to respond as are they victims.
3. This latter group attributes to the same extent the cause of both the minor and the serious accident to the perpetrator; that is to say, they hold him equally responsible for both accidents.

To some extent, these results confirm a general finding of the attribution research, called "the fundamental attribution error" (see e.g., Harvey, Ickes and Kidd, 1976/1978). By this we mean that people - particularly when observing somebody else's instead of their own behavior - are inclined to over-attribute causality to personal factors like dispositions, and underestimate the impact of situational or environmental factors. One possible explanation of this bias is given in Gestalt-terms of Figure: Background. To the victim-like perceiver, the perpetrator is Figure, thus salient; to the perpetrator-like perceiver, the environment is Figure.

However, this theory would at least need one extra premise to explain the differential impact of varying seriousness on attribution by perpetrator-like versus victim-like perceivers (see Chaikin and Darley's 2nd and 3rd finding). As already mentioned, perpetrator-like subjects may make defensive attributions to avoid that they in such situation would be blamed for their "careless" act. Such defensive avoidance behavior is more likely to occur with more serious accidents. Apparently, a similar reasoning cannot be applied to victim-like subjects (the more threatening the situation is appraised by victim-like subjects, the more they may want to protect themselves from the knowledge that they, too, could be at the mercy of such an event). Vidmar and Crinklaw (1974) therefore suggest that only when subjects feel a perpetrator ought to have foreseen the consequences of his behavior, they assign more responsibility for a serious rather than a minor outcome of his act.[17] Presumably, their victim-like subjects did not feel that way (their data indeed showed that they considered the perpetrator equally careful in both conditions). The fact that they hold the perpetrator responsible for his act, no matter how serious its consequences are, may be based on their belief that yet he <u>could</u> have done otherwise. The judgment that is being made is essentially a sanctioning judgment, not an explanatory judgment, and also not - as it is for perpetrator-like subjects - a judgment based on a value-maintenance set.

Hamilton (1980) argues that the use of such a moral-legal decision rule for attributing responsibility instead of, for example, applying the scientific principle of co-variation, may also provide an alternative explanation of the "fundamental attribution error."

Fincham and Jaspars (1980), too, have stressed the importance of the distinction between causal and moral responsibility. Holding someone responsible by demanding that he rebut an accusation does not explain his actions but simply indicates liability for punishment or compensation (p. 104). It is not simply that people use a notion of causality different from the one advanced in general attribution theory. However, in attribution theories the causal questions involved deal with the relation between intention/disposition and behavior (actor-act chain), whereas in assigning responsibility the central relation is between act and outcome. Many conflicting interpretations of research results stem from the conceptual confusion about causality and responsibility and from differences regarding the part of the actor-act-outcome chain that is given attention to.

## 3.  CONCLUDING REMARK

From a psychological point of view for both layman and politician alike, technological risk assessment - i.e., the assessment of aversive consequences due to a certain degree of exposure to potentially harmful events - most often is a complex process.[18] In this paper we have tried to give some idea of how people assess threatening events and how judgmental, attributional and motivational processes together lead to a (intermediary) response to threat. This we called stress or, specifically with respect to our research interest, feelings of insecurity concerning technological activities. One frequently raised argument in the debate about "How safe is safe enough?" is that there exists a gap between

the factual, actuarial or objective risks and the way in which the
public perceives the risks of technology, the so-called subjective
risks. With the foregoing analysis we hope to have convincingly
shown that there is no (psycho-) logic according to which people's
assessment of technological safety is proportionally related to
(can be predicted from) observed relative frequencies or statistically calculated probabilities of negative consequences of the
technology concerned. Even disregarding the fact that for most
activities only human death is the best defined and quantifiable
of all hazard consequences and impacts to the ecosystem are at
best difficult to judge (Harris et al., 1978, p. 9), the individual is not so much concerned with estimating uncertainty parameters of a physical or material system as he is with estimating
the uncertainty involved in his exposure to the threatening event
and in opportunities to influence or control his exposure.
Statistics are only marginally relevant to specific circumstances.

## NOTES

1) It appears to us that this area of essentially social choice
   processes is much in need of systematic investigation. For
   one thing, it would be important to study how qualitative
   descriptions (e.g., Wynne, 1980) could benefit from formal
   mathematical analysis of social choice mechanisms. One central
   conclusion that both approaches seem to have in common is
   that social choice processes should essentially be about
   alternative options. In their remarkable effort to critically
   review existing social choice mechanisms concerning the
   question "how safe is safe enough" (like revealed preference,
   cost benefit analysis), Fischhoff et al. (1980) conclude:
   "Acceptable risk problems are decision problems, that is,
   they require a choice between alternatives" (ii). Bezembinder
   and Van Acker (1979) from their point of view suggest that
   "the evaluation underlying intransitive choice and the alleged
   interpersonal incomparability of utility come together in
   calling for a notion of preference in which the basic object
   of evaluation is a pair of alternatives rather than a single
   alternative," (emphasis ours).

2) Many people fear the possible consequences of complex modern
   technologies. Those who support the technological status quo
   often call this response too emotional and - therefore -
   irrational. Stallen (1980; see also Wynne, 1980), has argued
   that such a derogation of opposition can be seen as based
   upon an ideological generalization of the applicability of
   the dominant scientific notion of rationality to all kinds
   of human decisions. This, of course, is not what we mean by
   emotional: it is affect, value, but not as opposed to fact.

3) The symbols of this figure are taken from system dynamics in an attempt to stimulate our thinking about the dynamics of personal risk assessment. They usually symbolize rates ($\bowtie$), levels ($\square$), infinitively large (re)sources ($\oslash$), auxiliary variables ($\bigcirc$), and information or energy flows ($\rightarrow$).

4) We do not distinguish between attention and perception because there appears to be no sound theoretical base anymore (see Keele and Trammel-Neill, 1978).

5) Note the similarities of these conditions with the first three conditions for a decision problem to occur: "A decision problem occurs when a decision maker (a) notices a discrepancy between an existing state and a desired state, (b) has the motivation, as well as (c) the potential to reduce this discrepancy, whereby (d) there is more than one possible course of action which may not be immediately available, (e) the implementation of a course of action demands an irreversible allocation of his resources, and (f) the utilities (of the consequences) associated with each choice-alternative are partly or entirely uncertain," (Vlek and Wagenaar, 1979, p. 257).

6) Sometimes the entire internal process is labelled as coping, sometimes only the positive, healthful way of responding to stress is called coping. For reasons of distinction, we use two separate terms, coping and acting, where others, e.g., Folkman and Lazarus (in press), use one, i.e., coping.

7) This divergence of consequences is a general effect with all prospective choices that have important consequences (and, thus, are more or less stressful in itself). Only when evaluations of alternative options involve public commitment to the final choice, evaluations converge (see Brownstein, Ostrove and Mills, 1979).

8) If so, it would be primarily because of an enormous growth of interest in nuclear power.

9) Social scientists are also no big business yet for technology. Fischhoff et al. (1980) state that "Given the enormous stakes riding on acceptable-risk decisions, our investment in research seems very small. Considering the cost of a day's delay in returning a nuclear facility to service or in approving a pipeline proposal, a research project that offered a 0.1 chance of responsibly shortening the decision-making period would have an enormous expected return on investment. Similar bargains would be found in studies that might improve public involvement in project planning (so as to avoid mid-

construction surprises), identify generic categories of new
chemicals (so as to reduce testing costs), decrease the uncertainty in drug licensing (so as to encourage innovative
research and development), or inform workers about occupational
risks (so as to enable them to make better decisions on their
own behalf). Such research could be a good place to invest
society's venture capital."

10) Even when staying within the area of psychology, there is
still more to risk assessment than merely risk perception,
as Figure 3 shows. However, risk assessment is often designated as "risk perception." The Society for Risk Analysis,
for example, has organized at its first annual meeting a
"Workshop on the Analysis of Real versus Perceived Risks."

11) Sociologists, for example, seem to concentrate more on natural
hazards (cf. Mileti, 1980).

12) The analysis used is based upon a point-vector unfolding model
for interpreting the structure of the observations (rank
order data).

13) The major differences are that Slovic c.s. 1. did not formu-
late all their stimuli as specific risky activities but as
global activities, technical systems of devices; 2. they subjected their data to a type of external analysis. Considering
the number of variables for classification of their stimuli,
this is a rather vulnerable procedure.

14) Arguments justifying the logarithmic transformation of response
are given by both Green and Brown (1980) and Stallen and
Vlek (Apeldoorn: TNO, mimeo).

15) Of these, the latter two influences seem to reflect self-
serving tendencies, whereas the other determinants reflect
mainly information-processing and perceptual tendencies.

16) The classification into these three perceptual sets was
originally developed by Jones and Thibaut in their analysis
of interpersonal perception.

17) This is one of Heider's five possibilities of associating behavioral outcomes with individuals: 1. global association
(= mere coincidence), 2. extended commission (= outcomes not
foreseeable), 3. careless commission (= outcomes could have
been foreseen), 4. purposive commission (= outcomes foreseen),
5. justified commission (= outcomes intended, see Shaw and
Sulzer, 1964).

18) Neither one should claim the exclusive use of the adjective rational to characterize his own risk assessment. Rationality is a quality of conflict resolution along the lines of prespecified rules for the allocation of one's analytic and synthetic resources in dealing with the various aspects of what one perceives as a problem (see e.g., Simon, 1978). Decision analysis, for example, offers different criteria, for evaluating such allocation of resources than psycho-analysis. Thus, one should be very careful calling someone else's response to risk "irrational."

## REFERENCES

Averill, J., 1973, Personal control over aversive stimuli and its relationship to stress, Psychological Bulletin, 80 (4), 286-303.

Bezembinder, Th., and van Acker, P., 1979, Intransitivity in individual and social choice, in: Lanterman, E. D. and Feger, H. (eds.), Similarity and Choice, New York: Wiley.

Bowers, I. S., 1968, Pain, Anxiety and Perceived Control, Journal of Consulting and Clinical Psychology, 9, 205-209.

Brehm, J. W., 1972, Responses to the loss of freedom: a theory of psychological reactance, Morristown: General Learning Press.

Brounstein, P. J., Ostrove, N., and Mills, J., 1979, Divergence of private evaluations of alternatives prior to choice, Journal of Personality and Social Psychology, 37 (11), p. 1957-1965.

Burger, J. M., and Arkin, R. M., 1980, Prediction, Control and Learned Helplessness, Journal of Personality and Social Psychology, 38 (3), 482-491.

Burton, I., and Kates, R. W., 1964, The perception of natural hazard in resource management, in: Natural Resource Journal, 3, 412-441.

Chaikin, A. L., and Darley, I. M., 1973, Victim or perpetrator: defensive attribution of responsibility and the need for order and justice, Journal of Personality and Social Psychology, 25, 268-275.

Crinklaw, L. D., and Vidmar, N., 1971, Inferential sets, locus of control and attribution of responsibility for an accident, Ontario: University of Western Ontario (Res. Bull. 203).

Fincham, F. D., and Jaspars, J. M., 1980, Attribution of responsi-
     bility, in: Berkowitz, L. (ed.), Advances in experimental
     social psychology, vol. 13, New York: Academic Press.

Fischhoff, B., 1975, Hindsight and Foresight: The effect of outcome
     knowledge on judgment under uncertainty, Journal of Experimen-
     tal Psychology: Human Perception and Performance, 1, 228-299.

Fischhoff, B., 1976, Attribution theory and judgment under uncertain-
     ty, in: Harvey, J. H., Ickes, W. J., and Kidd, R. F. (eds.),
     New directions in attribution research, Hillsdale:
     Erlbaum.

Fischhoff, B., 1977, Perceived informativeness of facts, Journal of
     Experimental Psychology: Human Perception and Performance, 3
     (2), 349-358.

Fischhoff, B., Lichtenstein, S., Slovic, P., Keeney, R., and
     Derby, S., 1980, Approaches to acceptable risk: a critical
     guide, Oak Ridge, Tenn.: Oak Ridge National Laboratory.

Folkman, S., and Lazarus, R. S. (in press), An analysis of coping
     in a middle-aged community sample, in: Journal of Health and
     Behavior.

Green, C., and Brown, D., 1980, Through a glass darkly: perceiving
     perceived risks to health and society, Paper prepared for the
     workshop on Perceived Risk, Eugene, Or.

Hamill, R., DeCamp Wilson, T., and Nisbett, R. E., 1980, Insensi-
     tivity to sample bias: generalizing from a typical case,
     Journal of Personality and Social Psychology, 39 (4), 578-589.

Hamilton, V. L., 1980, Intuitive psychologist or intuitive lawyer?
     Alternative models of the attribution process, Journal of
     Personality and Social Psychology, 39 (5), 767-772.

Harriss, R. C., Hohenemser, C., and Kates, R. W., 1978, Our Haz-
     ardous Environment. Environment, 20 (7).

Harvey, J. H., 1976, Attribution of freedom, in: Harvey, J. H.,
     Ickes, W. I., and Kidd, R. F. (eds.), New directions in
     attribution research, vol. 1., Hillsdale: Erlbaum.

Harvey, J. H., Ickes, W. J., and Kidd, R. F. (eds.), 1976/1978,
     New directions in attribution research, vol. 1/vol. 11,
     Hillsdale: Erlbaum.

Ittelson, W. H., 1973, Environment perception and contemporary
     perceptual theory, in: Ittelson, W. H. (ed.), Environment
     and Cognition, New York: Seminar Press.

Janis, I. L., and Mann, L., 1977, Decision Making: a psychological analysis of conflict, choice, and commitment, New York: Free Press.

Jordan, N., 1968, Themes in speculative psychology, London: Tavistock.

Kanner, A., Coyne, J. C., Schaefer, C., and Lazarus, R. S. (in press), Comparison of two modes of stress measurement: daily hassles and uplifts versus major life events, Journal of Behavioral Medicine.

Kates, R. W., 1962, Hazard and choice perception in flood plain management, Chicago: University of Chicago (Dept. of Geography, Research Paper 78).

Kates, R. W., 1979, Summary Report, in: R. W. Kates (ed.), Managing Technological Hazard: Research needs and opportunities, University of Colorado (Monograph no. 25, Institute of Behavioral Science).

Keele, S. W., and Trammell-Neill, W., 1978, Mechanisms of attention, in: Carterette, E. C., and Friedman, M. P. (eds.), Handbook of Perception: Perceptual Processing, New York: Academic Press.

Krohne, H. W., 1978, Individual differences in coping with stress and anxiety, in: Spielberger, C. D., and Sarason, I. G. (eds.), Stress and Anxiety, vol. 5, New York: Wiley.

Lazarus, R. S., 1966, Psychological stress and the coping process, New York: McGraw-Hill.

Lazarus, R. S., 1976, Patterns of adjustment, New York: McGraw-Hill.

Lazarus, R. S., 1980, The stress and coping paradigm, in: Eisdorfer, C., Cohen, D., and Kleinman, A. (eds.), Theoretical bases for psychopathology, New York: Spectrum.

Mileti, D. S., 1980, Human adjustment to the risk of environmental extremes, Sociology and Social Research, 64 (3), 237-347.

Monat, A., Averill, J. R., and Lazarus, R. S., 1972, Anticipatory stress and coping reactions under various conditions of uncertainty, Journal of Personality and Social Psychology, 24 (2), 237-253.

Nisbett, R. E., and Borgida, E., 1975, Attribution and the psychology of prediction, Journal of Personality and Social Psychology, 32, 532-543.

Perlmutter, L. C., and Monty, R. A. (eds.), 1980, Choice and Per-
    ceived Control, Hillsdale: Lawrence Erlbaum.

Pryor, J. B., and Kriss, M., 1977, The cognitive dynamics of
    salience in the attribution process, Journal of Personality
    and Social Psychology, 35, 49-55.

Rowe, W. D., 1977, An Anatomy of Risk, New York: Wiley.

Shaver, K. G., 1970, Defensive attribution: effects of severity
    and relevance on the responsibility assigned to an accident,
    Journal of Personality and Social Psychology, 14, 101-113.

Shaw, M. E., and Sulzer, J. L., 1964, An empirical test of Heider's
    levels in attribution of responsibility, Journal of Abnormal
    and Social Psychology, 69, 39-46.

Shephard, R. N., On subjectively optimum selection among multi-
    attribute alternatives, in: Shelley, M. W., and Bryan, G. L.
    (eds.), Human judgments and optimality, New York: Wiley.

Simon, H. A., 1978, Rationality as process and as product of
    thought, American Economic Review, 68, 1-16.

Slovic, P., and Lichtenstein, S., 1971, Comparison of Bayesian
    and regression approaches to the study of human information
    processing in judgment, Organizational Behavior and Human
    Performance, 6, 649-744.

Slovic, P., Fischhoff, B., and Lichtenstein, S., 1976, Cognitive
    Processes and Societal Risk Taking, in: Carroll, J. S., and
    Payne, J. W. (eds.), Cognitive and Social Behaviour,
    Hillsdale: Lawrence Erlbaum.

Solvic, P., Fischhoff, B., and Licntenstein, S., 1980, Perceived
    Risk, in: Schwing, R. C., and Albers, W. A. (eds), Societal
    Risk Assessment: How safe is safe enough?, New York: Plenum
    Press.

Stallen, P. J. M., 1980, Risk of Science of Science of Risk?, in:
    Conrad, J. (ed.), Society, Technology and Risk Assessment,
    New York: Academic Press.

Stallen, P. J. M., and Vlek, C. A. J., Probabilities, conditional
    probabilities and judged frequencies of fatal accidents,
    Apeldoorn: TNO (mimeo).

Starr, Ch., 1969, Social benefit versus technological risk
    Science, 165, 1232 -1238.

Tomas, A., and Stallen, P. J. M., 1981, Risk, Stress and Decision Making, Paper prepared for the Conference on Subjective Utility and Decision Making, Budapest.

Tuller, J., 1978, The scope of hazard management expenditure in the U. S., Worcester, Mass.: Clark University (mimeo CENTED).

Tversky, A., and Kahneman, D., 1971, Judgment under uncertainty: heuristics and biases, Science, 185, 1124-1131.

Vidmar, N., and Crinklaw, L. D., 1974, Attributing responsibility for an accident: a methodological and conceptual critique, Canadian Journal of Behavioral Science, 6, 112-130.

Vlek, C. A. J., and Wagenaar, W. A., 1979, Judgment and decision under uncertainty, in: Michon, J. A., Eijkman, E. G., and de Klerk, L. F. W. (eds.), Handbook of Psychonomics, vol. 2, Amsterdam: North-Holland Publishers.

Vlek, C. A. J., and Stallen, P. J. M., 1980, Rational and Personal Aspects of Risk, Acta Psychologica, 45, p. 273-300.

Vlek, C. A. J. and Stallen, P. J. M., 1981, Risk Perception in the Small and in the Large, Organizational Behavior and Human Performance, 28, 235-271

Walster, E., 1967, "Second-guessing" important events, Human Relations, 20, 239-250.

Wynne, B., 1980, Technology, Risk and Participation: on the social treatment of uncertainty, in: Conrad, J. (ed.), Society, Technology and Risk Assessment, New York: Academic Press.

Zajonc, R. B., Feeling and Thinking: Preferences need no Inferences, American Psychologist, 35 (2), p. 151-175.

RISK ANALYSIS AND TECHNOLOGICAL HAZARDS:   A POLICY-RELATED
BIBLIOGRAPHY


Vincent Covello and Mark Abernathy

U.S. National Science Foundation


## 1.  INTRODUCTION

During the past decade, the risk analysis literature has
grown from a handful of articles and books to a formidable collection of material.  The objective of this bibliography is to
provide a listing of this material, with emphasis on works of
generic interest to researchers and policymakers.  Aside from
inadvertent omissions and some minor exceptions, the following
types of material were intentionally excluded from the bibliography:  works dealing narrowly with a specific technology or
hazard; technical or methodological works dealing exclusively
with risk estimation or the measurement of probabilities and
uncertainties; articles from mass circulation magazines; book
reviews and foreign language works; unpublished dissertations
and theses; and unpublished symposium, conference, or workshop
papers.  Because numerous sources were consulted in compiling
the bibliography, it was not possible to provide complete reference information (such as page numbers) in all cases.

Ackerman, B., Rose-Ackerman, S., Sawyer, J.W., and Henderson,
    D.W.  The Uncertain Search for Environmental Quality.  New
    York: Free Press, 1974.

Acton, J.P.  Measuring the Monetary Value of Lifesaving Programs.
    Publication 5675.  Santa Monica, California: Rand Corporation,
    1976.

Adams, J.G.  "... and How Much for Your Grandmother?"  Environ-
    ment and Planning 6 (1974).

Adomeit, H.  "Risk and Risk-Taking."  Soviet Risk-Taking and
    Crises Behavior:  A Theoretical and Empirical Analysis.
    London: George Allen & Unwin, 1982, pp. 9-15.

Aharoni, A.  The No Risk Society.  Old Greenwich, Connecticut:
    Chatham Press, 1981.

Allera, E.J.  "An Overview of How the FDA Regulates Carcinogens
    Under the Federal Food, Drug, and Cosmetic Act."  Food,
    Drug, and Cosmetic Law Journal 33 (1978):59-70.

Allias, M.  "The Behavior of Rational Man in the Face of Risk:
    Critique of the Postulates and Axioms of the American
    School."  Econometrica 23 (April 1953).

Altzinger, E., Brook, M., Wilbert, J., Chernick, M.R., Elsner,
    B., and Foster, W.V.  Special Publication No. 4.  Compen-
    dium of Risk Analysis Techniques.  Aberdeen, Maryland:
    U.S. Army Material Systems Agency, 1972.

Anderson, D.R.  "All Risks Rating Within a Catastrophe Insurance
    System."  Journal of Risk and Insurance 43 (December 1976):
    629-651.

Anderson, J.E.  "Decision Analysis in Environmental Decision
    Making:  Improving the Concorde Balance."  Columbia Journal
    of Environmental Law 5 (Fall 1978).

Anderson, J.E.  "Economic Regulatory and Consumer Protection
    Policies."  Nationalizing Government:  Public Policies
    in America.  Edited by T. Lowi and A. Stone.  Beverly
    Hills, California: Sage Publications, 1978.

Andrews, R.  "Cost-Benefit Analysis as Regulatory Reform."
    Cost-Benefit Analysis and Environmental Regulations.
    Edited by D. Swartzman, R. Liroff, and K. Croke.
    Washington, D.C.: The Conservation Foundation, 1982.

Anthony, R. "Accountability and Credibility in the Management of Complex Hazardous Technology." Policy Studies Review 1:4 (May 1982):705-715.

Apostolakis, G.E. Mathematical Methods of Probabilistic Safety Analysis (UCLA-ENG-7464). Los Angeles: University of California, September 1974.

Apostolakis, G.E. "Probability and Risk Assessment: The Subjectivistic Viewpoint and Some Suggestions." Nuclear Safety 22 (January-February 1981):1.

Apostolakis, G.E., Garribba, S., and Volta, G., eds. Synthesis and Analysis Methods for Safety and Reliability Studies. NATO Advanced Study Institute Series. New York: Plenum Press, 1980.

Apostolakis, G.E., and Kaplan, S. "Pitfalls in Risk Calculations." Reliability Engineering 2 (1981):135-145.

Arrow, K.J. Essays in the Theory of Risk Bearing. Chicago: Markham Publishing Company, 1971.

Arrow, K.J., and Lind, R.C. "Uncertainty and the Evaluation of Public Investment Decisions." American Economic Review 60 (June 1970).

Arthur, W.B. The Economics of Risks to Life (RR-79-16). Laxenburg, Austria: International Institute for Applied Systems Analysis, December 1979.

Ashby, E. "Protection of the Environment: The Human Dimension." Proceedings of the Royal Society of Medicine 69 (1976): 721-730.

Ashby, E. "The Risk Equations: The Subjective Side of Assessing Risks." New Scientist 74 (May 19, 1977):398-400.

Ashford, N.A. "Alternatives to Cost-Benefit Analysis in Regulatory Decisions." Annals of the New York Academy of Sciences (July 1980):129-137.

Ashford, N.A. Crisis in the Workplace: Occupational Injury and Disease. Cambridge, Massachusetts: MIT Press, 1976.

Ashford, N.A. "The Limits of Cost-Benefit Analysis in Regulatory Decisions." Technology Review 82 (May 1980):70-72.

Ashford, N.A., Hattis, D., Heaton, G.R., Katz, J., Priest, C.
    and Volt, E.M.  Evaluating Chemical Regulations:  Trade-off
    Analysis and Impact Assessment for Environmental Decision-
    making.  Council on Environmental Quality and the Environ-
    mental Protection Agency, No. EQ4A-CA35, Washington, D.C.,
    March 1979.

Ashford, N.A., Hattis, D., Priest, C., Frankel, R., Mendez, W.,
    Andrews, R., Katz, C. and Mitcel, C.S. The Benefits of
    Environmental, Health, and Safety Regulations.  Committee
    on Governmental Affairs, Washington, D.C.: United States
    Senate Committee Print, March 25, 1980.

Atiyah, P.S.  Accidents, Compensation, and the Law.  London:
    Weidenfeld and Nicolson, 1970.

Atkinson, J.W.  "Motivational Determinants of Risk-Taking
    Behavior."  Psychology Review 64 (1957):359-372.

Ayers, F.T.  "The Management of Technological Risk."  Research
    Management (November 1977):24-28.

Bailey, M.J.  Reducing Risks to Life and Limb:  Measurement of
    the Benefits.  Washington, D.C.:  American Enterprise
    Institute, 1980.

Bailey, M.J., and Jensen, M.C.  "Risk and the Discount Rate for
    Public Investment."  Studies in the Theory of Capital
    Markets.  Edited by M.C. Jensen.  New York: Praeger Pub-
    lishers, 1972.

Baldweicz, W., Haddock, G., Lee, Y., Prajoto, Whitley, R., and
    Denny, V.  Historical Perspective on Risk for Large Scale
    Technological Systems (UCLA-ENG-7485).  Los Angeles:
    University of California, December 1974.

Baram, M.S.  "Cost-Benefit Analysis:  An Inadequate Basis for
    Health, Safety, and Environmental Regulatory Decisionmaking."
    Ecology Law Quarterly 8 (1980):473-531.

Baram, M.S.  "Radiation from Nuclear Power Plants:  The Need
    for Congressional Directives."  Harvard Journal on Legis-
    lation 14 (June 1977):905.

Baram, M.S.  "Regulation of Environmental Carcinogens:  Why Cost-
    Benefit Analysis May Be Harmful to Your Health."
    Technology Review 78 (July/August 1976):40-42.

Baram, M.S.  "Social Control of Science and Technology."
    Science 172 (May 7, 1971).

Baram, M.S. "Technology Assessment and Social Control." Science 180 (May 4, 1973):465-473.

Baram, M.S. "Testimony on H.R. 3441: The Risk Analysis Research and Demonstration Act of 1981." Environmental Professional 3:3/4 (1981):201-208.

Baram, M.S., et al. Alternatives to Regulation for Managing Risks to Health, Safety and Environment. Lexington, Massachusetts: Lexington Books, 1981.

Baram, M.S., and Miyares, J.R. "Managing Flood Risk: Technical Uncertainty in the National Flood Insurance Program." Columbia Journal of Environmental Law 7:2 (1982):129.

Baram, M.S., Sandberg, D., Dufault, L., and McAllister, K. "Managing Risks to Health, Safety and Environment by the Use of Alternatives to Regulation." New England Law Review 16 (1981-1982):657-687.

Barber, R. "Risk Factors in Social Change." Disasters 2:4 (1979):251-253.

Battelle Northwest Laboratories. Review of Decision Methodologies for Evaluating Regulatory Actions Affecting Public Health and Safety (BNWL-2158). Seattle: Battelle Northwest Laboratories, December 1976.

Baumol, W.J. "Environmental Protection and Income Distribution." Benefit-Cost & Policy Analysis 1974. Edited by R. Zeckhauser et al. Chicago: Aldine Publishing Company, 1975.

Baumol, W.J. "On the Social Rate of Discount." American Economic Review 58 (June-September 1968):785.

Baumol, W.J., and Oates, W.E. The Theory of Environmental Policy. Englewood Cliffs, New Jersey: Prentice Hall, 1975.

Baxter, W.F. People or Penguins: The Case for Optimal Pollution. New York: Columbia University Press, 1974.

Bazelon, D.L. "Coping with Technology Through the Legal Process." Cornell Law Review 62 (1977):817.

Bazelon, D.L. "The Judiciary: What Role in Health Improvement?" Science 211 (February 20, 1981):792-793.

Bazelon, D.L. "Risk and Responsibility." American Bar Association Journal 65 (1979):1066.

Bazelon, D.L. "Risk and Responsibility." Science 205 (July 20, 1979): 277-280.

Bazelon, D.L. "Science, Technology and the Court." Science 208 (1980): 661.

Beacher, G., Gros, J., and McCusker, K. Balancing Apples and Oranges: Methodologies for Facility Siting Decisions (RR-75-33). Laxenburg, Austria: International Institute for Applied Systems Analysis, 1975.

Beauchamp, D.E. "Public Health and Individual Liberty." Annual Review of Public Health 1 (1980):121-136.

Becker, G.M., and McClintock, C.G. "Value: Behavioral Decision Theory." Annual Review of Psychology 18 (1967):239-286.

Bem, D.J., Wallach, M., and Kogan, N. "Group Decision Making Under Risk of Aversive Consequences." Journal of Personality and Social Psychology 1 (1965):453-460.

Berg, G.G., and Maillie, H.D., eds. Measurement of Risks. New York and London: Plenum Press, 1981.

Berger, J.L., and Riskin, S.D. "Economic and Technological Feasibility in Regulating Toxic Substances Under the Occupational Safety and Health Act." Ecology Law Quarterly 7 (1978):285-358.

Bergstrom, T.C. "Living Dangerously." Risk-Benefit Methodology and Application (UCLA-ENG-7598). Edited by D. Okrent. Los Angeles: University of California, December 1975.

Berkowitz, M. "Occupational Safety and Health." Annals of the American Academy of Political and Social Science (May 1979):41-53.

Berman, D.M. "How Cheap is Life?" International Journal of Health Sciences 8 (1978):79-99.

Bertin, M. "Studies of Risks from Different Energy Sources: Methodological Comments." International Atomic Energy Agency Bulletin 22 (October 1980):72.

Besuner, P.M., Tetelman, A.S., Eagen, G.R., and Rau, C.A.  "The
    Combined Use of Engineering and Reliability Analysis in
    Risk Assessment of Mechanical and Structural Systems."
    Risk-Benefit Methodology and Application (UCLA-ENG-7598).
    Edited by D. Okrent.  Los Angeles: University of California,
    1975.

Bettman, O.L.  The Good Old Days--They Were Terrible!  New York:
    Random House, 1974.

Bick, T., and Kasperson, R.  "Pitfalls of Hazard Management:
    The CPSC Experiment."  Environment 20 (October 1978):
    30-42.

Bick, T., Hohenemser, C., and Kates, R.  "Target:  Highway
    Risks."  Environment 21:2 (1979):7-15, 29-38.

Black, S.C., and Niehaus, F.  "How Safe is 'Too' Safe?"  Inter-
    national Atomic Energy Agency Bulletin 22:1 (1980):40-50.

Boffey, P.  "Radiation Standards: Are the Right People Making
    the Decisions?"  Science 171 (February 1971):780-783.

Bogen, K.T.  Coordination of Regulatory Risk Analysis:  Current
    Framework and Legislative Proposals.  Washington, D.C.:
    Congressional Research Service, 1981.

Bogen, K.T.  "Public Policy and Technological Risk."  IDEA:
    The Journal of Law and Technology 21 (1980):37-74.

Bowen, J.  "The Choice of Criteria for Individual Risk."  Risk-
    Benefit Methodology and Application (UCLA-ENG-7598).
    Edited by D. Okrent.  Los Angeles: University of California,
    December 1975.

Bowman, C.H., et al.  The Prediction of Voting Behavior on
    a Nuclear Energy Referendum (RM-78-8).  Laxenburg, Austria:
    International Institute for Applied Systems Analysis
    Research, February 1978.

Boyer, B.B.  "Alternatives to Administrative Trial-Type Hearings
    for Resolving Complex Scientific, Economic, and Social
    Issues."  Michigan Law Review 71 (1972):111.

Brimblecombe, P., and Ogden, C.  "Air Pollution in Art and
    Literature."  Weather 32 (1977):285-291.

Broome, J.  "Trying to Value a Life."  Journal of Public
    Economics 9 (1978):91-100.

Broussalian, V.L.  "Risk Measurement and Safety Standards in
    Consumer Products," Household Production and Consumption.
    Edited by N. Terleckyj.  New York: Columbia University
    Press, 1975.

Brown, B.V.  "Projected Environmental Harm: Judicial Acceptance
    of a Concept of Uncertain Risk."  Journal of Urban Law 53
    (February 1976):497-531.

Brown, J.M.  "Probing the Law and Beyond: A Quest for Public
    Protection from Hazardous Product Catastrophes."  George
    Washington Law Review 38 (March 1970):431-462.

Brown, J.P.  "Toward an Economic Theory of Liability."  Journal
    of Legal Studies (1973):431-462.

Bunker, J.P., Barnes, B.A., and Mosteller, F., eds.  Costs,
    Risks, and Benefits of Surgery.  New York: Oxford Univer-
    sity Press, 1977.

Burger, E.J.  Protecting the Nation's Health:  The Problems of
    Regulation.  1976.

Burton, I., Fowle, C.D., and McCullough, R.S., eds.  Living With
    Risk:  Environmental Risk Management in Canada.  Environ-
    mental Monograph No. 3.  Toronto: Institute for Environ-
    mental Studies, University of Toronto, 1982.

Burton, I., Kates, R., and White, G.  The Environment as Hazard.
    New York:  Oxford, 1978.

Buttel, F., and Flinn, W.  "The Politics of Environmental Con-
    cern:  The Impacts of Party Identification and Political
    Ideology on Environmental Attitudes."  Environment and
    Behavior 10 (March 1978):17-35.

Butzel, A.K.  "Legal Mechanisms for Risk-Benefit Analysis:  Some
    Thoughts on the Significance of the Storm-King Case."
    Perspectives on Benefit-Risk Decision Making.  Edited by
    the Committee on Public Engineering Policy.  Washington,
    D.C.: National Academy of Engineering, 1972.

Cairns, J.  Cancer:  Science and Society.  San Francisco: W.H.
    Freeman, 1978.

Cairns, J.  "Estimating Hazard."  BioScience 30 (February
    1980):101-107.

Calabresi, G.  The Costs of Accidents:  A Legal and Economic
    Analysis.  New Haven, Connecticut: Yale University Press,
    1970.

Calabresi, G., and Bobbitt, P.  Tragic Choices.  New York: W.W.
    Norton and Company, 1978.

Callahan, D.J.  The Tyranny of Survival and Other Pathologies of
    Civilized Life.  New York: Macmillan Publishing Company,
    1973.

Carson, R.  Silent Spring.  Boston: Houghton Mifflin, 1962.

Carter, L.  "Dispute Over Cancer Risk Quantification."  Science
    203 (March 30, 1979):1324-1325.

Carter, L.  "How to Assess Cancer Risks."  Science 204 (May 25,
    1979):811.

Cederlof, R., et al.  "Air Pollution and Cancer:  Risk Assess-
    ment Methodology and Epidemiological Evidence."  Environ-
    mental Health Perspectives 22 (1978):1-13.

Chiang, C.L.  "A Stochastic Model of Competing Risks of Illness
    and Competing Risks of Death."  Biology.  Edited by J.
    Gurland.  Madison, Wisconsin: University of Wisconsin
    Press, 1964.

Chicken, J.C.  Hazard Control Policy in Britain.  New York:
    Pergamon Press, 1975.

Clark, E.M., and Van Horn, A.J.  "Risk-Benefit Analysis and
    Public Policy: A Bibliography."  Updated and extended by
    L. Hedal and E.A.C. Crouch.  Cambridge, Massachusetts:
    Energy and Environmental Policy Center, Harvard University,
    1978.

Clark, W.  "Managing the Unknown."  Managing Technological
    Hazard:  Research Needs and Opportunities.  Edited by R.
    Kates.  Boulder: Institute of Behavioral Science, Univer-
    sity of Colorado, 1977, pp. 109-142.

Clark, W.  "Witches, Floods and Wonder Drugs: Historical Per-
    spectives on Risk Management."  Societal Risk Assessment:
    How Safe is Safe Enough?  Edited by R. Schwing and W.
    Albers.  New York: Plenum Press, 1980, pp. 287-312.

Coase, R.H.  "The Problem of Social Cost."  The Journal of Law
     Economics  3 (1960):1-44.

Coates, J.F.  "Some Methods and Techniques for Comprehensive
     Impact Assessment."  Technological Forecasting and Social
     Change 6 (1974):341-357.

Coates, J.F.  "Technology Assessment: The Benefits, the Costs,
     the Consequences."  The Futurist 5 (December 1971):225.

Cohen, A.V.  "The Nature of Decisions in Risk Management."
     Dealing with Risk.  Edited by R.F. Griffiths.  New York:
     Halsted Press, 1981.

Cohen, A.V., and Pritchard, D.K.  Comparative Risks of Elec-
     tricity Production Systems:  A Critical Survey of the
     Literature.  Health and Safety Executive, Research Paper
     11.  London: HMSO, 1980.

Cohen, B.L.  "Society's Valuation of Life Saving in Radiation
     Protection and Other Contexts."  Health Physics 38
     (January 1980): 33-51.

Cohen, B.L., and Sing Lee, I.  "A Catalog of Risks."  Health
     Physics 36 (June 1979):707-722.

Cole, G., and Withey, S.  "Perspectives on Risk Perceptions."
     Risk Analysis:  An International Journal 1:2 (1982).

Comar, C.L.  "Risk, a Pragmatic de Minimis Approach".  Science
     203 (January 26, 1979).

Combs, B., and Slovic, P.  "Causes of Death:  Biased Newspaper
     Coverage and Biased Judgments."  Journalism Quarterly 56
     (1979): 837-843.

Commission of the European Communities.  Nuclear and Non-Nuclear
     Risk--An Exercise in Comparability (EUR 6417EN).  Brussels,
     Belgium: Commission of the European Communities, 1980.

Committee for a Study on Saccharin and Food Safety Policy,
     Institute of Medicine and the National Research Council/
     Assembly of Life Sciences.  Food Safety Policy.  Washington,
     D.C.: National Academy of Sciences, 1979.

Commoner, B.  "The Environmental Cost of Economic Growth."
     Economics of the Environment.  Edited by R. Dorfman and
     N.S. Dorfman.  New York: W.W. Norton & Company, 1972.

Conley, B.C. "The Value of Human Life in the Demand for Safety."
American Economic Review 66 (March 1976):45-55.

Connolly, T.H., and Mazur, A. "The Risk of Benefit-Risk Analysis." Proceedings of the Sixth Annual Health Physics
Society Midyear Symposium. Richmond, Washington, November
1970.

Conrad, J. "Society and Risk Assessment: An Attempt at Interpretation." Society, Technology and Risk Assessment.
Edited by J. Conrad. London: Academic Press, 1980, pp.
241-276.

Conrad, J., ed. Society, Technology, and Risk Assessment.
London: Academic Press, 1980.

Conservation Foundation. "Cost-Benefit Analysis: A Tricky
Game." Conservation Foundation Letter (December 1980).

Conway, R.A., ed. Environmental Risk Analysis for Chemicals.
New York: Van Nostrand Reinhold Company, 1982.

Cooper, R.M. "The Role of Regulatory Agencies in Risk-Benefit
Decision Making." Food, Drug, and Cosmetic Law Journal 33
(December 1978):755-773.

Coppola, A., and Hall, R.E. A Risk Comparison (NUREG/CR-1916).
Upton, New York: Brookhaven National Laboratory, prepared
for the Nuclear Regulatory Commission, February 1981.

Cornell, N.W., Noll, R.G., and Weingast, B. Safety Regulation
Washington, D.C.: Brookings Institution, 1976.

Costle, D.M. "Dollars and Sense: Putting A Price Tag on Pollution." Environment 21:8 (October 1979):25-27.

Council for Science and Society. The Acceptability of Risk.
London: Barry Rose, Ltd., 1977.

Covello, V. "The Perception of Technological Risks." Technological Forecasting and Social Change 22 (In press).

Covello, V. "Technological Hazards, Risk, and Society: A
Perspective on Risk Analysis Research." Risk Benefit
Analysis in Water Resources Planning and Management.
Edited by Y. Haimes. New York: Plenum Press, 1981.

Covello, V., and Abernathy, M.  "Actual vs. Perceived Risk:  A
    Policy Related Bibliography."  The Analysis of Actual vs.
    Perceived Risks.  Edited by V. Covello, G. Flamm, J.
    Rodricks, and R. Tardiff.  New York: Plenum, 1983.

Covello, V., and Menkes, J.  "Issues in Risk Analysis."  Risk
    in the Technological Society.  Edited by C. Hohenemser and
    J. Kasperson.  Boulder, Colorado: Westview Press, 1981.

Covello, V., Flamm, G., Rodricks, J., and Tardiff, R., ed.  The
    Analysis of Actual vs. Perceived Risk.  New York:  Plenum
    Press, 1983.

Covello, V., Menkes, J., and Nehnevajsa, J.  "Risk, Philosophy,
    and the Social and Behavioral Sciences."  Risk Analysis
    2:2 (1983).

Covello, V., Waller, R., Abramson, L., Bryson, M., Flanagan,
    G., and Uppuluri, R., ed.  Low Probability/High Consequence
    Risk Analysis.  New York: Plenum Press (In press).

Cox, D.C., and Baybutt, P.  "Methods for Uncertainty Analysis:
    A Comparative Survey."  Risk Analysis 1:4 (December 1981):
    251-258.

Craik, K.H.  "Environmental Psychology."  New Directions in
    Psychology.  Edited by T.M. Newcomb.  New York: Holt,
    Rinehart, and Winston, 1970.

Crandall, R. and Lave, L., eds.  The Scientific Basis of Health
    and Safety Regulations.  Washington, D.C.: Brookings Insti-
    tution, 1981.

Crenson, M.  The Un-Politics of Air Pollution:  A Study of Non-
    Decision Making in the Cities.  Baltimore, Maryland: Johns
    Hopkins University Press, 1971.

Crouch, E.A.C., and Wilson, R.  "Regulation of Carcinogens."  Risk
    Analysis:  An International Journal 1 (March 1981):47-58.

Crouch, E.A.C., and Wilson, R.  Risk/Benefit Analysis.
    Cambridge, Massachusetts: Ballinger Publishing Company,
    1982.

Crowe, M.J.  "Toward a 'Definitional Model' of Public Percep-
    tions of Air Pollution."  Journal of the Air Pollution
    Control Association 18 (March 1968):154-157.

Cumming, R.B.  "Is Risk Assessment a Science?"  Risk Analysis:
    An International Journal 1 (March 1981):1-4.

Darby, W.J.   "Acceptable Risk and Practical Safety." Journal of the American Medical Association 224 (1973):1165-1168.

Darby, W.J.   "Benefit Risk Decision Making and Food Safety." Food, Man and Society. Edited by D.N. Walcher, N. Kretchmer, and H. Barnett. New York: Plenum Press, 1976.

Dasgupta, A.K., and Pierce, D.W.   Cost-Benefit Analysis:   Theory and Practice. New York: Barnes and Noble, 1972.

Davies, J.C., Gusman, S., and Irwin, F.   Determining Unreasonable Risk Under the Toxic Substances Control Act. Washington, D.C.: The Conservation Foundation, 1979.

de Boer, C.   "The Polls:   Nuclear Energy." Public Opinion Quarterly (Fall 1977):402-411.

de Heer, H.J.   "Calculating How Much Safety is Enough." Chemical Engineering 80 (February 19, 1973):121-128.

Delcoigne, G.   "Education and Public Acceptance of Nuclear Power Plants." Nuclear Safety 20 (November-December 1979):655-664.

Del Sesto, S.L.   "Nuclear Reactor Safety and the Role of the Congressman:   A Content Analysis of Congressional Hearings." Journal of Politics 42 (February 1980):227-241.

Del Sesto, S.L.   Science, Politics, and Controversy:   Civilian Nuclear Power in the United States. Boulder, Colorado: Westview Press, 1979.

DeMuth, C.C.   "Constraining Regulatory Costs.   Part 1: The White House Review Programs." Regulation 4 (January-February 1980):13-26.

DeMuth, C.C.   "Constraining Regulatory Costs.   Part 2:   The Regulatory Budget." Regulation 4 (March-April 1980):29-39, 42-44.

de Neufville, R., and Pate, M.E.   "A Conceptual Risk Assessment Procedure." Two Conceptual Approaches to Health Risk Assessment for Alternative National Ambient Air Quality Standards. Washington, D.C.: U.S. Environmental Protection Agency, September 1980.

Derby, S., and Keeney, R.   "Risk Analysis:   Understanding 'How Safe is Safe Enough?'" Risk Analysis:   An International Journal 1:3 (1982):217-224.

Derr, P., Goble, R., Kasperson, R., and Kates, R. "Worker/
    Public Protection:  The Double Standard." Environment 23
    (September 1981):6-15, 31-36.

DeSchmukh, S.S. "Risk Analysis." Chemical Engineering 81
    (June 24, 1974):141-144.

Diamond, P. "Economic Factors in Benefit-Risk Decision Making."
    Perspectives on Benefit Risk Decision Making. Edited by
    the Committee on Public Engineering Policy. Washington,
    D.C.: National Academy of Engineering, 1972, pp. 115-120.

Dickstein, H.L. "National Environmental Hazards and Inter-
    national Law." International and Comparative Law Quarterly
    23 (1974):426-446.

Dierkes, M. "Assessing Technological Risks and Benefits."
    Technological Risk:  Its Perception and Handling in the
    European Community. Edited by M. Dierkes, S. Edwards, and
    R. Coppock. Cambridge, Massachusetts: Oelgeschlager, Gunn,
    and Hain, Publishers, Inc., 1980, pp. 21-30.

Dierkes, M., Edwards, S., and Coppock, R., eds. Technological
    Risk:  Its Perception and Handling in the European
    Community. Cambridge, Massachusetts: Oelgeschlager, Gunn
    & Hain, Publishers, Inc., 1980.

Ditton, R.B. NEPA of 1969: Bibliography on Impact Assessment
    Methods and Legal Considerations (CPL Exchange Bibliog-
    raphy, 415). Monticello, Illinois: Council of Planning
    Librarians, 1973.

Doniger, D. "Federal Regulation of Vinyl Chloride:  A Short
    Course in the Law and Policy of Toxic Substances Control."
    Ecology Law Quarterly 7 (1978):497-677.

Doniger, D. The Law and Policy of Toxic Substances Control:
    A Case Study of Vinyl Chloride. Baltimore, Maryland:
    Johns Hopkins University Press, 1978.

Dorfman, N.S. "The Social Value of Saving a Life." Health:
    What is it Worth? Measures of Health Benefits. Edited by
    S.E. Mushkin and D.W. Dunlop. New York: Pergamon Press,
    1979, pp. 61-68.

Dorfman, R., and Dorfman, N.S., eds. Economics of the Environ-
    ment. New York: W.W. Norton & Company, 1972.

Douglas, J. "Toward Better Methods of Risk Assessment." EPRI
    Journal 7 (March 1982):22-28.

Douglas, M., and Wildavsky, A.  Can We Know the Risks We Face?
    An Essay on Technological and Environmental Dangers.  A
    Report to the National Science Foundation.  Washington,
    D.C.: Technology Assessment and Risk Analysis Program,
    U.S. National Science Foundation, 1981.

Douglas, M., and Wildavsky, A.  Risk and Culture.  Berkeley,
    California: University of California Press, 1982.

Dowie, J., and Lefrere, P.  Risk and Chance:  Selected Readings.
    Milton Keynes, United Kingdom: Open University Press, 1980.

Downs, A.  "Up and Down with Ecology--The 'Issue-Attention
    Cycle'."  The Public Interest 28 (Summer 1972):38-50.

Drake, A., Keeney, R.L., and Morse, P., eds.  Analysis of Public
    Systems.  Cambridge, Massachusetts: MIT Press, 1972.

Dubos, R.  Man Adapting.  New Haven, Connecticut: Yale Univer-
    sity Press, 1965.

Dunster, H.J.  "The Approach of a Regulatory Authority to the
    Concept of Risk."  International Atomic Energy Agency
    Bulletin 22 (October 1980):123.

Dunster, H.J.  "The Assessment of Risk--Its Value and Limita-
    tions."  Nuclear Engineering International 24 (August
    1979):23-25.

Dunster, H.J.  "The Risk Equations:  Virtue in Compromise."
    New Scientist 74 (May 26, 1977):454-456.

Dunster, H.J., and McLean, A.S.  "The Use of Risk Estimates in
    Setting and Using Basic Radiation Protection Standards."
    Health Physics (July 1970).

Ebbin, S., and Kasper, R.  Citizen Groups and the Nuclear Power
    Controversy.  Cambridge, Massachusetts: MIT Press, 1975.

Edwards, C.A.  "Safety and Health Regulation of the Transporta-
    tion Industry:  Can the Industry Serve Two Masters?"
    ICC Practice Journal 43 (1976):614.

Edwards, W.  "Behavioral Decision Theory."  Annual Review of
    Psychology 12 (1961):473-498.

Edwards, W., and Tversky, A.  Decision Making, Selected Readings.
    Middlesex: Penguin Books, 1967.

Eichholz, G.G.  "Cost-Benefit and Risk-Benefit Assessment for
    Nuclear Power Plants."  Nuclear Safety 17 (September-
    October 1976):525-539.

Einhorn, H.J.  "Decision Errors and Fallible Judgment:  Impli-
    cations for Social Policy."  Judgment and Decision in
    Public Policy Formation.  Edited by K.R. Hammond.  Boulder,
    Colorado: Westview Press, 1978.

Ellis, H.M., and Keeney, R.L.  "A Rational Approach for Govern-
    ment Decisions Concerning Air Pollution."  Analysis of
    Public Systems.  Edited by A.W. Drake, R.L. Keeney, and
    P.M. Morse.  Cambridge, Massachusetts: MIT Press, 1972.

Elster, J.  "Risk, Uncertainty, and Nuclear Power."  Social
    Science Information.  London and Beverly Hills: Sage
    Publications, 1979, pp. 371-400.

Engelman, P.A., and Renn, O.  "On the Methodology of Cost-
    Benefit Analysis and Risk Perception."  Directions in
    Energy Policy.  Edited by B. Kursunoglo and A. Perlmutter.
    Cambridge, Massachusetts: Ballinger Publishing Company,
    1979, pp. 357-364.

Environmental Law Institute.  Cost-Benefit Analysis and Environ-
    mental, Health and Safety Regulation:  An Overview of the
    Agencies and Legislation.  Washington, D.C.:  Environmental
    Law Institute, 1980.

Epstein, R.  "A Theory of Strict Liability."  Journal of Legal
    Studies 2 (1973):151.

Epstein, S.S.  "Cancer, Inflation, and the Failure to Regulate."
    Technology Review 82 (December-January 1980):42-53.

Epstein, S.S.  "Information Requirements for Determining the
    Benefit-Risk Spectrum."  Perspective on Benefit-Risk
    Decision Making.  Edited by the Committee on Public
    Engineering Policy.  Washington, D.C.: National Academy of
    Engineering, 1972.

Epstein, S.S.  The Politics of Cancer.  Garden City, New York:
    Anchor Books, 1979.

Epstein, S.S., and Grundy, R.D., eds.  The Legislation of
    Product Safety.  2 vols.  Cambridge, Massachusetts: MIT
    Press, 1974.

Erickson, L.E. <u>Issues and Experiences in Applying Benefit Cost Analysis to Health and Safety Standards</u>. Report to the U.S. Nuclear Regulatory Commission. Richland, Washington: Battelle Pacific Northwest Laboratories, September 1977.

Etzioni, A. "How Much is a Life Worth?" <u>Social Policy</u> 9 (March-April, 1979):4-8.

Executive Office of the President, Office of Science and Technology Policy. <u>Identification, Characterization and Control of Potential Human Carcinogens: A Framework for Federal Decision Making</u>. February 1, 1979.

Fagnani, F. "Role and Function of Risk Assessment." <u>Society, Technology and Risk Assessment</u>. Edited by J. Conrad. London: Academic Press, 1980, pp. 165-172.

Fairley, W. "Assessment of Catastrophic Risks." <u>Risk Analysis: An International Journal</u> 1:3 (1982):197-204.

Falk, H. "The Effect of Personal Characteristics on Attitudes Toward Risk." <u>Journal of Risk and Insurance</u> 43 (June 1976):215-241.

Farmer, F.R. "Experience in the Reduction of Risk." <u>Proceedings of the Symposium on Major Loss Prevention in the Process Industries</u>. London: The Institution of Chemical Engineers, 1971.

Farmer, F.R. "Methodology of Energy Risk Comparisons." <u>International Atomic Energy Agency Bulletin</u> 22 (October 1980): 120.

Farmer, F.R. "Some Considerations of Major Non-Nuclear Hazards." <u>International Atomic Energy Agency Bulletin</u> 20 (December 1978): 13-20.

Faron, R.S. "Risk Assessment: New Legislation Clouds the Picture." <u>Environmental Professional</u> 3:3/4 (1981): 213-216.

Feinberg, J. <u>Doing and Deserving: Essays in the Theory of Responsibility</u>. Princeton, New Jersey: Princeton University Press, 1974.

Feingold, E. "The Great Cranberry Crisis." <u>Government Regulation of Business: A Casebook</u>. Edited by E.A. Bock. Englewood Cliffs, New Jersey: Prentice-Hall, 1965.

Feldstein, M.S.   "The Social Time Preference Discount Rate in
     Cost Benefit Analysis."  Economic Journal 74 (June
     1964):360.

Ferguson, A., ed.  The Benefits of Health and Safety Regulation.
     Cambridge, Massachusetts: Ballinger Press, 1981.

Ferkiss, V.C.  Technological Man:  The Myth and the Reality.
     New York: New American Library, 1969.

Ferreira, J., and Slesin, L.  Observations on the Social Impact
     of Large Accidents (Technical Report No. 122).   Cambridge,
     Massachusetts: Operations Research Center, Massachusetts
     Institute of Technology, October 1976.

Field, R.I.  "Patterns in the Laws on Health Risks."  Policy
     Analysis and Management 1:2 (Winter 1982):257-260.

Fiksel, J., and Rosenfield, D.B.  "Probabilistic Models for
     Risk Assessment."  Risk Analysis 2:1 (March 1982):1-8.

Fischer, D.W., and Von Winterfeldt, D.  "Setting Standards
     Against Oil Pollution in the North Sea."  Journal of
     Environmental Management 7 (1978):177-199.

Fischer, G.W.  "Willingness to Pay for Probabilistic Improve-
     ments in Functional Health Status:  A Psychological Per-
     spective."  Health:  What Is it Worth?  Measures of Health
     Benefits.  Edited by S.J. Mushkin and D.W. Dunlop.  New
     York: Pergamon Press, 1979, pp. 167-200.

Fischhoff, B.  "Behavioural Aspects of Cost-Benefit Analysis."
     Energy Risk Management.  Edited by G. Goodman and W. Rowe.
     London: Academic Press, 1979.

Fischhoff, B.  "Cost Benefit Analysis and the Art of Motorcycle
     Maintenance."  Policy Science 8 (June 1977):177-202.

Fischhoff, B.  "Hindsight/Foresight:  The Effect of Outcome
     Knowledge on Judgment Under Uncertainty."  Journal of
     Experimental Psychology:  Human Perception and Performance
     1 (1975):288-299.

Fischhoff, B.  "Informed Consent in Societal Risk--Benefit
     Decisions."  Technological Forecasting and Social Change 13
     (May 1979):347-357.

Fischhoff, B., Goitein, B. and Shapira, Z. "The Experienced Utility of Expected Utility Approaches." <u>Expectancy, Incentive and Action</u>. Edited by N. Feather. Hillsdale, New Jersey: Erlbaum, 1982.

Fischhoff, B., Hohenemser, C., Kasperson, R., and Kates, R. "Handling Hazards." <u>Environment</u> 20 (September 1978):16-37.

Fischhoff, B., Lichtenstein, S., Slovic, P., Derby, S., and Keeney, R. <u>Acceptable Risk</u>. New York: Cambridge University Press, 1981.

Fischhoff, B., Lichtenstein, S., Slovic, P., Keeney, R., and Derby, S. <u>Approaches to Acceptable Risk: A Critical Guide</u> (NUREG/CR-1614). Oak Ridge, Tennessee: Oak Ridge National Laboratories, 1980.

Fischhoff, B., Slovic, P., and Lichtenstein, S. "Fault Trees: Sensitivity of Estimated Failure Probabilities to Problem Representation." <u>Journal of Experimental Psychology: Human Perception and Performance</u> 4 (1978):330-344.

Fischhoff, B., Slovic, P., and Lichtenstein, S. "Labile Values: A Challenge for Risk Assessment." <u>Society, Technology and Risk Assessment</u>. Edited by J. Conrad. London: Academic Press, 1980, pp. 57-66.

Fischhoff, B., Slovic, P. and Lichtenstein, S. "Lay Foibles and Expert Fables in Judgments about Risk." <u>Progress in Resource Management and Environmental Planning, Vol. 3</u>. Edited by T. O'Riordan and R.K. Turner. Chichester, England: Wiley, 1981.

Fischhoff, B., Slovic, P. and Lichtenstein, S. "'The Public vs. The Experts: Perceived vs. Actual Disagreements about the Risks of Nuclear Power." <u>Risk Analysis</u> (In press.)

Fischhoff, B., Slovic P., and Lichtenstein, S. "Weighing the Risks." <u>Environment</u> 21 (May 1979):17-38.

Fischhoff, B., Slovic, P., Lichtenstein, S., Combs, B., and Read, S. "How Safe is Safe Enough? A Psychometric Study of Attitudes Toward Technological Risks and Benefits." <u>Policy Sciences</u> 8 (1978):127.

Fischhoff, B., and Whipple, C. "Assessing Health Risks Associated with Ambient Air Quality Standards." <u>Four Conceptual Approaches to Health Risk Assessment</u>. Washington, D.C.: U.S. Environmental Protection Agency, August 1980.

Fischhoff, B., and Whipple, C. "Risk Assessment: Evaluating Error in Subjective Estimates." Environmental Professional 3:3/4 (1981): 277-292.

Fishbein, M. Readings in Attitude Theory and Measurement. New York: John Wiley & Sons, 1967.

Fishbein, M., and Ajezen, I. Belief, Attitude, Intention and Behavior: An Introduction to Theory and Research. Reading, Massachusetts: Addison-Wesley, 1975.

Fishburn, P.C. Utility Theory for Decision Making. New York: John Wiley & Sons, 1970.

Flanders, J.P., and Thistlewaite, D.L. "Effects of Familiarization and Group Discussion Upon Risk Taking." Journal of Personality and Social Psychology 5 (1967):91-97.

Floriman, S. Blaming Technology: The Irrational Search for Scapegoats. New York: St. Martins, 1981.

Foreman, H., ed. Nuclear Power and the Public. Minneapolis: University of Minnesota Press, 1970.

Frederichs, G. "Risk Research--A 'Problem Community' and Its Role in Society." Society, Technology and Risk Assessment. Edited by J. Conrad. London: Academic Press, 1980, pp. 123-130.

Freeman, A. Industry Response to Health Risk (Rpt. 811). New York: The Conference Board, 1981.

Freeman, A.M., III. The Benefits of Environmental Improvement: Theory and Practice. Baltimore: John Hopkins University Press, 1979.

Freeman, A.M., III, and Havemen, R.H. "Residuals Charges for Pollution Control: A Policy Evaluation." Science 177 (July 28, 1972).

Friedman, M., and Savage, L.J. "The Utility Analysis of Choices Involving Risks." Journal of Political Economy 56 (1948):279-304.

Fuchs, V.R. Who Shall Live? New York: Basic Books, 1974.

Gardener, G.T., Tiemann, A.R., Gould, L.C., DeLuca, D.R., Doob, L.W., and Stolwijk, J.A. "Risk and Benefit Perceptions, Acceptability Judgments, and Self Reported Actions Toward Nuclear Power." *Journal of Social Psychology* 116 (1982): 179-197.

Gardiner, D.J., and Edwards, W. "Public Values: Multiattribute-Utility Measurement for Social Decision Making." *Human Judgment and Decision Processes*. Edited by M.F. Kaplan and S. Schwartz. New York: Academic Press, 1975.

Garfield, E. "Risk Analysis, Part 1. How We Rate the Risks of New Technologies." *Current Contents* 33 (August 16, 1982): 5-13.

Gelpe, M.R., and Tarlock, A.D. "The Use of Scientific Information in Environmental Decisionmaking." *Southern California Law Review* 48 (1974):371-427.

Gibson, S.B. "The Use of Quantitative Risk Criteria in Hazard Analysis." *Risk Benefit Methodology and Application* (UCLA-ENG-7598). Edited by D. Okrent. Los Angeles: University of California, December 1975.

Goetz, A.A. "Health Risk Appraisal: The Estimation of Risk." *Public Health Reports* 95 (1980):119-126.

Golay, M.W. "How Prometheus Came to Be Bound: Nuclear Regulation in America." *Technology Review* 83 (June-July 1980): 29-39.

Goldberg, L. "The Risks of Predicting Risks." *The Future of Risk*. New York: Risk Studies Foundation, 1978.

Goldman, M., ed. *Controlling Pollution: The Economics of a Cleaner America*. Englewood Cliffs, New Jersey: Prentice-Hall, 1967.

Goodin, R.E. "No Moral Nukes." *Ethics* 90 (April 1980):417-449.

Goodman, G.T., and Rowe, W.D., eds. *Energy Risk Management*. New York: Academic Press, 1979.

Goodwin, R. "Uncertainty as an Excuse for Cheating Our Children: The Case of Nuclear Wastes." *Policy Sciences* 10 (1978): 25-43.

Gori, G.B. "The Regulation of Carcinogenic Hazards." *Science* 208 (April 18, 1980):256-261.

Gould, L., and Walker, C.A., eds. Too Hot to Handle:  Public
    Policy Issues in Nuclear Waste Management. New Haven,
    Connecticut: Yale University Press, 1981.

Graham, J. "Some Explanations for Disparities in Lifesaving
    Investments." Policy Studies Review 1 (May 1982):692-704.

Graham, J., and Shakow, D. "Risk and Reward:  Hazard Pay for
    Workers." Environment 23:8 (1981):14-20, 44-45.

Graham, J., Shakow, D., and Cyr, C. "Are Workers Compensated
    for Hazardous Work?" Environment (1982. In press).

Graham, J.D., and Vaupel, J.W. "Value of a Life: What Differ-
    ence Does It Make?" Risk Analysis: 1:1 (March 1981): 89-95.

Green, A. "Comments on Legal Mechanisms." Perspectives on
    Benefit-Risk Decision Making. Edited by the Committee on
    Public Engineering Policy. Washington, D.C.: National
    Academy of Engineering, 1972.

Green, C.H. "Revealed Preference Theory:  Assumptions and Pre-
    sumptions." Society, Technology and Risk Assessment.
    Edited by J. Conrad. London: Academic Press, 1980, pp.
    49-56.

Green, C.H. "Risk:  Attitudes and Beliefs." Behaviour in Fires.
    Edited by D.V. Canter. Chichester: Wiley, 1980.

Green, C.H., and Brown, R.A. "Counting Lives." Journal of
    Occupational Accidents 2 (1978):55-70.

Green, C.H., and Brown, R.A. Life Safety:  What Is It and How
    Much Is It Worth? (CP52/78).  Borehamwood, Hertfordshire,
    U.K.: Department of the Environment, Building Research
    Establishment, 1978.

Green, C.H., and Brown, R.A. Metrics for Societal Safety (Note
    N 144/78).  Borehamwood, Hertfordshire, U.K.: Department
    of the Environment, Building Research Establishment, 1978.

Green, C.H., and Brown, R.A. Perceived Safety as an Indiffer-
    ence Function (Note N 156/78).  Borehamwood, Hertfordshire,
    U.K.: Department of the Environment, Building Research
    Establishment, 1978.

Green, C.H., and Brown, R.A.  The Perception of, and Attitudes
    Towards, Risk:  Final Report:  Vol. 2. Measure of Safety
    (FRO/028/68).  Dundee, Scotland: School of Architecture,
    Duncan of Jordanstone College of Art, University of Dundee,
    April 1977.

Green, C.H., and Brown, R.A.  The Perception of, and Attitudes
    Towards, Risk, Final Report:  Vol. 3. Stability of Percep-
    tion under Time and Data (FRO/028/68).  Dundee, Scotland:
    School of Architecture, Duncan of Jordanstone College of
    Art, University of Dundee, April 1977.

Green, C.H.; and Brown, R.A.  The Perception of, and Attitudes
    Towards, Risk, Final Report:  Vol. 4. Initial Experiments
    on Determining Satisfaction With Safety Levels (FRO/028/68).
    Dundee, Scotland: School of Architecture, Duncan of
    Jordanstone College of Art, University of Dundee, April
    1977.

Green, C.H., and Brown, R.A.  Problems of Valuing Safety  (Note
    N 70/78).  Borehamwood, Hertfordshire, U.K.: Department of
    the Environment, Building Research Establishment, 1978.

Green, H.P.  "Cost-Risk-Benefit Assessment and the Law: Intro-
    duction and Perspective."  George Washington Law Review 45
    (August 1977): 901-910.

Green, H. P.  "The New Technological Area: A View from the Law."
    Bulletin of the Atomic Scientists 23 (November 1967):12-18.

Green, H.P.  "Nuclear Power, Risk, Liability, and Indemnity."
    Michigan Law Review 71 (1973):470-510.

Green, H.P.  "Nuclear Safety and the Public Interest."  Nuclear
    News 15 (1972):75.

Green, H.P.  "The Risk-Benefit Calculus in Safety Determinations."
    George Washington Law Review 43 (March 1975):791-808.

Green, H.P.  "The Role of Law in Determining Acceptability of
    Risk."  Societal Risk Assessment:  How Safe is Safe Enough?
    Edited by R. Schwing and W. Albers.  New York: Plenum
    Press, 1980, pp. 225-267.

Greenberg, P.F.  "The Thrill Seekers."  Human Behavior 6 (April
    1977): 17-21.

Greene, M.  "The Government as an Insurer."  Journal of Risk
    and Insurance 43:3 (September 1976).

Greene, M. "A Review and Evaluation of Selected Government Programs to Handle Risk." The Annals of the American Academy of Political and Social Science (May 1979):129-144.

Greene, M. Risk and Insurance. Cincinnati, Ohio: Southwestern Publishing, 1977.

Greenwood, D., Kingsbury, G., and Cleland, J. A Handbook of Key Federal Regulations and Criteria for Multimedia Environmental Control (EPA-600/7-79-175). Washington, D.C.: U.S. Environmental Protection Agency, 1979.

Greer-Wooten, B. "Context, Concept and Consequence in Risk Assessment Research: A Comparative Overview of North American and European Approaches in the Social Sciences." Society, Technology and Risk Assessment. Edited by J. Conrad. London: Academic Press, 1980, pp. 67-104.

Griesmeyer, J., Simpson, M., and Okrent, D. The Use of Risk Aversion in Risk Acceptance Criteria (UCLA-ENG-7970). Los Angeles: University of California, 1980.

Grieves, F.L. International Law, Organization, and the Environment: A Bibliography and Research Guide. Tucson, Arizona: University of Arizona Press, 1974.

Griffiths, R.F., ed. Dealing with Risk: The Planning, Management and Acceptability of Technological Risk. New York: Halsted Press, 1981.

Griffiths, R.F. "Problems in the Use of Risk Criteria." Dealing with Risk: The Planning, Management and Acceptability of Technological Risk. Edited by R.F. Griffiths. New York: Halsted Press, 1981.

Guzzardi, W., Jr. "The Mindless Pursuit of Safety." Fortune 99 (April 9, 1979):54-64.

Hadden, S., and Hazelton, J. "Public Policy Toward Risk." Policy Studies Journal 9:1 (1980):109-116.

Haddon, W.H., Jr., Suchman, E.A., and Klein, D. Accident Research: Methods and Approaches. New York: Harper and Row, 1964.

Hafele, W. "Hypotheticality and the New Challenges: The Pathfinder Role of Nuclear Energy." Minerva 12 (1974):302-322.

Haimes, Y., ed. Risk Benefit Analysis in Water Resources Plan- ning and Management. New York: Plenum Press, 1981.

Halbett, R. "Why Do We Still Dice With Death?" New Society 53 (September 1980):502-504.

Hall, W.K. "Why Risk Analysis Isn't Working." Long Range Planning (December 1975):25-29.

Halliwell, L., Oahes, F., and Slater, D.H. "Hazards, Risks, and Social Responsibilities." Consulting Engineer (June 1979):75-80.

Hamilton, L.D. "Comparative Risks from Different Energy Systems: Evolution of the Methods of Studies." International Atomic Energy Agency Bulletin 22 (October 1980):35.

Hamilton, R. The Role of Nongovernmental Standards in the Devel- opment of Mandatory Federal Standards Relating to Safety or Health. Washington, D.C.: Administrative Conference of the United States, 1978.

Hammer, W. Product Safety Management and Engineering. Englewood Cliffs, New Jersey: Prentice-Hall, 1980.

Hammond, E. C., and Selikoff, I.J., eds. "Public Control of Environmental Health Hazards." Annals of the New York Academy of Sciences 329 (October 26, 1979):1-405.

Hammond, J. "Risk Spreading Through Underwriting and the Insurance Mechanism." Societal Risk Assessment: How Safe is Safe Enough? Edited by R. Schwing and W. Albers. New York: Plenum Press, 1980, pp. 147-176.

Hammond, K.R., and Adleman, L. "Science, Values and Human Judgment." Science 194 (October 22, 1976):389-396.

Handler, P. "A Rebuttal: The Need for a Sufficient Scientific Base for Government Regulation." George Washington Law Review 43 (March 1975):808-813.

Handler, P. "Some Comments on Risk Assessment." 1979--Current Issues and Studies. Washington, D.C.: National Research Council, National Academy of Sciences, 1979.

Hanke, S.H., and Walker, R.A. "Benefit-Cost Analysis Recon- sidered: An Evaluation of the Mid-State Project." Benefit-Cost and Policy Analysis 1974. Chicago: Aldine Publishing Company, 1975.

Hapgood, F. "Risk-Benefit Analysis: Putting a Price on Life."
    Atlantic 243 (January 1979):33-38.

Harberger, A.C., et al., eds. Benefit-Cost Analysis, 1971.
    Chicago: Aldine, 1972.

Hardin, C.M. "The Effects of Over-Regulation." Food, Drug, and
    Cosmetic Law Journal 34 (January 1979):50-57.

Harris, Louis and Associates, Inc. Harris Perspective 1979:
    A Survey of the Public and Environmental Activists on the
    Environment (59). New York: Louis Harris and Associates,
    1979.

Harris, Louis and Associates, Inc. Risk in a Complex Society:
    A Marsh and McLennan Public Opinion Survey. New York:
    Marsh and McLennan, 1980.

Harris, Louis and Associates, Inc. A Second Survey of Public
    and Leadership Attitudes Toward Nuclear Power Development
    in the United States. New York: EBASCO, 1976.

Harris, R.J. "Why the Pinto Jury Felt Ford Deserved $125 Million
    Penalty." Wall Street Journal, February 14, 1978.

Harrison, D.J., Jr. Who Pays for Clean Air? The Cost and Bene-
    fit Distribution of Federal Auto Emission Standards.
    Cambridge, Massachusetts: Ballinger Publishing Company,
    1975.

Harriss, R., Hohenemser, C., and Kates, R. "Our Hazardous
    Environment." Environment 20 (September 1978)6-15.

Hartley, H.O., Manton, K.G., and Woodbury, M.A. "Estimation of
    Risk of Adverse Health Effects Associated with Air Quality
    Standards for Pollutants." Four Conceptual Approaches to
    Health Risk Assessment. Prepared for U.S. Environmental
    Protection Agency, August 1980.

Haveman, R.H. "Common Property, Congestion and Environmental
    Pollution." Quarterly Journal of Economics 87 (May 1973):
    278-287.

Haveman, R.H. "Economic Evaluation of Long Run Uncertainties."
    Futures 9 (1977):383-403.

Havemen, R.H., et al., eds. Benefit-Cost and Policy Analysis,
    1973. Chicago: Aldine Publishing Company, 1974.

Havender, W. "Assessing and Controlling Risks." Social Regulation: Strategies for Reform. Edited by E. Bardach and R. Kagan. Contemporary Institute, 1982.

Head, G. The Risk Management Process. New York: Risk and Insurance Management Society, Inc., 1978.

Health and Safety Executive. Canvey: An Investigation of Potential Hazards from Operations in the Canvey Island/ Thurrock Area. London: H.M.S.O., 1978.

Healy, J.W. "What is Hazardous? What is Safe?" Environmental Health Perspectives 27 (1978):317-321.

Henderson, G.F. "Insurance Underwriting for Environmental Risk." Environmental Professional 3:3/4 (1981):209-212.

Hensley, G. "Safety and Reliability Assessment for Major Hazard Plants." Health and Safety in the Workplace 2:5 (January 1980):62-67.

Herbert, J.H., Swanson, C., and Reddy, P. "A Risky Business: Energy Production and the Inhaber Report." Environment 20 (1979):28-33.

Herfindahl, O.C., and Kneese, A.V. Quality of the Environment: An Economic Approach to Some Problems in Using Land, Water and Air. Baltimore: Johns Hopkins University Press, 1965.

Hershey, J.C., and Schoemaker, P.J.H. "Risk Taking and Problem Context in the Domain of Losses: An Expected Utility Analysis." Journal of Risk and Insurance 57 (1980):11-13.

Hetman, F. Society and the Assessment of Technology. Paris: Organization for Economic Cooperation and Development, 1973.

Heuser, F.W., and Homke, P. "Reliability Analysis and Its Application for Safety Assessment of Nuclear Plants." Risk Benefit Methodology and Application (UCLA-ENG-7598). Edited by D. Okrent. Los Angeles: University of California, December 1975.

Hickman, L., and Al-Hibri, eds. Technology and Human Affairs. St. Louis, Missouri: Mosby, 1981.

Hirshleifer, J. "The Economic Approach to Risk-Benefit Analysis." Risk-Benefit Methodology and Application (UCLA-ENG-7578). Edited by D. Okrent. Los Angeles: University of California, November 1974.

Hirshleifer, J., Berstrom, T., and Rappaport, E. Applying Cost-Benefit Concepts to Projects Which Alter Human Mortality (UCLA-ENG-7578). Los Angeles: University of California, November 1974.

Hirshleifer, J., and Shapiro, D.L. "The Treatment of Risk and Uncertainty." Public Expenditures and Policy Analysis. Edited by R. Haveman and J. Margolis. Chicago: Markham Publishing Company, 1970.

Hoel, D. "A Representation of Mortality Data by Competing Risks." Biometrics 28 (1972):475-488.

Hoerger, F.O.. et al. "Risk Benefit and Cost Effectiveness Methodologies in Setting Priorities and Making Decisions Under the TSCA." Toxic Substances Journal 1 (Summer 1979).

Hoffman, M.E. "The Consumer Product Safety Commission: In Search of a Regulatory Pattern." Columbia Journal of Law and Social Problems 12 (1976):393-450.

Hoffman, S.D. Unreasonable Risk of Injury Revisited. Chicago: Underwriters Laboratories, 1976.

Hohenemser, C. "The Failsafe Risk." Environment 17:1 (1975): 6-10.

Hohenemser, C., and Kasperson, J., eds. Risk in the Technological Society. Boulder, Colorado: Westview Press, 1981.

Holdren, J.P. "Energy Hazards: What to Measure, What to Compare." Technology Review 85 (April 1982).

Holdren, J.P. "The Nuclear Controversy and the Limitations of Decision Making by Experts." Bulletin of the Atomic Scientists 32 (1976): 20-22.

Holdren, J.P., Anderson, K., Gleick, P., Mintzer, I., Morris, G. and Smith, K. Risk of Renewable Energy Sources: A Critique of the Inhaber Report (ERG 79-3). Berkeley: Energy and Resources Group, University of California, and Honolulu: Resource Systems Institute, East-West Center, June 1979.

Holdren, J.P., Smith, K., and Morris, G. "Energy: Calculating the Risks." Science 204 (May 11, 1979):564-568.

Holifield, C. "A Political Point of View: Risk-Benefit Methodology and Application." Risk-Benefit Methodology and Application (UCLA-ENG-7598). Edited by D. Okrent. Los Angeles: University of California, December 1975.

Hood, S., ed. Human Values and Economic Policy. New York: New York University Press, 1967.

Horden, J. Accident Risks in Norway: How Do We Perceive and Handle Risks? Oslo: Risk Research Committee, Royal Norwegian Council for Scientific and Industrial Research, 1980.

Howard, N. "What Price Safety? The 'Zero-Risk' Debate." Dun's Review 114 (September 1979):49-51, 53, 57.

Howard, R. "An Assessment of Decision Analysis." Special Issue on Decision Analysis, Operations Research 28 (January-February 1980):1.

Howard, R. "On Making Life and Death Decisions." Societal Risk Assessment: How Safe is Safe Enough? Edited by R. Schwing and W. Albers. New York: Plenum Press, 1980, pp. 89-106.

Howard, R., Matheson, J.E., and North, D.W. "The Decision to Seed Hurricanes." Science 176 (1972):1191-1202.

Hutt, P.B. "Public Policy Issues in Regulating Carcinogens in Food." Food, Drug, and Cosmetic Law Journal 33 (October 1978):541-557.

Hutt, P.B. "Unresolved Issues in the Conflict Between Individual Freedom and Government Control of Food Safety." Food, Drug, and Cosmetic Law Journal 33 (October 1978):558-589.

Ingeles, O.G. "Safety in Civil Engineering." Reliability Engineering 1 (1980):15-27.

Ingram, H.M., and Mann, D.E. "Environmental Policy: From Innovation to Implementation." Nationalizing Government: Public Policies in America. Edited by T.J. Lowi and A. Stone. Beverly Hills, California: Sage Publications, 1979.

Inhaber, H. Energy Risk Management. New York: Gordon and Breach, 1982.

Inhaber, H. "Modelling the Risks of Energy Systems." Simula-
    tion, Modelling and Decision in Energy Systems. Edited by
    M.B. Carvey and M.H. Hamza. Anaheim, California: Acta
    Press, 1978, pp.56-69.

Inhaber, H. "Risk and Consequences in Energy Production."
    Interdisciplinary Science Reviews 5:4 (1980):304-311.

Inhaber, H. "Risk from Conventional and Nonconventional Energy
    Sources." Science 203 (February 23, 1979):718-723.

Inhaber, H. Risk of Energy Production (Report AECB-rev. 3).
    Ottawa: Atomic Energy Control Board, 1979.

Inhaber, H. "The Risk of Producing Energy." Philosophical
    Transactions of the Royal Society, A376 (1981):121-132.

Interagency Regulatory Liaison Group. Scientific Basis for
    Identification of Potential Carcinogens and Estimation of
    Risk. Washington, D.C.: Interagency Regulatory Liaison
    Group, February 1979.

Irwin, G.A. "Discussion Paper on A. Mazur: Societal and Scien-
    tific Causes of the Historical Development of Risk Assess-
    ment." Society, Technology and Risk Assessment. Edited
    by J. Conrad. London: Academic Press, 1980, pp. 159-162.

Irwin, J., and Stoner, G.D. "Facet of Biohazard Control
    Program--Agent Registration, Risk Assessment and Computer-
    ization of Data." American Journal of Public Health 66
    (1976):372-374.

Jackel, H. "Technological Risk: A Problem of Growing Concern
    to Government." Technological Risk: Its Perception and
    Handling in the European Community. Edited by M. Dierkes,
    S. Edwards, and R. Coppack. Cambridge, Massachusetts:
    Oelgeschlager, Gunn, and Hain, Publishers, Inc., 1980, pp.
    7-11.

Jackson, J., and Kunreuther, H. Low Probability Events and
    Determining Acceptable Risk: The Case of Nuclear
    Regulation (WP-81-007). Laxenburg, Austria: International
    Institute for Applied Systems Analysis, 1981.

Jacobs, P. "Analyzing Environmental Health Hazards." Environ-
    mental Science and Technology 13 (May 1979):526-529.

Jacoby, H.D., et al. Clearing the Air: Federal Policy on Auto-
    motive Emissions Control. Cambridge, Massachusetts:
    Ballinger Publishing Company, 1973.

Jarret, H., ed. Environmental Quality in a Growing Economy. Baltimore: Johns Hopkins University Press, 1966.

Jasonoff, S., and Nelkin, D. "Science, Technology and the Limits of Judicial Competence." Science 214 (December 11, 1981):1211-1215.

Jennergren, L.P., and Keeney, R.L. "Risk Assessment." Handbook of Applied Systems Analysis. Laxenburg, Austria: International Institute of Applied Systems Analysis (In press.)

Johnson, P. "The Perils of Risk Avoidance." Regulation 4 (May-June 1980):15-19.

Johnson, R. "The Characteristics of Risk Assessment Research." Society, Technology and Risk Assessment. Edited by J. Conrad. London: Academic Press, 1980, pp. 105-122.

Johnson, R., and Gummett, P. Directing Technology: Policies for Promotion and Control. London: Redwood Burn Ltd., 1979.

Johnson, W.G. "MORT: The Management Oversight and Risk Tree." Journal of Safety Research 7 (March 1975):4-15.

Jones-Lee, M.W. "Maximum Acceptable Physical Risk and a New Measure of Financial Risk-Aversion." Economic Journal 90 (September 1980): 550-568.

Jones-Lee, M.W. "The Value of Changes in the Probability of Death or Injury." Journal of Political Economy 99 (July/August 1974): 835-849.

Jones-Lee, M.W. The Value of Life-An Economic Analysis. Chicago: University of Chicago Press, 1976.

Jones-Lee, M.W., ed. Valuing of Life and Safety. Amsterdam: North Holland, 1982.

Jordan, R.E., III. "Alternatives Under NEPA: Toward an Accommodation." Ecology Law Quarterly 3 (Fall 1973):705-757.

Joskow, P.L. "Approving Nuclear Power Plants: Scientific Decisionmaking or Administrative Charade?" Bell Journal of Economics and Management Science 5 (Spring 1974).

Junger, P.D. "The Inapplicability of Cost Benefit Analysis to Environmental Policies." Ekistics 46 (1979):61, 184-194.

Kahan, J.P. How Psychologists Talk About Risk (P-6403). Santa Monica, California: Rand Corporation, October 1979.

Kahan, J.S. "Reporting Substantial Risks Under Section 8(e) of the Toxic Substances Control Act." Boston College Law Review 19 (July 1978):859-879.

Kahneman, D., Slovic, P., and Tversky, A. Judgment Under Uncertainty: Heuristics and Biases. New York: Cambridge University Press, 1981.

Kahneman, D., and Tversky, A. "Intuitive Predictions: Biases and Corrective Procedures." TIMS Studies in Management Science 12 (1979):313-327.

Kahneman, D., and Tversky, A. "On the Psychology of Prediction." Psychological Review 80 (July 1973):237-251.

Kahneman, D., and Tversky, A. "Prospect Theory: An Analysis of Decision Under Risk." Econometrica 47:2 (March 1979): 263-291.

Kalelkar, A., and Drake, E. "Handle With Care: Using Risk Analysis for Hazardous Material Facilities." Risk Management (March 1981):44-50.

Kantrowitz, A. "Controlling Technology Democratically." American Scientist 63 (September/October 1975):505-509.

Kantrowitz, A. "The Science Court Experiment." Trial 13 (1977):48.

Kaplan, S., and Garrick, B.J. "On the Quantitative Definition of Risk." Risk Analysis: An International Journal 1 (March 1981):11-28.

Kasper, R.G. "Cost-Benefit Analysis in Environmental Decision-making." George Washington Law Review 45 (August 1977): 1013-1024.

Kasper, R.G. "Perceived Risk: Implications for Policy." Impacts and Risks of Energy Strategies: Their Analysis and Role in Management. London: Academic Press, 1979.

Kasper, R.G. "Perceptions of Risk and Their Effects on Decision-Making." Societal Risk Assessment: How Safe is Safe Enough? Edited by R. Schwing and W. Albers. New York: Plenum Press, 1980, pp. 71-80.

Kasperson, R.E., ed. Equity Issues in Radioactive Waste Management. Cambridge, Massachusetts: Oelgeschlager, Gunn and Hain (In press).

Kasperson, R.E. "Societal Management of Technological Hazards." Managing Technological Hazard: Research Needs and Opportunities. Edited by R. Kates. Boulder: Institute of Behavioral Science, University of Colorado, 1977, pp. 49-80.

Kasperson, R.E., Berk, G., Pijawka, D., Sharaf, A., and Wood, J. "Public Opposition to Nuclear Energy: Retrospect and Prospect." Science, Technology and Human Values 5 (Spring 1980):11-23.

Kastenberg, W., McKone, T., and Okrent, D. On Risk Assessment in the Absence of Complete Data (UCLA-ENG-7677). Los Angeles: University of California, 1976.

Kates, R.W. "Assessing the Assessors: The Art and the Ideology of Risk Assessment." Ambio 6 (1977):247-252.

Kates, R.W. Risk Assessment of Environmental Hazard. New York: John Wiley & Sons, 1978.

Kates, R.W., ed. Managing Technological Hazard: Research Needs and Opportunities. Boulder: Institute of Behavioral Science, University of Colorado, 1977.

Kates, R.W., Hohenemser, C., and Kasperson, J.X., eds. Technology as Hazard. Cambridge, Massachusetts: Oelgeschlager, Gunn & Hain Publishers, Inc., 1982.

Katz, M. "The Function of Tort Liability in Technology Assessment." University of Cincinnati Law Review 38 (1969): 587-592.

Katz, M. "Legal Mechanisms." Perspectives on Benefit-Risk Decision Making. Washington, D.C.: National Academy of Engineering, 1972.

Keeler, E., Spence, M., and Zeckhauser, R. "The Optimal Control of Pollution." Journal of Economic Theory 4 (February 1972):19-24.

Keeler, E., and Zeckhauser, R.  Another Type of Risk Aversion (RM-5996-PR).  Santa Monica, California: Rand Corporation, May 1969.

Keeney, R.L.  Energy Policy and Value Tradeoffs (RM-75-76). Laxenburg, Austria: International Institute for Applied Systems Analysis, December 1975.

Keeney, R.L.  "Equity and Public Risk."  Operations Research 28:3 (May-June 1980):527-534.

Keeney, R.L.  Siting Energy Facilities.  New York: Academic Press, 1980.

Keeney, R.L.  "Utility Functions for Equity and Public Risk." Management Science 26:4 (April 1980).

Keeney, R.L., and Kirkwood, C.W.  "Group Decision Making Using Cardinal Social Welfare Functions."  Management Science 22 (1975):  430-437.

Keeney, R.L., and Nair, K.  Evaluating Potential Nuclear Power Plant Sites in the Pacific Northwest Using Decision Analysis (PP-76-81).  Laxenburg, Austria: International Institute for Applied Systems Analysis, January 1976.

Keeney, R.L., and Raiffa, H.  "A Critique of Formal Analysis in Public Decision Making."  Analysis of Public Systems. Edited by A.W. Drake, R.L. Keeney, and P.M. Morse. Cambridge, Massachusetts: MIT Press, 1972.

Keeney, R.L., and Raiffa, H.  Decisions with Multiple Objectives: Preferences and Value Tradeoffs.  New York: John Wiley & Sons, 1976.

Keeney, R.L., and Robilliard, G.A.  Assessing and Evaluating Environmental Impacts at Proposed Nuclear Power Plant Sites (PP-76-3).  Laxenburg, Austria: International Institute for Applied Systems Analysis, February 1976.

Kelman, S.  "Cost-Benefit Analysis:  An Ethical Critique." Regulation 5 (January-February 1980):33-40.

Kelman, S.  "Cost-Benefit and Environmental, Safety, and Health Regulation:  Ethical and Philosophical Considerations." Cost-Benefit Analysis and Environmental Regulations. Edited by D. Swartzman, R. Liroff, and K. Croke.  Washington, D.C.: The Conservation Foundation, 1982.

Kemeny, J.G.  "Saving American Democracy:  The Lessons of Three
    Mile Island."  Technology Review 82 (June/July 1980):65-75.

Kidner, R., and Richards, K.  "Compensation to Dependents of
    Accident Victims."  Economic Journal 84 (March 1974).

Kirschman, J.C.  "Toxicology--The Exact Use of an Inexact
    Science."  Food, Drug, and Cosmetic Law Journal 31 (1976):
    455-461.

Kirschten, D.  "Can Government Place a Value on Saving a Human
    Life?"  National Journal 11 (February 17, 1979):252-255.

Klausner, S., ed.  Why Man Takes Chances:  Studies in Stress-
    Seeking.  Garden City, New York: Doubleday, 1968.

Kleindorfer, P., and Kunreuther, H.  "Descriptive and Prescrip-
    tive Aspects of Health and Safety Regulation."  The
    Benefits of Health and Safety Regulation.  Edited by A.
    Ferguson.  Cambridge, Massachusetts: Ballinger, 1981.

Kletz, T.  "Benefits and Risks: Their Assessment in Relation to
    Human Needs."  Dealing with Risk.  Edited by R.F. Griffiths.
    New York: Halsted Press, 1981.

Kletz, T.  "The Risk Equations--What Risks Should We Run?"  New
    Scientist 74:1051 (May 12, 1977).

Kneese, A.V.  "Benefit-Cost Analysis and Unscheduled Events in
    the Nuclear Fuel Cycle."  Resources 44 (September 1973):
    1-8.

Kneese, A.V., et al.  Managing the Environment:  International
    Economic Cooperation for Pollution Control.  New York:
    Praeger Publishers, 1972.

Kneese, A.V., and Bower, B.T.  Environmental Quality and
    Residuals Management.  Baltimore: Johns Hopkins University
    Press, 1979.

Kneese, A., Schulze, W.D., Dorfman, R., Nordhaus, W., Muskin,
    J., and Sorrentino, J.  "Environmental Problems."
    American Economic Review 67 (February 1977): 326-350.

Kogan, N., and Wallach, M.A. "Risk Taking as a Function of the Situation, the Person, and the Group." New Directions in Psychology III. New York: Holt, Rinehart, and Winston, 1967.

Kraft, M. "The Use of Risk Analysis in Federal Regulatory Agencies: An Exploration." Policy Studies Review 1:4 (May 1982):666-675.

Krantz, D.H., Luce, D., Suppes, P., and Tversky, A. Foundations of Measurement, Vol. I. New York: Academic Press, 1971.

Kranzberg, M. "Prospects for Change." Societal Risk Assessment: How Safe is Safe Enough? Edited by R. Schwing and W. Albers. New York: Plenum Press, 1980, pp. 319-332.

Kraus, R. "Environmental Carcinogenesis: Regulations on the Frontiers of Science." Environmental Law 7 (Fall 1976).

Krier, J.E. "The Pollution Problem and Legal Institutions: A Conceptual Overview." UCLA Law Review 18 (1971):429-435.

Krutilla, J.V. "Some Environmental Effects of Economic Development." Daedalus 96 (Fall 1967):1058.

Krutilla, J.V., and Fisher, A.C. The Economics of National Environments. Baltimore, Maryland: Johns Hopkins University Press, 1975.

Kunreuther, H. "Changing Societal Consequences of Risks From Natural Hazards." Annals of the American Academy of Political and Social Science (May 1979):104-116.

Kunreuther, H. Decision Making for Low Probability Events: A Conceptual Framework (WP-80-169). Laxenburg, Austria: International Institute for Applied Systems Analysis, 1980.

Kunreuther, H. The Economics of Protection Against Low Probability Events (WP-81-3). Laxenburg, Austria: International Institute for Applied Systems Analysis, 1981.

Kunreuther, H. "Limited Knowledge and Insurance Protection." Public Policy 24 (1976):227-61.

Kunreuther, H. Recovery From Natural Disasters: Insurance or Federal Aid. Washington, D.C.: American Enterprise Institute, 1973.

Kunreuther, H. Societal Decision Making for Low Probability
   Events: Descriptive and Prescriptive Aspects (IIASA
   WP-80-164). Laxenburg, Austria: International Institute
   for Applied Systems Analysis, 1980.

Kunreuther, H., Ginsberg, R., Miller, L., Sagi, P., Slovic, P.,
   Borkan, B., and Katz, N. Disaster Insurance Protection:
   Public Policy Lessons. New York: John Wiley & Sons, 1978.

Kunreuther, H., and Kleindorfer, P. Guidelines for Coping with
   Natural Hazards and Climatic Change (PP-80-013). Laxenburg,
   Austria: International Institute for Applied Systems
   Analysis, 1980.

Kunreuther, H., and Kleindorfer, P. Insuring Against International Hazards: Descriptive and Prescriptive Aspects
   (CP-81-031). Laxenburg, Austria: International Institute
   for Applied Systems Analysis, 1981.

Kunreuther, H., and Lathrop, J.W. "Siting Hazardous Facilities:
   Lessons from LNG." Risk Analysis 1:4 (December 1981):289-302.

Kunreuther, H., and Ley, E., eds. Decision Processes and
   Institutional Aspects of Risk. Laxenburg, Austria: International Institute for Applied Systems Analysis (In press).

Kunreuther, H., and Ley, E. The Risk Analysis Controversy.
   New York: Springer Verlag, 1982.

Lagadec, P. Major Technological Risk: An Assessment of
   Industrial Disasters. New York: Pergamon Press, 1982.

Lagadec, P. "Societal Challenges in Risk Assessment. Discussion
   Paper on J. Conrad: Society and Risk Assessment--An
   Attempt at Interpretation." Society, Technology and Risk
   Assessment. Edited by J. Conrad. London: Academic Press,
   1980, pp. 277-280.

Lancianese, F.W. "Survey Probes Risk in a Complex Society."
   Occupational Hazards 42 (October 1980):109-113.

Landefeld, J.S., and Seskin, E.P. "The Economic Value of Life:
   Linking Theory to Practice." American Journal of Public
   Health 72 (1982):555-566.

Landau, J.L. "Economic Dream or Environmental Nightmare? The
   Legality of the 'Bubble Concept' in Air and Water Pollution
   Control." Boston College Environmental Affairs Law Review
   8(4):741-783.

LaPorte, T.R.  "Public Attitudes Toward Present and Future
    Technology."  Social Studies of Science 5 (1975):373-391.

LaPorte, T.R., and Metlay, D.  "Technology Observed:  Attitudes
    of a Wary Public."  Science 188 (April 11, 1975):121-127.

LaPorte, T.R., and Metlay, D.  They Watch and Wonder:  Public
    Attitudes Toward Advanced Technology.  University of
    California Institute of Governmental Studies, Berkeley,
    1975.

Lathrop, J.  "Measuring Social Risk and Determining Its Accept-
    ability."  Two Blowouts in the North Sea:  Managing Tech-
    nological Disaster.  Edited by D.W. Fischer.  Oxford:
    Pergamon Press, 1980.

Lathrop, J.  The Role of Risk Assessment in Facility Siting:
    An Example from California (WP-80-150).  Laxenburg,
    Austria: International Institute for Applied Systems
    Analysis, 1980.

Lathrop, J., and Linnerooth, J.  The Role of Risk Assessment
    in a Political Decision Process (WP-81-119).  Laxenburg,
    Austria: International Institute for Applied Systems
    Analysis, 1981.

Lathrop, J., and Linnerooth, J.  "The Role of Risk Assessment
    in a Political Decision Process."  Analyzing and Aiding
    Decision Processes.  Edited by P. Humphreys and A. Vari.
    Amsterdam: North Holland, 1982.

Lave, L.  "Economic Tools for Risk Reduction."  Societal Risk
    Assessment:  How Safe is Safe Enough?.  Edited by R. Schwing
    and W. Albers.  New York: Plenum Press, 1980, pp. 115-128.

Lave, L.  "Health, Safety, and Environmental Regulations."
    Setting National Priorities:  Agenda for the 1980s.  Edited
    by J.A. Pechman.  Washington, D.C.: Brookings Institution,
    1980.

Lave, L.  "Risk, Safety and the Role of Government."  Perspec-
    tives on Benefit-Risk Decision Making.  Edited by the
    Committee on Public Engineering Policy.  National Academy
    of Engineering, 1972.

Lave, L.  "Safety in Transportation: The Role of Government."
    Law Contemporary Problems 33 (Summer 1968):512-535.

Lave, L. The Strategy of Social Regulation: Decision Frame-works for Policy. Washington, D.C.: Brookings Institution, 1981.

Lave, L., and Seskin, E.P. Air Pollution and Human Health. Baltimore: John Hopkins University Press for Resources for the Future, 1977.

Lave, L., and Seskin, E. "Epidemiology, Causality, and Public Policy." American Scientist 67 (March-April 1979):178-186.

Lave, L., and Silverman, L. "Economic Costs of Energy Related Environmental Pollution." Annual Review of Energy 1 (1976):601-626.

Lawless, E.W. Technology and Social Shock. New Brunswick, New Jersey: Rutgers University Press, 1977.

Lawless, E.W. "Anticipating Technologically-Derived Risk." Impact Assessment Bulletin 1:3 (1982):54-66.

Lawless, E.W. "The Nature of Controversy Over Technology." Weed Control in Forest Management: Proceedings of the 1981 John S. Wright Forestry Conference. Edited by H.A. Holt and B.C. Fischer. West Lafayette, Indiana: Purdue Research Foundation, 1981.

Lawless, E.W. "Technology and Social Shock." Futurific 4:3/4 (1980): 243-256.

Leape, J.P. "Quantitative Risk Assessment in Regulation of Environmental Carcinogens." Harvard Environmental Law Review 4 (1980):86.

Lederberg, J. "The Freedom and the Control of Science: Notes From the Ivory Tower." Southern California Law Review 45 (1972):609.

Lee, K.N. "A Federalist Strategy for Nuclear Waste Management." Science 208 (May 16, 1980):679-684.

Lehr, G. "Aspects of Technological Risk from the Perspective of Policymaking." Technological Risk: Its Perception and Handling in the European Community. Edited by M. Dierkes, S. Edwards, and R. Coppack. Cambridge, Massachusetts: Oelgeschlager, Gunn, and Hain, Publishers, Inc., 1980, pp. 3-6.

Lerch, I. "Risk and Fear." New Scientist 185 (January 3, 1980):8-11.

Leventhal, H. "Environmental Decisionmaking and the Role of the Courts." University of Pennsylvania Law Review 122 (1974):509.

Levine, S. "The Role of Risk Assessment in the Nuclear Regulatory Process." Nuclear Safety 19 (September-October 1978):556-564.

Lewis, H.A., Budnitz, R.J., Kouts, H.J., Loewenstein, W.B., Rowe, W.D., Von Hippel, S., and Zachariasen, S. Risk Assessment Review Group Report to the U.S. Nuclear Regulatory Commission (NUREG/CR-0400). Washington, D.C., 1978.

Lewis, H.W. "The Safety of Fission Reactors." Scientific American 242:3 (March 1980).

Libby, L.M. Technological Risk Versus Natural Catastrophes (P-4602). Santa Monica, California: Rand Corporation, March 1971.

Lichtenstein, S., Fischhoff, B., and Phillips, L.D. "Calibration of Probabilities: The State of the Art to 1980." Judgment Under Uncertainty: Heuristics and Biases. Edited by D. Kahneman, P. Slovic, and A. Tversky. New York: Cambridge University Press, 1982.

Lichtenstein, S., Slovic, P., Fischhoff, B., Layman, M., and Combs, B. "Judged Frequency of Lethal Events." Journal of Experimental Psychology: Human Learning and Memory 4 (1978):551-578.

Lifton, R.J., and Olson, E. Living and Dying. New York: Praeger Publishers, 1974.

Lijinsky, W., Coulston, F., Wolfe, M., and Martin, J. "Should the Delaney Clause be Changed?" Chemical and Engineering News 55:26 (June 1977):24-42.

Lind, R.C. "The Analysis of Benefit-Risk Relationships: Unresolved Issues and Areas for Future Research." Perspectives on Benefit-Risk Decision Making. Washington, D.C.: National Academy of Engineering, 1972, pp. 109-114.

Lindblom, C. The Policy-Making Process, 2nd ed. Englewood Cliffs, New Jersey: Prentice-Hall, 1980.

Lindblom, C. "The Science of Muddling Through." The Making of Decisions. Edited by W.J. Gore and J.W. Byson. New York: Free Press, 1964.

Linnerooth, J. The Evaluation of Life-Saving: A Survey (RR-75-21). Laxenburg, Austria: International Institute for Applied Systems Analysis, July 1975.

Linnerooth, J. "Methods for Evaluating Mortality Risk." Futures 8 (August 1976):293-304.

Linnerooth, J. "Models for Determining Life Values: Some Comments." Risk-Benefit Methodology and Application (UCLA-ENG-7598). Edited by D. Okrent. Los Angeles: University of California, December 1975.

Linnerooth, J. A Review of Recent Modelling Efforts to Determine the Value of Human Life (RM-75-67). Laxenburg, Austria: International Institute for Applied Systems Analysis, 1975.

Linnerooth, J. "The Value of Human Life: A Review of the Models." Economic Inquiry 17 (January 1979).

Liroff, R. "Cost-Benefit Analysis in Environmental Regulation: Will It Clear the Air or Muddy the Water?" Cost-Benefit Analysis and Environmental Regulations. Edited by D. Swartzman, R. Liroff, and K. Croke. Washington, D.C.: The Conservation Foundation, 1982.

Liroff, R. "Cost-Benefit Analysis in Federal Environmental Programs." Cost-Benefit Analysis and Environmental Regulations. Edited by D. Swartzman, R. Liroff, and K. Croke. Washington, D.C.: The Conservation Foundation, 1982.

Litai, D., Lanning, D.K., and Rasmussen, N.C. "The Public Perception of Risk." The Analysis of Actual vs. Perceived Risks. Edited by V. Covello, G. Flamm, J. Rodricks, and R. Tardiff. New York: Plenum, 1983.

Lovins, A.B. "Cost-Risk-Benefit Assessments in Energy Policy." George Washington Law Review 45 (August 1977):911-943.

Lowenstein, R. "Price-Anderson Act: An Imaginative Approach to Public Liability Concerns." Forum 12:2 (Winter 1977): 594-604.

Lowrance, W.  "Choosing Our Pleasures and Our Poisons:  Risk
    Assessment for the '80's."  Science, Technology and The
    Issues of the Eighties:  Policy Outlook.  Edited by A.
    Teich and R. Thornton.  Boulder, Colorado: Westview Press,
    1982.

Lowrance, W.  "The Nature of Risk."  Societal Risk Assessment:
    How Safe is Safe Enough?  Edited by R. Schwing and W.
    Albers.  New York: Plenum Press, 1980, pp. 5-14.

Lowrance, W.  Of Acceptable Risk:  Science and the Determina-
    tion of Safety.  Los Altos, California: W. Kaufman, Inc.,
    1976.

Lowrance, W. "Probing Societal Risks."  Chemical and Engineering
    News 59 (July 6, 1981):13-20.

Luce, R.D., and Raiffa, H.  Games and Decisions.  New York:
    John Wiley & Sons, 1975.

MacLean, D.E.  "Risk and Consent:  Philosophical Issues for
    Centralized Decisions."  Risk Analysis 2:2 (1983).

Maderthaner, R., Guttman, G., Swaton, E., and Otway, H.J.
    "Effect of Distance on Risk Perception."  Journal of
    Applied Psychology 63:3 (1978):380-382.

Maderthaner, R., Pahner, P., Guttman, G., and Otway, H.J.
    "Perceptions of Technological Risks:  The Effect of Con-
    frontation" (RM-76-53).  Laxenburg, Austria: International
    Institute for Applied Systems Analysis, 1976.

Magnuson, H.J.  "Soviet and American Standards for Industrial
    Health."  Archives of Environmental Health 10 (1965):542-
    545.

Majone, G.  "On the Logic of Standard Setting in Health and
    Related Fields."  Systems Aspects of Health Planning.
    Edited by N.T.J. Bailey and M. Thompson.  New York: North
    Holland Publishing Company, 1975.

Majone, G.  "Process and Outcome in Regulatory Decisions."
    American Behavioral Scientist 22 (1979):561-583.

Majone, G., and Quade, E.S., eds. Pitfalls of Analysis.
    Chichester, England: Wiley, 1980.

Mann, N.R., Shafer, R.E., and Singpurwalla, N.O.  Methods for
    Statistical Analysis of Reliability and Life Data.  New
    York: John Wiley & Sons, 1974.

Marcus, G. "Quantitative Risk Assessment: Why the Controversy?" Congressional Research Service Review (April 1981):23-24.

Mark, R.K., and Stuart-Alexander, D.E. "Disasters as a Necessary Part of Benefit-Cost Analysis." Science 197 (September 16, 1977):1160-1162.

Marquis, D.G., and Reitz, H.J. "Effects of Uncertainty on Risk Taking in Individual and Group Decisions." Behavioral Science 14 (July 1969):281-288.

Marrone, J. "Price-Anderson Act: The Insurance Industry's View." Forum 12:2 (Winter 1977):605-611.

Marshall, J.M. "Optimum Safety and Production of Information When Risks Are Insured." Risk-Benefit Methodology and Application (UCLA-ENG-7598). Edited by D. Okrent. Los Angeles: University of California, December 1975.

Martin, J.G. "Note: Procedures for Decisionmaking Under Conditions of Scientific Uncertainty: The Science Court Proposal." Harvard Journal on Legislation 16 (1979):445.

Martin, J.G. "The Delaney Clause and Zero Risk Tolerance." Food, Drug and Cosmetic Law Journal 34 (January 1979): 43-49.

Martin, J.G. "The Proposed Science Court." Michigan Law Review 75 (1977):1058.

Masten, J. "Comment: Epistemic Ambiguity and the Calculus of Risk: Ethyl Corporation v. Environmental Protection Agency." South Dakota Law Review 21 (1976):425.

Mattson, R., Ernst, M., Minners, W., and Spangler, M. "Concepts, Problems, and Issues in Developing Safety Goals and Objectives for Commercial Nuclear Power." Nuclear Safety 21:6 (November-December 1980):703-716.

Maugh, T.H. "How Safe is Safe?" Science 22 (October 6, 1978):39.

Maxey, M.N. "Bio-ethical Perspective on Radiation Protection and Safety." Directions in Energy Policy. Edited by B. Kursunoglu and A. Perlmutter. Cambridge, Massachusetts: Ballinger Publishing Company, 1979.

May, W.W. "Dollars for Lives: Ethical Considerations in the Use of Cost/Benefit Analysis by For-Profit Firms." Risk Analysis 2:1 (March 1982):35-46.

Maynard, W.S., Nealey, S.M., Hebart, J.A., and Lindell, M.K.
Public Values Associated with Nuclear Waste Disposal
(BNWL-1997). Seattle: Battelle Human Affairs Research
Centers, 1976.

Mazur, A. "Disputes Between Experts." Minerva 2 (1973):243-
262.

Mazur, A. The Dynamics of Technical Controversy. Washington,
D.C.: Communications Press, 1981.

Mazur, A. "Opposition to Technological Innovation." Minerva
13 (1975): 58-81

Mazur, A. "Societal and Scientific Causes of the Historical
Development of Risk Assessment." Society, Technology and
Risk Assessment. Edited by J. Conrad. London: Academic
Press, 1980, pp. 151-158.

McClay, J. Federal Regulation: Roads to Reform. Washington,
D.C.: Commission on Law and the Economy, American Bar
Association, 1979.

McCormick, N.J. Reliability and Risk Analysis. New York:
Academic Press, 1981.

McCray, W.R. "Pesticide Regulation: Risk Assessment and Burden
of Proof." George Washington Law Review 45:5, 1066-1094.

McEnvoy, J. "The American Concern with the Environment."
Natural Resources and the Environment. Edited by W.R.
Burch, Jr., et al. New York: Harper and Row, 1972.

McGarity, T.O. "The Courts, the Agencies, and NEPA Threshold
Issues." Texas Law Review 55 (1977):801.

McGarity, T.O. "Substantive and Procedural Discretion in
Administrative Resolution of Science Policy Questions:
Regulating Carcinogens in EPA and OSHA." Georgetown Law
Review 67 (1979):729-795.

McGinty, L. and Atherley, G. "Acceptability vs. Democracy."
New Scientist 74:1051 (May 12, 1977).

McKnight, A.D., et al. Environmental Pollution Control: Technical, Economic, and Legal Aspects. London: Allen &
Unwin, 1974.

McLoughlin, J. "Risk and Legal Liability." <u>Dealing with Risk:</u>
<u>The Planning, Management and Acceptability of Technological</u>
<u>Risk</u>. Edited by R.F. Griffiths. New York: Halsted Press,
1981.

Meadows, D.H., Meadows, D.L., Randers, J., and Behrens, W.W.
<u>The Limits to Growth</u>. New York: Signet, 1972.

Mehlman, M.A., Shapiro, R.E., and Blumenthal, H., eds. <u>Advances</u>
<u>in Modern Toxicology</u>. Vol. 1: <u>New Concepts in Safety</u>
<u>Evaluation</u>. New York: John Wiley & Sons, 1979.

Mehr, R. "Changing Responsibility for Personal Risks and
Societal Consequences: Premature Death and Old Age."
<u>Annals of the American Academy of Political and Social</u>
<u>Science</u> (May 1979):1-11.

Melber, B.D., Nealey, S.M., Hammersla, J, and Rankin, W.L.
<u>Nuclear Power and the Public: Analysis of Collected</u>
<u>Survey Research</u> (PNL-2430). Seattle: Battelle Human
Affairs Research Centers, 1977.

Melville, M. "Risks on the Job: The Worker's Right to Know."
<u>Environment</u> 23:9 (November 1981):12-20, 42-45.

Mendeloff, J. <u>Regulating Safety: An Economic and Political</u>
<u>Analysis of Occupational Safety and Health Policy</u>.
Cambridge, Massachusetts: MIT Press, 1979.

Menkes, J. "Risk or Angst?" <u>Risk Analysis</u> 1:4 (December 1981):
237-240.

Merkhofer, M.W. "A Procedure Based on Decision Analyses for
Assessing the Health Risks Associated with Alternative
Ambient Air Quality Standards." <u>Four Conceptual Approaches</u>
<u>to Health Risk Assessment</u>. Washington, D.C.: U.S. Environ-
mental Protection Agency, August 1980.

Merkhofer, M.W. "Risk Assessment: Quantifying Uncertainty in
Health Effects." <u>Environmental Professional</u> 3:3/4 (1981):
249-264.

Merrill, R.A. <u>Federal Regulation of Cancer Causing Chemicals</u>.
Washington, D.C.: Administrative Conference of the United
States, 1980.

Merrill, R.A. "Regulating Carcinogens in Food: A Legislator's
Guide to the Food Safety Provisions of the Federal Food,
Drug, and Cosmetic Act." <u>Michigan Law Review</u> 77 (December
1978):171.

Merrill, R.A.   "Risk-Benefit Decisionmaking by the Food and Drug
    Administration."   George Washington Law Review 45 (August
    1977):994-1012.

Metropolitan Life Insurance Company.   Statistical Bulletin--
    Catastrophic Accidents 58 (March 1977).

Metzger, H.P.   "Risk-Benefit Analysis is a Dream."   Risk vs.
    Benefit:   Solution or Dream?   Edited by H.J. Otway.   Los
    Alamos, New Mexico: Los Alamos National Laboratory, 1971,
    pp. 14-21.

Michelman, F.   "Pollution as a Tort:   A Non-Accidental Perspec-
    tive on Calabresi's Costs."   Yale Law Journal 80 (1971):
    647.

Michelman, F.   "Property, Utility and Fairness: Comments on the
    Ethical Foundations of 'Just Compensation' Law."   Harvard
    Law Review 80 (1967):1165.

Mileti, D.S.   "Human Adjustment to the Risk of Environmental
    Extremes."   Sociology and Social Research 64 (April 1980):
    327-347.

Miller, B., and Hall, R.E.   Index of Risk Exposure and Risk
    Acceptance Criteria   (NUREG/CR-1930).   Prepared for the
    U.S. Nuclear Regulatory Commission.   Upton, New York:
    Brookhaven National Laboratory, February 1981.

Miller, J.C., III, and Yandle, B., eds.   Benefit-Cost Analyses
    of Social Regulation:   Case Studies from the Council on
    Wage and Price Stability.   Washington, D.C.: American
    Enterprise Institute for Public Policy Research, 1979.

Mishan, E.   Cost-Benefit Analysis:   An Introduction.   New York:
    Praeger Publishers, 1976.

Mishan, E.   "Evaluation of Life and Limb."   Journal of Political
    Economy   79 (July/August 1971):687-705.

Mishan, E.   "Pareto Optimality and the Law."   Oxford Economic
    Papers 19 (1967):255.

Mitchell, R.C.   "National Environmental Lobbies and the Apparent
    Illogic of Collective Action."   Collective Decision Making:
    Applications from Public Choice Theory.   Edited by C.S.
    Russell.   Baltimore and London: Johns Hopkins University
    Press, 1979.

Mitchell, R.C.  Public Opinion on Environmental Issues:  Results
    of a National Public Opinion Survey.  Washington, D.C.:
    Council on Environmental Quality, Department of Agriculture,
    Department of Energy, and Environmental Protection Agency,
    1980.

Mitchell, R.C.  "Silent Spring/Solid Majorities."  Public
    Opinion, 2 (August-September 1979).

Mitnick, B.M., and Weiss, C.  "The Siting Impasse and a Rational
    Choice Model of Regulatory Behavior: An Agency for Power
    Plant Siting."  Journal of Environmental Economics and
    Management 1 (1974):150-171.

MITRE Corporation.  Risk Assessment and Governmental Decision-
    making:  Symposium/Workshop on Nuclear and Nonnuclear
    Energy Systems (MTR-79W-00335).  McLean, Virginia: MITRE
    Corporation, 1979.

Moll, K.D., and Tihansky, D.P.  "Risk Benefit Analysis for
    Industrial and Social Needs."  American Industrial Hygiene
    Association Journal 38 (1977):153-161.

Montague, S., and Beardsworth, E.  Benefit-Risk--A Critique and
    Cultural Analysis for Non-Quantitative Risk Assessments.
    Upton, New York: Brookhaven National Laboratory, December
    1974.

Mooney, G.H.  The Valuation of Human Life.  New York: MacMillan
    Press Ltd., 1977.

Moreau, D.  Quantitative Assessment of Health Risks by Selected
    Federal Regulatory Agencies.  Washington, D.C.: U.S.
    Environmental Protection Agency, Office of Air Quality
    Planning and Standards, October 1980.

Morgan, M.G.  "Choosing and Managing Technology-Induced Risk."
    IEEE Spectrum 18 (December 1981):53-60.

Morgan, M.G.  "Probing the Question of Technology-Induced Risk."
    IEEE Spectrum 18 (November 1981):58-64.

Morgan, M.G.  Some Methodological Issues in Estimating the
    Social Costs of the Energy System (BEAG-HE/EE19/75).
    Upton, New York: Brookhaven National Laboratory, Depart-
    ments of Applied Science and Medicine, June 1975.

Morgan, M.G., Morris, S.C., Meier, A.K., and Shenk, D.L.  "A
    Probabilistic Methodology for Estimating Air Pollution
    Health Effects from Coal-Fired Power Plants."  Energy
    Systems and Policy 2 (1978):287-310.

Morgan, M.G., Morris, S.C., Rish, W.R., and Meier, A.K. "Sulfur
    Control in Coal-Fired Power Plants: A Probabilistic
    Approach to Policy Analysis." Journal of the Air Pollu-
    tion Control Association 28 (1978):993-997.

Morgan, T.E. "Who Pays for Clean Air--Cost and Benefit Distri-
    bution of Federal Automobile Emission Standards." Ecology
    Law Quarterly 6 (1978):859-868.

Morton, D.R. A Selected, Partially Annotated Bibliography of
    Recent (1980-1981) Natural Hazards Publications. Boulder,
    Colorado: University of Colorado Natural Hazards Research
    and Applications Information Center, 1982.

Moss, T.H. "Environmental Versus Emission Control Costs--A
    Legislative Perspective." Atmospheric Sulfur Deposition:
    Environmental Impact and Health Effects. Edited by D.S.
    Shriner, C.R. Richmond, and S.E. Lindberg. Ann Arbor,
    Michigan: Ann Arbor Science Publishers, 1980.

Moss, T.H. "Risk Analysis in Public Policy Decisions."
    Environmental Professional 3:3/4 (1981):197-200.

Muelhause, C.O. Costs, Real and Perceived, Examined in a Risk-
    Benefit Framework. Paris: Organization for Economic
    Cooperation and Development, 1972.

Muelhause, C.O. "Risk-Benefit Analysis." ASTM Standardization
    News (February 1973):8-27.

Muller, F.G. "Benefit Cost Analysis: A Questionable Part of
    Environmental Decisionmaking." Journal of Environmental
    Systems 4 (Winter 1974).

Muller, F.G. "Benefit Cost Analysis--Necessary Part of Environ-
    mental Decisionmaking--Reply." Journal of Environmental
    Systems 7 (1978):293-295.

Muntzing, L.M. "Siting and Environment: Essentials in an
    Effective Nuclear Siting Policy." Energy Policy 4 (March
    1976):3-11.

Murch, A., ed. Environmental Concern: Reasoned Attitudes and
    Behavior Toward Environmental Problems. New York: MSS
    Information Corporation, 1974.

Muschett, F.D. "Risk Assessment: Developing Standards Within
    a Geographic Perspective." Environmental Professional
    3:3/4 (1981):293-296.

Mushkin, S.J., and Dunlop, D.W. Health: What is it Worth?
   Measures of Health Benefits. New York: Pergamon Press,
   1979.

National Academy of Engineering. Perspectives on Benefit Risk
   Decision-Making. Washington, D.C.: National Academy of
   Engineering, Committee on Public Engineering Policy, 1972.

National Academy of Engineering. Public Safety: A Growing
   Factor in Modern Design. Washington, D.C.: National
   Academy of Engineering, 1970.

National Academy of Engineering. A Study of Technology Assessment; Report of the Committee on Public Engineering Policy.
   Washington, D.C.: National Academy of Engineering, Committee on Public Engineering Policy, 1969.

National Academy of Sciences. Energy in Transition 1985-2010:
   Final Report of the Committee on Nuclear and Alternative
   Energy Systems. San Francisco: W.H. Freeman, 1979.

National Academy of Sciences. Food Safety Policy: Scientific
   and Social Considerations, Part 2. Washington, D.C.:
   National Academy of Sciences, March 1979.

National Academy of Sciences. How Safe is Safe? The Design of
   Policy on Drugs and Food Additives. Washington, D.C.:
   National Academy of Sciences, 1974.

National Academy of Sciences. A Review of the State of the Art
   in Mass Media Disaster Reporting. Washington, D.C.:
   National Academy of Sciences, National Research Council,
   1980.

National Academy of Sciences, Committee on the Biological
   Effects of Ionizing Radiations. The Effects on Popula-
   tions of Exposure to Low Levels of Ionizing Radiation.
   Washington, D.C.: National Academy of Sciences, 1980.

National Academy of Sciences, Committee on Environmental
   Decision Making. Decision Making in the Environmental
   Protection Agency. Washington, D.C.: National Academy of
   Sciences, 1977.

National Academy of Sciences, Committee on Principles of
   Decision Making for Chemicals in the Environment.
   Decision Making for Regulating Chemicals in the Environment.
   Washington, D.C.: National Academy of Sciences, 1975.

National Academy of Sciences, Committee on Prototype Explicit
    Analyses for Pesticides. "Risk Assessment and Benefit
    Assessment." Regulating Pesticides. Washington, D.C.:
    National Academy of Sciences, 1980, pp. 65-130.

National Academy of Sciences, Committee on Risk and Decision
    Making. Risk and Decision Making: Perspectives and
    Research. Washington, D.C.: National Academy of Sciences,
    1982.

National Academy of Sciences, Committee for a Study on Saccharin
    and Food Safety Policy. Food Safety Policy: Scientific
    and Societal Considerations. Washington, D.C.: National
    Academy of Sciences, 1979.

National Academy of Sciences, Environmental Studies Board.
    Regulating Pesticides. Washington, D.C.: National Academy
    of Sciences, 1980.

National Academy of Sciences/National Research Council, Advisory
    Committee on Biological Effects of Ionizing Radiation.
    Considerations of Health Benefit Cost Analysis for Activi-
    ties Involving Ionizing Radiation and Alternatives.
    Washington, D.C.: National Academy of Sciences, 1977.

National Academy of Sciences, Safe Water Drinking Committee.
    "Problems of Risk Estimation." Drinking Water and Health
    3 (1980):25-65.

National Air Pollution Control Administration. Instances, If
    Any, Where Cost-Benefit Analysis Has Been Applied to
    Environmental Problems--United States. Paris: Organiza-
    tion for Economic Cooperation and Development, 1972.

National Council on Radiation Protection and Measurements.
    Perceptions of Risk: Proceedings of the Fifteenth Annual
    Meeting, March 14-15, 1979. Washington, D.C.: National
    Council on Radiation Protection and Measurements, March
    1980.

National Planning Association/Environmental Protection Agency.
    Social Decisionmaking for High Consequence, Low Probability
    Occurrences. Corvallis, Oregon: National Safety Council,
    October 1978.

National Safety Council. Accident Facts, 1980 Edition.
    Chicago: National Safety Council, 1980.

National Science Foundation. The Five-Year Outlook: Problems, Opportunities and Constraints in Science and Technology, (2 volumes.) Washington, D.C.: U.S. Government Printing Office, 1980.

National Science Foundation. Risk Assessment and Catastrophe Management Policies: The Role of Science and Technology. Papers Commissioned as Inputs to Annual Science and Technology Report (ASTR), 2nd ed. Washington, D.C.: National Science Foundation, 1980.

Nehnevajsa, J., and Menkes, J. "Technology Assessment and Risk Analysis." Technological Forecasting and Social Change 19 (1981): 245-255.

Nelkin, D. Jetport. New Brunswick, New Jersey: Transactional Books, 1975.

Nelkin, D. Nuclear Power and Its Critics. Ithaca, New York: Cornell University Press, 1971.

Nelkin, D. "The Political Impact of Technical Expertise." Social Studies of Science 5 (1975):35-54.

Nelkin, D. "The Role of Experts in a Nuclear Siting Controversy." Bulletin of the Atomic Scientists 30 (November 1974):29-36.

Nelkin, D. "Scientists in an Environmental Controversy." Science Studies 1 (1971):245.

Nelkin, D. "Some Social and Political Dimensions of Nuclear Power: Examples from Three Mile Island." American Political Science Review 75 (March 1981):132-145.

Nelkin, D. Technological Decisions and Democracy. Beverly Hills, California: Sage Publications, 1977.

Nelkin, D., ed. Controversy: Politics of Technical Decisions. Beverly Hills: Sage Publications, 1979.

Nelkin, D., and Pollak, M. "Consensus and Conflict Resolution: The Politics of Assessing Risk." Science and Public Policy (October 1979).

Nelkin, D., and Pollack, M. "Political Parties and the Nuclear Energy Debate in France and Germany." Comparative Politics (January 1980).

Nelkin, D., and Pollack, M. "Problems and Procedures in the Regulation of Technological Risk." Societal Risk Assessment: How Safe is Safe Enough? Edited by R. Schwing and W. Albers. New York: Plenum Press, 1980, pp. 233-248.

Nelkin, D., and Pollack, M. "Public Participation in Technological Decisions: Reality or Grand Illusion?" Technology Review 81 (August-September 1979):55-64.

Newill, Y.A. "Regulatory Decision Making: The Scientists' Role." Journal of Washington Academy of Science 64 (1974).

Nicholson, W.J., ed. "Management of Assessed Risk for Carcinogens." Annals of New York Academy of Sciences 63 (1981). 229 p.

Niskanen, W.A., et al., eds. Benefit-Cost Policy Analysis, 1972. Chicago: Aldine Publishing Company, 1973.

Noll, R. "Information Decision Making Procedures and Energy Policy." American Behavioral Science 4 (March/April 1976).

Nordhause, W., and Litan, R. A Regulatory Budget for the United States. Washington, D.C.: Brookings Institution, 1981.

Nozick, R. "Prohibition, Compensation, and Risk." Anarchy, State and Utopia. New York: Basic Books, 1974.

O'Hare, M. "Not on My Block You Don't: Facility Siting and the Strategic Importance of Compensation." Public Policy 25 (Fall 1977):4.

Oi, W.Y. "Safety at Any Price?" Regulation 1 (November/ December 1977):1623.

Okrent, D. Alternative Risk Management Policies for State and Local Governments: Final Report (UCLA-ENG-8240). Los Angeles: University of California, School of Engineering and Applied Science, 1982.

Okrent, D. Alternative Risk Management Policies for State and Local Governments: Executive Summary (UCLA-ENG-8241). Los Angeles: University of California, School of Engineering and Applied Science, 1982.

Okrent, D., Meyer, M., and Solomon, K.A. Risk Management Practices in Local Communities: Five Alternatives (UCLA-ENG-8242). Los Angeles: University of California, School of Engineering and Applied Science, 1982.

Okrent, D., Solomon, K.A., Meyer, M., Nelson, P., Szabo, J., and Tsai, R. Management of Risks Associated with Drinking Water at the State and Local Levels (UCLA-ENG-8243). Los Angeles: University of California, School of Engineering and Applied Science, 1982.

Okrent, D., and Sarin, R.K. Risk Management Policy for Earthquake Hazard Reduction (UCLA-ENG-8244). Los Angeles: University of California, School of Engineering and Applied Science, 1982.

Okrent, D., Solomon, K.A., Meyer, M., Szabo, J., and Nelson, P. Classification of Risks (UCLA-ENG-8245). Los Angeles: University of California, School of Engineering and Applied Science, 1982.

Okrent, D., and Bordas, W. Problems of State and Local Risk Management: An Overview (UCLA-ENG-8246). Los Angeles: University of California, School of Engineering and Applied Science, 1982.

Okrent, D. "Comment on Societal Risk." Science 208 (April 25, 1980): 372-375.

Okrent, D. A General Evaluation Approach to Risk-Benefit for Large Technological Systems and Its Application to Nuclear Power (UCLA-ENG-7777). Los Angeles: University of California, December 1977.

Okrent, D. "Risk-Benefit Evaluation for Large Technological Systems." Nuclear Safety 20 (March-April 1979):148-164.

Okrent, D., ed. Risk-Benefit Methodology and Application: Some Papers Presented at the Engineering Foundation Workshop, September 22-26, 1975, Asilomar, California (UCLA-ENG-7598). Los Angeles: University of California, 1975.

Okrent, D., and Whipple, C. An Approach to Societal Risk Acceptance Criteria and Risk Management (UCLA-ENG-7746). Los Angeles: University of California, School of Engineering and Applied Science, 1977.

Okun, A.M. Equality and Efficiency: The Big Trade-Off. Washington, D.C.: Brookings Institution, 1975.

Olson, C.A. "An Analysis of Wage Differentials Received by Workers on Dangerous Jobs." Journal of Human Resources 16 (Spring 1981): 167-185.

Olson, S.H. "An Ecology of Work Place Hazards." Economic
    Geography 55 (1979):287-308.

Omenn, G.S., and Friedman, R.D. Individual Differences in
    Susceptibility and Regulation of Environmental Hazards.
    Washington, D.C.: Office of Science and Technology Policy,
    1980.

O'Neill, B. "A Danger-Theory Model of Danger Compensation."
    Accident Analysis & Prevention 9 (1977):157-165.

Open University. Risk: A Second Level University Course. 6
    vols. Milton Keynes, United Kingdom: Open University Press,
    1980.

Opinion Research Corporation. "Public Attitudes Toward Environ-
    mental Trade-Offs." ORC Public Opinion Index 33 (August
    1975):1-8.

Organization for Economic Cooperation and Development. Problems
    in Transfrontier Pollution. Paris: Organization for
    Economic Cooperation and Development, 1974.

O'Riordan, T. "Risk Perception Studies and Policy Priorities."
    Risk Analysis 2:2 (1983).

O'Riordan, T., and Turner, R.K., eds. Progress in Resource
    Management and Environmental Planning, Vol. 3. Chichester:
    John Wiley & Sons, 1981.

Orr, D.W. "Catastrophe and Social Order." Human Ecology 7
    (1979): 41-52.

Orr, D.W. "Social Risks and the Energy Option." Science and
    Public Policy 4:2 (April 1977):179-183.

Otway, H.J. "Discussion Paper on A. Mazur: Societal and
    Scientific Causes of the Historical Development of Risk
    Assessment." Society, Technology and Risk Assessment.
    Edited by J. Conrad. London: Academic Press, 1980, pp.
    163-164.

Otway, H.J. "The Perception of Technological Risks: A Psycho-
    logical Perspective." Technological Risk: Its Perception
    and Handling in the European Community. Edited by M.
    Dierkes, S. Edwards, and R. Coppock. Cambridge, Massa-
    chusetts: Oelgeschlager, Gunn and Hain, Publishers, Inc.,
    1980, pp. 35-45.

Otway, H.J.   A Review of Research on the Identification of
    Factors in Influencing Social Response to Technological
    Risks (IAEA-CN-36/4).  Proceedings of the International
    Atomic Energy Agency Conference on Nuclear Power and Its
    Fuel Cycle, Salzburg, Austria, 1977.

Otway, H.J.   Risk Assessment and Societal Choices (RM-75-XX).
    Laxenburg, Austria: International Institute for Applied
    Systems Analysis, February 1975.

Otway, H.J., and Cohen, I.J.   Revealed Preferences:  Comments
    on the Starr Benefit-Risk Relationships (RM 75-5).
    Laxenburg, Austria: International Institute for Applied
    Systems Analysis, 1975.

Otway, H.J., and Edwards, W.   Application of a Simple Multi-
    Attribute Rating Technique to Evaluation of Nuclear Waste
    Disposal Sites:  A Demonstration (RM-77-31).  Laxenburg,
    Austria: International Institute for Applied Systems
    Analysis, 1977.

Otway, H.J., and Fishbein, M.   The Determinants of Attitude
    Formation:  An Application to Nuclear Power (RM-76-80).
    Laxenburg, Austria: International Institute for Applied
    Systems Analysis, 1976.

Otway, H.J., and Fishbein, M.   Public Attitudes and Decision
    Making (RM-77-54).  Laxenburg, Austria: International
    Institute for Applied Systems Analysis, 1977.

Otway, H.J., Linnerooth, J., and Niehaus, F.   "Risk Estimation,
    Evaluation, and Management."  Transactions of the American
    Nuclear Society 26 (June 1977):263-264.

Otway, H.J., Maderthaner, R., and Guttman, G.   Avoidance
    Response to the Risk Environment:  A Cross-Cultural Com-
    parison (RR-75-14).  Laxenburg, Austria: International
    Institute for Applied Systems Analysis, 1975.

Otway, H.J., Maurer, D., and Thomas, K.   "Nuclear Power, The
    Question of Public Acceptance."  Futures 10 (April 1978):
    109-118.

Otway, H.J., and Pahner, P.D.   "Risk Assessment."  Futures 8
    (April 1976):122-134.

Otway, H.J., Pahner, P.D., and Linnerooth.   Social Values in
    Risk Acceptance (RM-75-54).  Laxenburg, Austria: Inter-
    national Institute for Applied Systems Analysis, November
    1975.

Otway, H., and Thomas, K. "Reflections on Risk Perception and Policy." Risk Analysis 2:2 (1983).

Otway, H., and Von Winterfeldt, D. "Beyond Acceptable Risk: On the Social Acceptability of Technologies." Policy Sciences 14 (1982):247-256.

Page, T. "Behind the Looking Glass: Administrative, Legislative and Private Approaches to Cosmetic Safety Substantiation." UCLA Law Review 24 (1977):795.

Page, T. "A Generic View of Toxic Chemicals and Similar Risks." Ecology Law Quarterly 7 (1978):207-243.

Pahner, P.D. "The Psychological Displacement of Anxiety: An Application to Nuclear Energy." Risk-Benefit Methodology and Application (UCLA-ENG-7598). Edited by D. Okrent. Los Angeles: University of California, December 1975.

Pahner, P.D. A Psychological Perspective of the Nuclear Energy Controversy (RM-76-67). Laxenburg, Austria: International Institute for Applied Systems Analysis, 1976.

Parry, G.W. and Winter, P. "Characterization and Evaluation of Uncertainty in Probabilistic Risk Analysis." Nuclear Safety 22 (January-February 1981):28-42.

Parzyck, D.C., and Inhaber, H. "Comparative Health Risks of Energy Production." IEEE Transactions on Nuclear Science, NS-28 1 (1981):226-230.

Pate, M.E. Risk and Public Policy (Report No. 37). Stanford, California: John A. Blume Earthquake Engineering Center, Stanford University, July 1979.

Pauly, M.V. "Risk and the Social Rate of Discount." American Economic Review 60 (1970):195.

Payne, J.W. "Relation of Perceived Risk to Preferences Among Gamblers." Journal of Experimental Psychology: Human Perception and Performance 104 (1975):86-94.

Pearce, D. "The Limits of Cost Benefit Analysis as a Guide to Environmental Policy." Kyklos 29 (1976):97-112.

Pearce, D.W.  "The Preconditions for Achieving Consensus in the
    Context of Technological Risk." Technological Risk:  Its
    Perception and Handling in the European Community.  Edited
    by M. Dierkes, S. Edwards, and R. Coppack.  Cambridge,
    Massachusetts: Oelgeschlager, Gunn, and Hain, Publishers,
    Inc., 1980, p. 58.

Pearce, E.G.  "The Nuclear Debate Is About Values." Nature 274
    (July 1978):200.

Peskin, H.M., and Seskin, E.P., eds.  Cost-Benefit Analysis and
    Water Pollution Policy.  Washington, D.C.: Urban Institute,
    1975.

Petak, W., and Atkisson, A.  Natural Hazard Risk Assessment and
    Public Policy.  New York: Springer-Verlag, 1982.

Peterson, J.M.  "Benefit Cost Analysis--A Necessary Part of
    Environmental Decisionmaking." Journal of Environmental
    Systems 5 (1975).

Pfenningsdorf, W.  "Environment, Damages and Compensation."
    American Bar Foundation 2 (1979):347-448.

Philipson, L.L.  Investigation of the Feasibility of the Delphi
    Technique for Estimating Risk Analysis Parameters
    (TES-20-74-5).  Prepared for the Office of Hazardous
    Materials, Department of Transportation.  Los Angeles:
    University of Southern California, 1974.

Pochin, E.E.  "The Acceptance of Risk." British Medical
    Bulletin 31:3 (1975):184-190.

Pochin, E.E.  "The Need to Estimate Risks." Physics in Medi-
    cine and Biology 25:1 (January 1980):1-12.

Poole, R.W., ed.  Instead of Regulation:  Alternatives to
    Federal Regulatory Agencies.  Lexington, Massachusetts:
    Lexington Books, 1982.

Porter, A.  A Guidebook for Technology Assessment and Impact
    Analysis.  New York: North Holland Publishing Company,
    1980.

Portney, P.  "Toxic Substance Policy and the Protection of Human
    Health." Current Issues in U.S. Environmental Policy.
    Edited by P. Portney.  Baltimore: Johns Hopkins University
    Press, 1978, pp. 105-143.

Poulinquen, L.Y.  Risk Analysis in Project Appraisal.  World
    Bank Staff Occasional Papers, No. 11.  International Bank
    for Reconstruction and Development.  Baltimore, Maryland:
    Johns Hopkins University Press, 1970.

Powers, W.T.  Behavior: The Control of Perception.  Chicago:
    Aldine, 1973.

Pratt, J., Raiffa, H., and Schlaifer, R.  "The Foundations of
    Decision Under Uncertainty."  The American Statistical
    Association Journal 59 (1964):353-376.

Pratt, J., and Zeckhauser, R.  "Inferences From Alarming Events."
    Journal of Policy Analysis and Management 1:3 (Spring
    1982).

Priest, W.C.  "Cost-Benefit Problems in Safety and Health Eval-
    uations."  Hazard Prevention (November/December 1979).

Primack, J., and von Hippel, F.  Advice and Dissent: Scientists
    in the Political Arena.  New York: Basic Books, 1974.

Quarantelli, E.L., and Dynes, R.R.  "Response to Social Crises
    and Disaster."  Annual Review of Sociology 3 (1979):23-49.

Quarantelli, E.L., ed.  Disasters: Theory and Research.
    Beverly Hills, California: Sage Publications, Inc., 1978.

Raiffa, H.  Decision Analysis: Introductory Lectures on Choice
    Under Uncertainty.  Reading, Massachusetts: Addison-Wesley,
    1968.

Raiffa, H., Schwartz, W.B., and Weinstein, M.C.  "Evaluating
    Health Effects of Societal Decisions and Programs."
    Decision Making in the Environmental Protection Agency:
    Selected Working Papers.  Washington, D.C.: National
    Academy of Sciences, 1977.

Raiffa, H., Schwartz, W., and Weinstein, M.  On Evaluating
    Health Effects of Societal Programs.  Cambridge, Massa-
    chusetts: Harvard University, October 1976.

Raiffa, H., and Zeckhauser, R.  "Reporting of Uncertainties in
    Risk Analysis."  Two Conceptual Approaches to Health Risk
    Assessment for Alternative National Ambient Air Quality
    Standards.  Washington, D.C.: U.S. Environmental Protec-
    tion Agency, September 1980.

Ramo, S.  "Regulation of Technological Activities:  A New
    Approach."  Science 213 (August 21, 1981):837-842.

Rapoport, A., and Wallsten, T.S. "Individual Decision Behavior."
Annual Review of Psychology 23 (1972):131-175.

Rappaport, E.B. "Remarks on the Economic Theory of Life Value."
Risk-Benefit Methodology and Application (UCLA-ENG-7598).
Edited by D. Okrent. Los Angeles: University of Cali-
fornia, December 1975.

Rasmussen, N., Kleitman, D.J., Stewart, R.B., and Yellin, J.
"Nuclear Power: Can We Live With It?" Technology Review
81 (1979):32-47.

Rasmussen, N., Lanning, D., and Litai, D. "Public Perception
of Risks." The Analysis of Actual vs. Perceived Risks.
Edited by V. Covello, G. Flamm, J. Rodricks, and R.
Tardiff. New York: Plenum Press, 1982 (In press).

Ravetz, J. "Public Perceptions of Acceptable Risks as Evidence
for Their Cognitive, Technical, and Social Structure."
Technological Risk: Its Perception and Handling in the
European Community. Edited by M. Dierkes, S. Edwards, and
R. Coppack. Cambridge, Massachusetts: Oelgeschlager, Gunn,
and Hain, Publishers, Inc., 1980, pp. 46-57.

Ravetz, J. "Public Perceptions of Acceptable Risks." Science
and Public Policy (October 1979).

Ravetz, J. "The Risk Equations: The Political Economy of Risks."
New Scientist 74 (September 8, 1977):598-599.

Rawls, J. "Concepts of Distributional Equity: Some Reasons
for the Maxi-Min Criterion." Benefit-Cost Analysis 1974.
Edited by R. Zeckhauser et al. Chicago: Aldine Publishing
Company, 1975.

Rawls, J. "A Symposium on Rawls' Theory of Justice." University of Chicago Law Review 40 (1973):486-555.

Rawls, J. Theory of Justice. Cambridge, Massachusetts:
Harvard University Press, 1971.

Reissland, J., and Harries, V. "Scale for Measuring Risks."
New Scientist 83:1172 (September 13, 1979):809-811.

Renn, O. Man, Technology and Risk: A Study on Intuitive Risk
Assessment and Attitudes Towards Nuclear Energy. Spezielle
Berichteder Kernforschunganleg. Julich, Nr. 115, Julich,
West Germany, Kernforschunganleg, 1981.

Renn, O., and Stichl, P.  What is Risk?:  An Interdisciplinary
    Approach to Risk Assessment and Risk Evaluation.  London:
    Ballinger, 1982.

Reutilinger, S.  Techniques for Project Appraisal Under
    Uncertainty.  World Bank Staff Occasional Papers No. 10.
    Baltimore, Maryland: Johns Hopkins Press, 1972.

Rheingold, P.D., and Birnbaum, S.L., eds.  Product Liability:
    Law, Practice, Science.  New York: Practicing Law Insti-
    tute, 1975.

Rhoads, S.E.  "How Much Should We Spend to Save a Life?"  The
    Public Interest 51 (Spring 1978):74-92.

Rhoads, S.E., ed.  Valuing Life:  Public Policy Dilemmas.
    Boulder, Colorado: Westview Press, 1980.

Ricci, P., and Molton, S.  "Risk and Benefit in Environmental
    Law."  Science 214 (December 4, 1981):1096-1100.

Rice, D., and Cooper, B.  "Economic Value of Human Life."
    American Journal of Public Health (November 1967).

Richmond, C.R., Walsh, P., and Copenhaver, E., eds.  Health Risk
    Analysis.  Philadelphia: Franklin Institute Press, 1981.

Richmond, C.R.  The Science of Risk Assessment (ORNL/PPA-80/2).
    Oak Ridge, Tennessee: Oak Ridge National Laboratory, 1980.

Ritch, J.B., Jr.  "Protecting Public Health from Toxic Chemi-
    cals."  Environmental Science and Technology 13 (August
    1979):922-926.

Rivard, J.B.  "Risk Minimizing by Optimum Allocation of
    Resources Available for Risk Reduction."  Nuclear Safety
    12 (July/August 1971):305-309.

Roback, H.  "Politics and Expertise in Policy Making."  Per-
    spectives on Benefit-Risk Decision Making.  Edited by the
    Committee on Public Engineering Policy.  Washington D.C.:
    National Academy of Engineering, 1972.

Rodgers, W.  "Benefits, Costs and Risks:  Oversight of Health
    and Environmental Decisionmaking."  Harvard Environmental
    Law Review 4 (1980):191-226.

Rodgers, W.  Environmental Law.  St. Paul, Minnesota: West
    Publishing Company, 1977.

Rodgers, W. "Judicial Review of Risk Assessments: The Role of Decision Theory in Unscrambling the Benzene Decision." Environmental Law 11 (Winter 1981):301-320.

Roschin, A.V., and Timofeevsaya, L.A. "Chemical Substance in the Work Environment: Some Comparative Aspects of U.S.S.R. and U.S. Hygienic Standards." AMBIO 4 (1975): 30-33.

Rosen, S.O. "Cost Benefit Analysis, Judicial Review, and the National Environmental Policy Act." Environmental Law 7 (1977):363-381.

Rosenbaum, W.A. The Politics of Environmental Concern. New York: Praeger publishers, 1977.

Rosenthal, A.J. "The Federal Power to Protect the Environment: Available Devices to Compel or Induce Desired Conduct." Southern California Law Review 45 (1972):397.

Ross, M. "Quantitative Decision Making." Risk vs. Benefit: Solution or Dream? Edited by H.J. Otway. Los Alamos, New Mexico: Los Alamos National Laboratory, 1971, pp. 33-36.

Rothschild, N.M.V. "Coming to Grips with Risk." The Wall Street Journal (March 13, 1979).

Rothschild, N.M.V., and Stiglitz, J. "Increasing Risk I: A Definition." Journal of Economic Theory 2 (1971):225-243.

Rothschild, N.M.V., and Stiglitz, J. "Increasing Risk II: Its Economic Consequences." Journal of Economic Theory 3 (1971):66-84.

Rowe, W.D. An Anatomy of Risk. New York: John Wiley & Sons, 1977.

Rowe, W.D. Corporate Risk Assessment: Strategies and Technologies. New York: Marcel Dekker, Inc., 1982.

Rowe, W.D. "Development of Approaches for Acceptable Levels of Risk." Risk-Benefit Methodology and Application (UCLA-ENG-7598). Edited by D. Okrent. Los Angeles: University of California, December 1975.

Rowe, W.D. "Governmental Regulation of Societal Risks." George Washington Law Review 45 (August 1977):944-968.

Rowe, W.D. "Risk." Chemtech 7 (August 1977):477-483.

Rowe, W.D.  "Risk Assessment Approaches and Methods."  Society,
     Technology and Risk Assessment.  Edited by J. Conrad.
     London: Academic Press, 1980, pp. 3-29.

Rowe, W.D., and Gordon, G.T., eds. Managing Societal Risks.
     London: Academic Press, 1979.

Royal Society of London.  The Assessment and Perception of Risk.
     London: The Royal Society, 1981.

Ruff, L.E.  "Federal Environmental Regulation."  Study on
     Federal Regulation.  Vol. 6:  Framework for Regulation.
     Appendix.  U.S. Senate Committee on Governmental Affairs,
     96th Congress, 1st Session.  Washington, D.C.: U.S.
     Government Printing Office, 1978.

Rushefsky, M.  "Technical Disputes:  Why Experts Disagree."
     Policy Studies Review 1 (May 1982):676-685.

Sabatier, P.A.  "Regulatory Policy-Making:  Toward a Framework
     of Analysis."  Natural Resources Journal 17 (July 1977):
     415-460.

Sagan, L.A.  "Human Cost of Nuclear Power."  Science 177
     (August 11, 1972):487-493.

Sage, A., and White, E.  "Methodologies for Risk and Hazard
     Assessment: A Survey and Status Report."  IEEE Transactions
     on Systems, Man and Cybernetics 6:10 (August 1980):425-445.

Salem, S.L., Solomon, K.A., and Yesley, M.S.  Issues and Prob-
     lems in Inferring a Level of Acceptable Risk (P6519).
     Santa Monica, California: Rand Corporation, August 1980.

Sampson, A.R., and Smitling, R.L.  Assessing Risks Through the
     Determination of Rare Event Probabilities."  Washington,
     D.C.: Bolling Air Force Base, Air Force Office of Scien-
     tific Research, 1980.  Available from National Technical
     Information Service, Springfield, Virginia, Order No.
     PB80-1106030.

Samuels, S.  "Role of Scientific Data in Health Decisions."
     Environmental Health Perspectives 32 (October 1979):
     301-307.

Sapolsky, H.M.  "Science, Voters, and the Fluoridation Con-
     troversy."  Science 162 (1968):427-433.

Sather, H.N. <u>Biostatistical Aspects of Risk Benefit: The Use of Competing Risk Analysis</u>. Prepared for the National Science Foundation. Los Angeles: University of California, School of Engineering and Applied Science, September 1974.

Schaefer, R. <u>What Are We Talking About When We Talk About Risk? A Critical Survey of Risk and Risk Preference Theories</u> (RM-78-69). Laxenburg, Austria: International Institute for Applied Systems Analysis, December 1978.

Schelling, T.C. "The Life You Save May Be Your Own." <u>Problems in Public Expenditures Analysis</u>. Edited by S. Chase. Washington, D.C.: Brookings Institution, 1968.

Schlaifer, R.O. <u>Analysis of Decisions Under Uncertainty</u>. New York: McGraw-Hill, 1969.

Schneider, S.H. "Comparative Risk Assessment of Energy Systems." <u>Energy</u> 4 (1979):919-932.

Schneiderman, M. "The Uncertain Risk We Run: Hazardous Materials." <u>Societal Risk Assessment: How Safe is Safe Enough?</u> Edited by R. Schwing and W. Albers. New York: Plenum Press, 1980, pp. 19-38.

Schoemaker, P.J.H., and Kunreuther, H.C. "An Experimental Study of Insurance Decisions." <u>Journal of Risk and Insurance</u> 46 (1980): 603.

Schulze, W.D. "Ethics, Economics and the Value of Safety." <u>Societal Risk Assessment: How Safe is Safe Enough?</u> Edited by R. Schwing and W. Albers. New York: Plenum Press, 1980.

Schulze, W.D., and Kneese, A.V. "Risk in Benefit-Cost Analysis." <u>Risk Analysis: An International Journal</u> 1:1 (March 1981): 81-88.

Schuster, G. "Technological Risk: Its Perception and Handling in the EEC." <u>Technological Risk: Its Perception and Handling in the European Community</u>. Edited by M. Dierkes, S. Edwards, and R. Coppack. Cambridge, Massachusetts: Oelgeschlager, Gunn, and Hain, Publishers, Inc., 1980, pp. 11-19.

Schweig, B.B. "Products Liability Problem." <u>Annals of the American Academy of Political and Social Science</u> (May 1970):94-103.

Schwing, R. Expenditure to Reduce Mortality Risk and Increase Longevity (GMR-2353-A). Warren, Michigan: Societal Analysis Department, General Motors Research Laboratories, February 1978.

Schwing, R. "Longevity Benefits and Costs of Reducing Various Risks." Technological Forecasting and Social Change 13 (May 1979):333-345.

Schwing, R. "Risks in Perspective-Safe Enough? Edited by R. Schwing and W. Albers. New York: Plenum Press, 1980, pp. 129-142.

Schwing, R., and Albers, W.A., Jr., eds. Societal Risk Assessment: How Safe is Safe Enough? New York: Plenum Press, 1980.

Scott, S., ed. What Decisionmakers Need to Know: Policy and Social Science Research on Seismic Safety. Institute of Governmental Studies, University of California, 1979.

Sell, R.G. "What Does Safety Propaganda Do for Safety: A Review." Applied Economics 8:4 (1977):203-214.

Shapo, M. A Nation of Guinea Pigs: The Unknown Risks of Chemical Technology. New York: Free Press, 1979.

Sheridan, T.B. "Human Error in Nuclear Power Plants." Technology Review 82:4 (1980):23-33.

Shue, H. "Exporting Hazards." Boundaries: National Autonomy and Its Limits. Edited by P. Brown and H. Shue. Totowa, New Jersey: Rowman and Littlefield, 1981.

Siddall, E. Risk, Fear, and Public Safety (Report AECL 7404). Mississaga, Ontario: Atomic Energy of Canada Limited, 1981.

Simon, J. The Ultimate Resource. New Jersey: Princeton University Press, 1981.

Sinclair, C. "Costing the Hazards of Technology." New Scientist 44 (October 16, 1979):120-122.

Sinclair, C., Marstrand, P., and Newick, P. Innovation and Human Risk: The Evaluation of Human Life and Safety in Relation to Technical Change. London: Centre for the Study of Industrial Innovation, 1977.

Singer, M.  "How to Reduce Risks Rationally."  Public Interest
     51 (1978):93-112.

Sive, D.  "Environmental Decisionmaking: Judicial and Policital
     Review."  CASE Western Reserve Law Review 28 (1978):827.

Sjoberg, L.  "Risk Generation and Risk Assessment in a Social
     Perspective."  Foresight, the Journal of Risk Management 3
     (1978):4-12.

Sjoberg, L.  "The Risks of Risk Analysis."  Acta Psychologica
     45 (1980):301-321.

Sjoberg, L.  "Strength of Belief and Risk."  Policy Sciences 2
     (August 1979):39-52.

Slesin, L., and Ferreria, J., Jr.  Social Values and Public
     Safety: Implied Preferences Between Accident Frequency
     and Severity.  Cambridge, Massachusetts: MIT Laboratory of
     Architecture and Planning, September 1976.

Slovic, P.  "Assessment of Risk-Taking Behavior."  Psychological
     Bulletin 61 (1964):220-233.

Slovic, P., and Fischhoff, B.  "Cognitive Process and Societal
     Risk Taking."  Cognition and Societal Behavior.  Edited by
     J.S. Carroll and J.W. Payne.  Potomac, Maryland: Lawrence
     Erlbaum Associates, 1976.

Slovic, P., Fischhoff, B., and Lichtenstein, S.  "Behavioral
     Decision Theory."  Annual Review of Psychology 28 (1977):
     1-39.

Slovic, P., Fischhoff, B., and Lichtenstein, S.  "Character-
     izing Perceived Risk."  Technological Hazard Management.
     Edited by R.W. Kates and C. Hohenemser.  Cambridge,
     Massachusetts: Oelgeschlager, Gunn & Hain, 1982.

Slovic, P., Fischhoff, B., and Lichentstein, S.  "Cognitive
     Processes and Societal Risk Taking."  Cognitive Processes
     and Social Behavior.  Edited by J.S. Carroll and J.W.
     Payne.  Potomac, Maryland: Lawrence Erlbaum, 1976.

Slovic, P., Fischhoff, B., and Lichtenstein S.  "Facts and
     Fears: Understanding Perceived Risk."  Societal Risk
     Assessment: How Safe is Safe Enough?  Edited by R.
     Schwing and W. Albers, Jr.  New York: Plenum, 1980, pp.
     181-216.

Slovic, P., Fischhoff, B., and Lichtenstein, S.  "Informing People about Risk."  Product Labeling and Health Risks (Banbury Report 6).  Edited by L. Morris, M. Marsis, and I. Barofksy.  Cold Spring Harbor, New York: Cold Spring Harbor Laboratory, 1980.

Slovic, P., Fischhoff, B., and Lichtenstein, S.  "Rating the Risks."  Environment 21:3 (April 1979):14-39.

Slovic, P., Fischhoff, B. and Lichtenstein, S.  "Response Mode, Framing, and Information Processing Effects in Risk Assessment."  New Directions for Methodology of Social and Behavioral Science:  The Framing of Questions and the Consistency of Response.  Edited by R.M. Hogarth.  San Francisco: Jossey-Bass.  (In press.)

Slovic, P., Fischhoff, B., and Lichtenstein, S.  "Risk Assessment:  Basic Issues."  Managing Technological Hazard: Research Needs and Opportunities.  Edited by R. Kates.  Boulder, Colorado: Institute of Behavioral Science, University of Colorado, 1977, pp. 81-108.

Slovic, P., Fischhoff, B., and Lichtenstein, S.  "Risky Assumptions."  Psychology Today 14 (June 1980):44-45, 47-48.

Slovic, P., Fischhoff, B., and Lichtenstein, S.  "Why Study Risk Perception?"  Risk Analysis 2:2 (1983).

Slovic, P., Fischhoff, B., Lichtenstein, S., Corrigan, B., and Combs, B.  "Preference for Insuring Against Probable Small Losses:  Insurance Implications."  Journal of Risk and Insurance 45 (June 1977): 237-258.

Slovic, P., Kunreuther, H., and White, G.  "Decision Processes, Rationality and Adjustments to Natural Hazards."  Natural Hazards:  Local, National, and Global.  Edited by G.F. White.  New York: Oxford University Press, 1974.

Slovic, P., Lichtenstein, S, and Fischhoff, B.  "Images of Disaster:  Perception and Acceptance of Risks from Nuclear Power."  Energy Risk Management.  Edited by G. Goodman and W. Rowe.  London: Academic Press, 1979.

Slovic, P., and Tversky A.  "Who Accepts Savage's Axiom?"  Behavioral Science 19 (1974):368-372.

Smalley, R.D.  "Risk Assessment:  An Introduction and Critique."  Coastal Zone Management Journal 7 (1980):133-162.

Smith, R.S. The Occupational Safety and Health Act: Its Goals and Achievements. Washington, D.C.: American Enterprise Institute for Public Policy Research, 1976.

Smith, R.S. "Protecting Workers' Health and Safety." Instead of Regulation. Edited by R. Poole. Lexington, Massachusetts: Lexington Books, 1981.

Soble, S.M. "A Proposal for the Administrative Compensation of Victims of Toxic Substance Pollution: A Model Act." Harvard Journal on Legislation 14 (June 1977):683-824.

Solomon, K.A., and Abraham, S.C. "The Index of Harm: A Useful Measure for Comparing Occupational Risk Across Industries." Health Physics 38 (March 1980):375-391.

Solomon, K.A., and Joksimovic, V.J. "Quantitative Safety Goals Through More Adequate Risk Management and Risk Assessment." Journal of Reliability Engineering (In press).

Solomon, K.A., and Okrent, D. Catastrophe Events Leading to De Facto Limits on Liability (UCLA-ENG-7732). Los Angeles: University of California, May 1977.

Solomon, K.A., and Okrent, D. De Facto Liability Limits (UCLA-ENG-7732). Los Angeles: University of California, 1977.

Solomon, K.A., and Okrent, D. "Liability Limits and Insurance." Hazard Prevention 14:5 (May/June 1978).

Solow, R. "The Economics of Resources or the Resources of Economics." Benefit-Cost and Policy Analysis. Edited by R. Zeckhauser and A. Harberger. Chicago: Aldine Publishing Company, 1975.

Solow, R. "An Economist's Approach to Pollution and Its Control." Science 173 (August 6, 1971):498.

Solow, R. "Intergenerational Equity and Exhaustible Resources." in Symposium on the Economics of Exhaustible Resources, Edinburgh, 1974 (Special number of the "Review of Economic Studies"):29-46.

Spangler, M.B. "Environmental and Social Issues of Site Choice for Nuclear Power Plants." Energy Plants 2 (March 1974): 18-32.

Spangler, M.B. "Risks and Psychic Costs of Alternative Energy Sources for Generating Electricity." The Energy Journal (January 1981).

Spangler, M.B. "The Role of Interdisciplinary Analysis in
    Bridging the Gap Between the Technical and Human Sides of
    Risk Assessment." Risk Analysis 2:2 (1983).

Spangler, M. "The Role of Syndrome Management and the Future
    of Nuclear Energy." Risk Analysis 1:2 (September 1981):
    179-188.

Stallen, P.M.J. "Risk of Science or Science of Risk?" Society,
    Technology and Risk Assessment. Edited by J. Conrad.
    London: Academic Press, 1980, pp. 131-148.

Starr, C. "Benefit-Cost Relationships to Socio-Technical
    Systems." Environmental Aspects of Nuclear Power Situa-
    tions (IAEA SM-146/47). Vienna: International Atomic
    Energy Agency, 1971.

Starr, C. "Benefit-Cost Studies in Sociotechnical Systems."
    Perspectives on Benefit-Risk Decision Making. Edited by
    the Committee on Public Engineering Policy. Washington,
    D.C.: National Academy of Engineering, 1972, pp. 17-42.

Starr, C. "Social Benefit Versus Technological Risk:  What Is
    Our Society Willing to Pay for Safety?" Science 165
    (September 19, 1969):1232-1238.

Starr, C. "Some Comments on the Public Perception of Personal
    Risk and Benefit." Risk vs. Benefit: Solution or Dream?
    Edited by H.J. Otway. Los Alamos, New Mexico: Los Alamos
    National Laboratory, 1971.

Starr, C., and Ritterbush, P., eds. Science, Technology and
    the Human Prospect. New York: Pergamon Press, 1980.

Starr, C., et al. Energy and the Environment:  A Risk-Benefit
    Analysis. New York: Pergamon Press, 1977.

Starr, C., Rudman, R., and Whipple, C. "Philosophical Basis
    for Risk Analysis." Annual Review of Energy (1976):629-
    662.

Starr, C., and Whipple, C. "Risks of Risk Decisions." Science
    208 (June 6, 1980):1114-1119.

Statts, E.B. Federal Policies for Regulating Carcinogenic
    Compounds (R 11). Washington, D.C.: General Accounting
    Office, Winter 1977.

Stein, J. "Assessing the Risk." Mosaic 13 (September/October
    1982):17-23.

Sterling Hobe Corporation. Development of Risk-Benefit-Cost
Analysis for Policy Formulation. Washington, D.C.: July
1979, NTIS Order No. PB-80-165103, 213 pp.

Stewart, R.B. "The Development of Administrative and Quasi-
Constitutional Law in Judicial Review of Environmental
Decisionmaking: Lessons from the Clean Air Act." Iowa
Law Review 62 (1977):713.

Stokey, E., and Zeckhauser, R. A Primer for Policy Analysis.
New York: W.W. Norton and Company, 1978.

Stone, C. "Should Trees Have Standing? Toward Legal Rights
for Natural Objects." Southern California Law Review 45
(1972):450.

Stone, J.M. "A Theory of Capacity, Insurance and Catastrophic
Risks." Journal of Risk and Insurance 40 (1973):339.

Stumpf, S.E. "Culture, Values, and Food Safety." BioScience
28 (March 1978):186-190.

Svenson, O. A Vulnerable or Resilient Society? Some Reflections on a Problem Area (Report No. 19). Stockholm:
Swedish Council for Social Science Research, 1979.

Swartzman, D., Liroff, R., and Croke, K., eds. Cost-Benefit
Analysis and Environmental Regulations. Washington, D.C.:
The Conservation Foundation, 1982.

Swaton, E., Maderthaner, R., Pahner, P.D., Guttman, G., and
Otway, H.J. The Determinants of Risk Perception: A Survey
(RM-76-XX). Laxenburg, Austria: International Institute
for Applied Systems Analysis, 1976.

Swelm, R.O. "Utility Theory Insights into Risk Taking."
Harvard Business Review (November-December 1966):123-136.

Tamerin, T., and Resnick, L.P. "Risk Taking by Individual
Option--Case Study: Cigarette Smoking." Perspectives on
Benefit-Risk Decision Making. Washington, D.C.: National
Academy of Engineering, 1972, pp. 73-84.

Tannenbaum, S.R. "Relative Risk Assessment." Advances in
Modern Toxicology Vol. 1: New Concepts in Safety
Evaluation. Edited by M.A. Mehlman, R.F. Shapiro, and H.
Blumenthal. New York: John Wiley & Sons, 1979.

Task Force of the Presidential Advisory Group on Anticipated
    Advances in Sciences and Technology.  "The Science Court
    Experiment:  An Interim Report."  Science 193 (1976):653.

Taylor, G.C.  Socioeconomic Analysis of Hazardous Waste Manage-
    ment Alternatives:  Methodology and Demonstration.  Denver,
    Colorado: University of Denver Research Institute, 1981.
    Available from National Technical Information Service,
    Springfield, Virginia, Order No. PB81-218968.

Thaler, R., and Gould, W.  "Public Policy Toward Life Saving:
    Should Consumer Preferences Rule?"  Policy Analysis and
    Management 1 (Winter 1982):223-242.

Thaler, R., and Rosen, S.  "The Value of Saving a Life:  Evi-
    dence from the Labor Market."  Household Production and
    Consumption.  Edited by N. Terleckyj.  New York: Columbia
    University Press, 1975.

Thayer, M.A.  "Contingent Valuation Techniques for Assessing
    Environmental Impacts."  Journal of Environmental Economics
    & Management 8 (1981):27-44.

Thomas, K., Maurer, D., Fishbein, M., Otway, H.; Hinkle, R.;
    and Simpson, D.  A Comparative Study of Public Beliefs
    About Five Energy Systems (RR 80-1).  Laxenburg, Austria:
    International Institute for Applied Systems Analysis,
    1979.

Thomas, K., Swaton, E., Fishbein, M., and Otway, H.  Nuclear
    Energy:  The Accuracy of Policy Maker's Perceptions of
    Public Beliefs (RR 80-2).  Laxenburg, Austria: Inter-
    national Institute for Applied Systems Analysis, 1979.

Thompson, K.H.  "Margin of Safety as a Risk-Management Concept
    in Environmental Legislation."  Columbia Journal of
    Environmental Law 6 (Fall 1979):1-29.

Thompson, M.  "Aesthetics of Risk: Culture or Context."
    Societal Risk Assessment.  Edited by R. Schwing and W.
    Albers, Jr.  Plenum Press, 1980.  pp. 273-286.

Thompson, M.  An Outline of the Cultural Theory of Risk (WP-80-
    177).  Laxenburg, Austria: International Institute for
    Applied Systems Analysis, 1980.

Thorngate, W.  "Efficient Decision Heuristics."  Behavioral
    Science 25 (1980):219-225.

Trauberman, J.  "An Overview of Risk-Benefit Analysis in
    Environmental Law."  Environmental Professional 3:3/4
    (1981):217-224.

Tribe, L.H.  Channeling Technology Through Law.  Chicago:
    Bracton.

Tribe, L.H.  "From Environmental Foundations to Constitutional
    Structures:  Learning from Nature's Future."  Yale Law
    Review 84 (1975):545-556.

Tribe, L.H.  "Technology Assessment and the Fourth Discon-
    tinuity:  The Limits of Instrumental Rationality."
    Southern California Law Review 46 (1973):617-660.

Tribe, L.H.  "Ways Not to Think About Plastic Trees:  New
    Foundations for Environmental Law."  Yale Law Review 83
    (1974):1315-1348.

Tribe, L.H., Schelling, C.S., and Voss, J.  When Values
    Conflict--Essays on Environmental Analysis, Discourse,
    and Decision.  Cambridge,Massachusetts: Ballinger Pub-
    lishing Company, 1976.

Trubeck, D.H.  "Allocating the Burden of Environmental Uncer-
    tainty:  The NRC Interprets NEPA's Substantive Mandate."
    Wisconsin Law Review (1977):747.

Tubiana, M.  "One Approach to the Study of Public Acceptance."
    Directions in Energy Policy.  Edited by B. Kursunoglu and
    A. Perlmutter.  Cambridge, Massachusetts: Ballinger Pub-
    lishing Company, 1979.

Tversky, A.  "Elimination by Aspects: A Theory of Choice."
    Psychological Review 79 (1972):281-299.

Tversky, A., and Kahneman, D.  "Availability:  A Heuristic View
    for Judging Frequency and Probability."  Cognitive
    Psychology 5 (1973):207-232.

Tversky, A., and Kahneman, D.  "The Framing of Decisions and
    the Psychology of Choice."  Science 211 (1981):453-458.

Tversky, A., and Kahneman, D.  "Judgment Under Uncertainty:
    Heuristics and Biases."  Science 185 (September 27,
    1974):1124-1131.

Tversky, A., and Sattath, S.  "Preferences Trees."  Psychological
    Review 86 (1979):542-573.

Union of Concerned Scientists. The Risks of Nuclear Power
    Reactors. A Review of the NRC Reactor Safety Study
    (WASH-1400, NUREG-75/ 014). Cambridge, Massachusetts:
    Union of Concerned Scientists, 1977.

U.S. Congress, General Accounting Office. Regulation of Cancer-
    Causing Food Additives--Time for a Change? (HRD-82-3).
    Washington, D.C.: U.S. General Accounting Office, 1981.

U.S. Congress, House Committee on Interstate and Foreign Com-
    merce, Subcommittee on Oversight and Investigations.
    Cost-Benefit Analysis: Wonder Tool or Mirage? 96th
    Congress, 2nd Session. Washington, D.C.: U.S. Government
    Printing Office, 1980.

U.S. Congress, House Committee on Interstate and Foreign
    Commerce. Use of Cost-Benefit Analysis by Regulatory
    Agencies. 96th Congress, 1st Session. Washington, D.C.:
    U.S. Government Printing Office, 1979.

U.S. Congress, House Committee on Science and Technology.
    Societal Risks of Energy Systems. 97th Congress, 1st
    Session. Washington, D.C.: U.S. Government Printing
    Office, 1981.

U.S. Congress, House Committee on Science and Technology, Sub-
    committee on Science, Research, and Technology. Compara-
    tive Risk Assessment. 96th Congress. Washington, D.C.:
    U.S. Government Printing Office, 1980.

U.S. Congress, House Committee on Science and Technology, Sub-
    committee on Science, Research and Technology. A Review
    of Risk Assessment Methodologies. 97th Congress, 2nd
    Session. Washington, D.C.: U.S. Government Printing
    Office, 1982.

U.S. Congress, House Committee on Science and Technology,
    Subcommittee on Science, Research and Technology. Risk/
    Benefit Analysis in the Legislative Process: Joint
    Hearings, July 24, 25, 1979. 96th Congress, 1st Session
    (No. 71). Washington, D.C.: U.S. Government Printing
    Office, 1980.

U.S. Congress, House Committee on Science and Technology,
    Subcommittee on Science, Research and Technology, and the
    American Association for the Advancement of Science.
    Congress/Science Forum: Risk/Benefit Analysis in the
    Legislative Process. Joint Hearings, 96th Congress, 1st
    Session. Washington, D.C.: U.S. Government Printing
    ·Office, 1980.

U.S. Congress, Office of Technology Assessment. <u>Assessment of Technologies for Determining Cancer Risks from the Environment</u>. Washington, D.C.: U.S. Government Printing Office, 1981.

U.S. Congress, Office of Technology Assessment. "Methods for Assessing Health Risks." <u>Environmental Contaminants in Food</u>. Washington, D.C.: U.S. Congressional Office of Technology Assessment, December 1979, pp. 59-70.

U.S. Congress, Senate Committee on Governmental Affairs. <u>Benefits of Environmental, Health, and Safety Regulation</u>. Prepared by the Center for Policy Alternatives at the Massachusetts Institute of Technology. Washington, D.C.: U.S. Government Printing Office, 1980.

U.S. Congress, Senate Committee on Governmental Affairs. <u>Study on Federal Regulation</u>. 96th Congress, 1st Session. Washington, D.C.: U.S. Government Printing Office, 1978.

U.S. Council on Environmental Quality, Toxic Substances Strategy Committee. <u>Toxic Chemicals and Public Protection: A Report to the President</u>. Washington, D.C.: U.S. Government Printing Office, May 1980.

U.S. Council on Environmental Quality, and the U.S. Department of State. <u>The Global 2000 Report to the President: Entering the Twenty-First Century</u>. Volume I, <u>The Summary Report</u>. Washington, D.C.: U.S. Government Printing Office, 1980.

U.S. Council on Environmental Quality, and the U.S. Department of State. <u>The Global 2000 Report to the President: Entering the Twenty-First Century</u>. Volume II, <u>The Technical Report</u>. Washington, D.C.: U.S. Government Printing Office, 1980.

U.S. Council on Environmental Quality, and the U.S. Department of State. <u>The Global 2000 Report to the President: Entering the Twenty-First Century</u>. Volume III, <u>The Government's Global Model</u>. Washington, D.C.: U.S. Government Printing Office, 1980.

U.S. Department of Health, Education, and Welfare, National Institute of Environmental Health Sciences. <u>Human Health and the Environment: Some Research Needs</u>. Washington, D.C.: U.S. Government Printing Office, 1977.

U.S. Environmental Protection Agency. Considerations of Health Benefit-Cost Analysis for Activities Involving Ionizing Radiation Exposure and Alternatives. Report of the Advisory Committee on the Biological Effects of Ionizing Radiation (EPA 520/4-77-003). Washington, D.C.: U.S. Environmental Protection Agency, 1977.

U.S. Environmental Protection Agency. The Quality of Life Concept, A Potential New Tool for Decision Makers. Washington, D.C.: U.S. Environmental Protection Agency and the National Bureau of Economic Research, 1973.

U.S. Environmental Protection Agency. Siting Hazardous Waste Management Facilities and Public Opposition (SW-809). Washington, D.C.: U.S. Environmental Protection Agency, 1979.

U.S. Interagency Regulatory Liaison Group, Work Group on Risk Assessment. Scientific Bases for Identifying Potential Carcinogens and Estimating Their Risks. Washington, D.C., February 1979.

U.S. Library of Congress, Congressional Research Service. Risk Assessment, Acceptability and Management. Washington, D.C.: U.S. Government Printing Office, 1981.

U.S. Nuclear Regulatory Commission. Comparative Risk-Cost-Benefit Study of Alternative Sources of Electrical Energy (WASH-1224). Washington, D.C.: U.S. Atomic Energy Commission, December 1974.

U.S. Nuclear Regulatory Commission. Executive Seminar on the Future Role of Risk Assessment and Reliability Engineering in Nuclear Regulation. Washington, D.C.: National Technical Information Service, 1981.

U.S. Nuclear Regulatory Commission. Fault Tree Handbook (NUREG-0492). Washington, DC: U.S. Nuclear Regulatory Commission. January 1981.

U.S. Nuclear Regulatory Commission. Plan for Developing a Safety Goal (NUREG-9735). Washington, D.C.: U.S. Nuclear Regulatory Commission, August 1980.

U.S. Nuclear Regulatory Commission. Reactor Safety Study: An Assessment of Accidental Risks in U.S. Commercial Nuclear Power Plants (WASH-1400, NUREG 75/014). Washington, D.C.: U.S. Atomic Energy Commission, Nuclear Regulatory Commission, October 1975.

U.S. President's Commission for a National Agenda for the Eighties. <u>Panel on Science and Technology: Promises and Dangers</u>. Washington, D.C.: U.S. Government Printing Office, 1980.

U.S. Supreme Court. <u>Industrial Union Department, AFL-CIO v. American Petroleum Institute (Benzene Case)</u>, 100 S. Ct. 2844-2877 (1980).

Upton, A.C. "The Biological Effects of Low Level Ionizing Radiation." <u>Scientific American</u> 246 (February 1982).

Urkowitz, A.G., and Laessig, R.E. "Assessing the Believability of Research Results Reported in the Environmental Health Matrix." <u>Public Administration Review</u> 42:5 (September/October 1982).

Usher, D. "An Imputation to the Measure of Economic Growth for Changes in Life Expectancy." <u>Measurement of Economic and Social Performance</u>. Edited by M. Moss. Washington, D.C.: National Bureau of Economic Research, 1973.

Van Horn, A., and Wilson, R. <u>The Status of Risk-Benefit Analysis</u>. Cambridge, Massachusetts: Harvard University Energy and Environmental Policy Center, November 1976.

Van Reijen, G., and Vinck, W. "Risk Analysis and Its Watch-dog Role." <u>Society, Technology and Risk Assessment</u>. Edited by J. Conrad. London: Academic Press, 1980, pp. 39-48.

Vaupel, J. "Early Death: An American Tragedy." <u>Law and Contemporary Problems</u> 40 (Autumn 1976):74-116.

Vaupel, J. "Truth and Consequences: Roles for Analysts and Scientists in Health, Safety and Environmental Policy Making." <u>Improving Environmental Regulation</u>. Edited by W.A. Magrat. Cambridge, Massachusetts: Ballinger, 1982.

Vaupel, J., and Graham, J.D. "Egg in Your Bier?" <u>The Public Interest</u> 58 (Winter 1980):3-17.

Velimirovic, H. <u>An Anthropological View of Risk Phenomena</u> (RM-75-XX). Laxenburg, Austria: International Institute for Applied Systems Analysis, 1975.

Vesely, W.E. "Estimating Common Cause Failure Probabilities in Reliability and Risk Analyses: Marshall-Olkin Specializations." <u>Journal of the Society for Industrial and Applied Mathematics</u> (1977):314-341.

Vig, N., and Bruer, P.  "The Courts and Risk Assessment."
    Policy Studies Review 1 (May 1982):716-727.

Viscusi, W.K.  Employment Hazards.  Cambridge, Massachusetts:
    Harvard University Press, 1979.

Viscusi, W.K.  "The Impact of Occupational Safety and Health
    Regulation."  Bell Journal of Economics 10 (Spring 1979):
    117-140.

Viscusi, W.K.  "The Informational Requirements for Effective
    Regulatory Review:  An Analysis of the EPA Lead Standards."
    Policy Studies Review 1 (May 1982):686-691.

Viscusi, W.K.  Social Regulation of Risk.  Cambridge, Massa-
    chusetts:  Harvard University Press (In press).

Viscusi, W.K., and Zeckhauser, R.  "Environmental Policy Choice
    Under Uncertainty."  Journal of Environmental Economics
    and Management (1976):97-112.

Vlek, C., and Stallen, P.J.  "Judging Risks and Benefits in the
    Small and in the Large."  Organizational Behavior and
    Human Performance 28 (October, 1981).

Vlek, C., and Stallen, P.J.  "Rational and Personal Aspects of
    Risk."  Acta Psychologica 45 (1980).

Von Neuman, J., and Morgenstern, O.  Theory of Games and
    Economic Behavior.  Princeton: Princeton University Press,
    1944.

Von Winterfeldt, D.  A Decision Theoretic Model for Standard
    Setting and Regulation (RM-78-7).  Laxenburg, Austria:
    International Institute for Applied Systems Analysis,
    February 1978.

Von Winterfeldt, D.  "Four Theses on the Application of Risk
    Assessment Methods.  Discussion Paper on W.D. Rowe:  Risk
    Assessment Approaches and Methods."  Society, Technology
    and Risk Assessment.  Edited by J. Conrad.  London:
    Academic Press, 1980, pp. 35-38.

Von Winterfeldt, D., Edwards, W., Anson, J., Stillwell, W., and
    Slovic, P.  Development of a Methodology to Evaluate Risks
    from Nuclear Electric Power Plants:  Phase I--Identifying
    Social Groups and Structuring Their Values and Concerns.
    Final Report to Sandia National Laboratories, Albuquerque,
    New Mexico, May 1980.

Von Winterfeldt, D., John, R.S., and Borcherding, K.  "Cognitive Components of Risk Ratings." Risk Analysis 1:4 (December 1981): 277-288.

Von Winterfeldt, D., and Rios, M.  "Conflicts about Nuclear Power Safety:  A Decision Theoretic Approach." Proceedings of the ANS/ENS Topical Meeting on Thermal Reactor Safety. Edited by M.H. Fontana and D.R. Patterson. Springfield, Virginia: National Technical Information Service, 1980, pp. 696-709.

Walker, R., and Bayley, S.  "Quantitive Assessment of Natural Values in Benefit-Cost-Analysis." Journal of Environmental Systems 7 (1978):131-147.

Walsh, J.R.  "Capital Concept Applied to Man." Quarterly Journal of Economics (February 1935):255-285.

Walter, I., and Ugleon, J.L.  "Environmental Policies in Developing Countries" AMBIO 8 (1979):102-109.

Ward, B., and Dubos, R.  Only One Earth. New York: W.W. Norton and Company, 1972.

Watson, S.R.  "Multi-attribute Utility Theory for Measuring Safety." European Journal of Operational Research 10 (1982):77-81.

Watson, S.R.  "On Risks and Acceptability." Journal of the Radiological Protection Society 1 (1981):21-25.

Weatherwax, R.K.  "Virtues and Limitations of Risk Analysis." Bulletin of the Atomic Scientists 31 (1975):29-32.

Weaver, S.  "Inhaber and the Limits of Cost-Benefit Analysis." Regulation (July/August, 1979):14.

Webre, A.L., and Liss, P.H.  The Age of Cataclysm. New York: Putnam, 1974.

Weidenbaum, M.L.  The Costs of Government Regulation of Business. A study prepared for the Subcommittee on Economic Growth and Stabilization of the Joint Economic Committee, Congress of the United States.  Washington, D.C.: U.S. Government Printing Office, 1978.

Weidenbaum, M.L.  The Impacts of Government Regulation. St. Louis: Center for the Study of American Business, 1978.

Weidenbaum, M.L., and Derina, R.  The Cost of Federal Regula-
    tion of Economic Activity (Reprint No. 88).  Washington,
    D.C.: American Enterprise Institute, 1978.

Weinberg, A.M.  "Reflections on Risk Assessment."  Risk
    Analysis: An International Journal 1 (March 1981):5-8.

Weinberg, A.M.  "Science and Trans-Science."  Minerva 10
    (1972):209-222.

Weinberg, A.M.  "Social Institutions and Nuclear Energy."
    Science 177 (1972):27.

Weinstein, M.C.  "Decision Making for Toxic Substances Control."
    Public Policy 27 (1979):333-383.

Weinstein, N.D.  "Seeking Reassuring or Threatening Information
    about Environmental Cancer."  Journal of Behavioral
    Medicine 16 (1979): 220-224.

Weisbrod, B.A.  "Income Redistribution Effects and Benefit-Cost
    Analysis."  Problems in Public Expenditure Analysis.
    Edited by S.B. Chase.  Washington, D.C.: Brookings
    Institution, 1968.

Weisbrod, B.A.  "The Value of Human Capital."  Journal of
    Political Economy (October 1961):425-436.

Wells, D.  "Site Control of Hazardous Waste Facilities."
    Policy Studies Review 1 (May 1982):728-735.

Wendt, D.; and Vlek, C.A.J., eds.  Subjective Probability,
    Utility and Human Decision Making.  Dordrecht: Reidel,
    1974.

Whipple, C., Ricci, P., and Sagan, L., eds.  Technological Risk
    Assessment.  Winchester, Massachusetts: Sijthoff and
    Nordhoff, 1982 (In press).

White, G.F.  "Formation and Role of Public Attitudes."
    Environmental Quality in a Growing Economy.  Edited by H.
    Jarret.  Baltimore: Johns Hopkins University Press, 1966.

White, G.  "Human Response to Natural Hazard."  Perspectives on
    Benefit-Risk Decision Making.  Edited by the Committee on
    Public Engineering Policy.  Washington, D.C.: National
    Academy of Engineering, 1972, pp. 43-49.

White, G., and Haas, J.E.  Assessment of Research on Natural
    Hazards.  Cambridge, Massachusetts: MIT Press, 1975.

White, L., Jr. "Technology Assessment from the Stance of a
Medieval Historian." American Historical Review 79
(February 1974).

Whorton, J.C. Before Silent Spring. Princeton, New Jersey:
Princeton University Press, 1974.

Whyte, A., and Burton, I., eds. Environmental Risk Assessment.
Chichester, England: John Wiley & Sons, 1980.

Wiggins, J.H. "Balanced Risk: An Approach to Reconciling Man's
Need with His Environment." Risk-Benefit Methodology and
Application (UCLA-ENG-7598). Edited by D. Okrent. Los
Angeles: University of California, December 1975.

Wiggins, J.H. "Earthquake Safety in the City of Long Beach
Based on the Concept of Balanced Risk." Perspectives on
Benefit-Risk Decision Making. Washington, D.C.: National
Academy of Engineering, 1972.

Wiggins, J.H. Generalized Description on the Concept of Risk,
Acceptable Risk/Balanced Risk, Socio-Economic Aspect of
Risk Taking/Decision Making. Berkeley, California:
Earthquake Engineering Research Institute, February 1978.

Wildavsky, A. "No Risk is the Highest Risk of All." American
Scientist 67 (January-February 1979):32-37.

Wildavsky, A. "Richer is Safer." The Public Interest 60
(Summer 1980):23-30.

Williams, A. "Cost-Benefit Analysis: Bastard Science or
Insidious Poison in the Body Politick?" Journal of Public
Economics 1 (July 1972).

Williams, C.A., Head, G.L., and Glendenning, G.W. Principles
of Risk Management and Insurance. Malvern, Pennsylvania:
American Institute for Property and Liability Underwriters,
1978.

Williams, D.L. "Benefit-Cost Analysis in Natural Resources
Decision Making." Natural Resources Law 11 (1979):761-796.

Williams, R. "Government Response to Man-Made Hazards."
Government and Opposition 12 (Winter 1977):3-19.

Wilson, J.Q., ed. The Politics of Regulation. New York: Basic
Books, 1980.

Wilson, R. "Analyzing the Daily Risks of Life." Technology Review 81 (February 1979):40-46.

Wilson, R. "The Concept of Risk." CTFA Cosmetic Journal 10 (1978).

Wilson, R. "The Costs of Safety." New Scientist 68 (October 30, 1975): 274-275.

Wilson, R. "Examples in Risk-Benefit Analysis." Chemical Technology 6 (October 1975):604-607.

Wilson, R. "A Rational Approach to Reducing Cancer Risks." Cancer News 32 (1978).

Wilson, R., and Jones, W. Energy, Ecology and the Environment. New York: Academic Press, 1974, 353 pp.

Winkler, R.L., and Sarin, R.K. "Risk Assessment: Consulting the Experts." Environmental Professional 3:3/4 (1981): 265-276.

Winkler, R.L., and Sarin, R.K. "A Risk Assessment Methodology for Environmental Pollutants." Four Conceptual Approaches to Health Risk Assessment. Washington, D.C.: U.S. Environmental Protection Agency, August 1980.

Winner, L. Autonomous Technology: Technologies Out of Control as a Theme in Political Thought. Cambridge, Massachusetts: MIT Press, 1977.

Winner, L. "On Criticizing Technology." Public Policy 20 (1972):35-39.

Wollan, M.J. The Process of Setting Safety Standards on the Courts, Congress, and Administrative Agencies (Discussion Paper 204). Washington, D.C.: George Washington University Program of Policy Studies in Science and Technology, 1968.

Wolpert, J. "The Dignity of Risk." Transactions, Institute of British Geographers 5 (November 4, 1980).

Wynne, B. "Discussion Paper on J. Conrad: Society and Risk Assessment--An Attempt at Interpretation." Society, Technology and Risk Assessment. Edited by J. Conrad. London: Academic Press, 1980, pp. 281-287.

Wynne, B. "Technology, Risk and Participation: On the Social Treatment of Uncertainty." Society, Technology and Risk Assessment. Edited by J. Conrad. London: Academic Press, 1980, pp. 173-208.

Yellin, J. "High Technology and the Courts: Nuclear Power and the Need for Institutional Reform." Harvard Law Review 94 (January 1981):489-560.

Yellin, J. "Judicial Review and Nuclear Power: Assessing the Risks of Environmental Catastrophe." George Washington Law Review 45 (August 1977):969.

Zebroski, E.L. "Attainment of Balance in Risk-Benefit-Perceptions." Risk-Benefit Methodology and Application. Some Papers Presented at the Foundation Workshop, Asilomar, California (UCLA-ENG-7598). Edited by D. Okrent. Los Angeles: University of California, 1975.

Zeckhauser, R. "Procedures for Valuing Lives." Public Policy 23:4 (Fall 1975):419-464.

Zeckhauser, R. "Risk Spreading and Distribution." Benefit-Cost and Policy Analysis 1974. Edited by R. Zeckhauser et al. Chicago: Aldine Publishing Company, 1975.

Zeckhauser, R. "The Risks of Growth." Daedalus (Fall 1973): 103-1181.

Zeckhauser, R. "Social and Economic Factors in Food Safety Decision-Making." Food Technology 33 (November 1979): 47-52.

Zeckhauser, R., et al., eds. Benefit-Cost and Policy Analysis, 1974. Chicago: Aldine, 1975.

Zeckhauser, R., and Shepard, D. "Where Now for Saving Lives?" Law and Contemporary Problems 40 (Autumn 1976):5-45.

Zimmerman, B.K. "Risk-Benefit Analysis: The Cop-out of Government Regulation." Trial 14 (February 1978):43-47.

Zimmerman, R. "Formation of New Organizations to Manage Risk." Policy Studies Review 1 (May 1982):736-747.